T0178415

Small Water Supplies

Other books in Clay's Library of Health and the Environment:

E Coli: Environmental Health Issues of VTEC 0157 – *Sharon Parry and Stephen Palmer*

Environmental Health and Housing – *Jill Stewart*

Air Quality Assessment and Management – *D. Owen Harrop*

Air Pollution, 2nd edition – *Jeremy Colls*

Environmental Health Procedures, 6th edition – *W.H. Bassett*

Also available from Spon Press:

Clay's Handbook of Environmental Health 18th Edition – *Edited by W.H. Bassett*

Decision-making in Environmental Health – *Edited by C. Corvalán, D. Briggs and G. Zielhuis*

Groundwater Quality Monitoring – *S. Foster, P. Chilton and R. Helmer*

Legal Competence in Environmental Health – *Terence Moran*

Monitoring Bathing Waters – *Edited by Jamie Bartram and Gareth Rees*

Statistical Methods in Environmental Health – *J. Pearson and A. Turton*

Water Pollution Control – *Richard Helmer and Ivanildo Hespanhol*

Toxic Cyanobacteria in Water – *Edited by Ingrid Chorus and Jamie Bartram*

Upgrading Water Treatment Plants – *Glen Wagner and Renato Pinheiro*

Water Quality Assessments – *Deborah Chapman*

Water Quality Monitoring – *Jamie Bartram and Richard Balance*

Urban Traffic Pollution – *D. Schwela and O. Zali*

Small Water Supplies

A practical guide

David Clapham

Routledge
Taylor & Francis Group

LONDON AND NEW YORK

First published 2004 by Spon Press

2 Park Square, Milton Park, Abingdon, Oxfordshire OX14 4RN
52 Vanderbilt Avenue, New York, NY 10017

Routledge is an imprint of the Taylor & Francis Group, an informa business

First issued in paperback 2019

Copyright © 2014 Taylor & Francis

Typeset in Sabon by
Newgen Imaging Systems (P) Ltd, Chennai, India

British Library Cataloguing in Publication Data
A catalogue record for this book is available from the British Library

Library of Congress Cataloging in Publication Data
A catalog record for this book has been requested

ISBN: 978-0-415-28282-6 (hbk)
ISBN: 978-0-367-39372-4 (pbk)

Clay's Library of Health and the Environment

An increasing breadth and depth of knowledge is required to tackle the health threats of the environment in the 21st century, and to accommodate the increasing sophistication and globalisation of policies and practices.

Clay's Library of Health and the Environment provides a focus for the publication of leading-edge knowledge in this field, tackling broad and detailed issues. The flagship publication *Clay's Handbook of Environmental Health*, now in its 19th edition, continues to serve environmental health officers and other professionals in over thirty countries.

Series Editor:
Bill Bassett: Honorary Fellow, School of Postgraduate Medicine and Health Sciences, University of Exeter, and formerly Director of Environmental Health and Housing, Exeter City Council, UK

Editorial Board:
Xavier Bonnefoy: Regional Adviser, European Centre for Environment and Health, World Health Organization, Bonn, Germany
Don Boon: Director of Environmental Health and Trading Standards, London Borough of Croydon, UK
David Chambers: Head of Law School, University of Greenwich, UK
Michael Cooke: Environmental Health and Sustainable Development Consultant, UK, formerly Chief Executive of the CIEH

For Val, Catherine, Joseph and Beatrice

Contents

Illustrations

Plates

Figures

Tables

Preface

When Bill Bassett the editor of Clay's, the environmental health officers' bible, rang to ask me if I would be interested in writing a book about exotic fruit and vegetables, I knew exactly what he wanted. What he wanted was a book on private water supplies. I have been teaching people about private water supplies for over ten years now and am often asked if there is a book that explains all the different aspects of this subject in a reader-friendly way. Not too scientific – but not too basic. I always had to reply that such a book did not exist, but one day I would try to write it. Well, it took slightly more than one day, but here it is. Hopefully, gentle reader, this book will be useful when locating, investigating, examining and trying to improve any small or private water supply you come into contact with. I have tried to explain the whole process in a logical fashion, from the rain dropping out of the sky until it comes out of the tap, older, wiser and chlorinated (again hopefully).

I have endeavoured to explain all you need to know, including the physics and chemistry of water, why private water supplies are a particular problem, what creatures and contaminants can be found in the water, what this means for the safety of the supply and what to do about them. I have even travelled around the world in a quest for your enlightenment, visiting Europe and both North and South America to find interesting examples of the small water supply maker's art, and have lived to tell the tale. Where possible, I have used examples from practical experience and I would like to thank all those owners of small water supplies who have amazed me with their ingenuity and artistry.

Acknowledgements

I would like to thank the following people for their help during the writing of this book. Dr Nigel Horan of the University of Leeds for his continued inspiration, help, advice, enthusiasm and relentless questioning of facts. Without him, I would never have been introduced to private water supplies in the first place. John Murphy of Marshall's Pumps for his wealth of expertise, good humour and technical drawings. Professor David Kay and John Watkins of the Centre for Research in Environmental Health for their support and expert knowledge and Gary O'Neill from Yorkshire Water who has often provided me with much useful information.

The people at the Chartered Institute of Environmental Health have always been most gentlemanly in their dealings with me, especially Howard Price and Graham Jukes. David Drury of the Drinking Water Inspectorate always has a lot of useful information. Finally thanks to the staff and management of the environmental protection section of the City of Bradford Metropolitan District Council, particularly John Major, who has always been supportive.

I would also like to thank all the people who helped me during my study tour of the US looking at small water supplies. In particular, Chuck and Sue Stirling, Clyde Carlson, Janelle Taylor, Ric Jensen, Norma Natividad, Jodi Miller, Phil and Lee Woods and Joan Rose. NSF International made the trip possible and should be commended on their support of international public health initiatives, particularly when they choose the recipients of their munificence so wisely. Bob Tanner and Jim Kendzel should be thanked in particular.

In Peru, my friends Marcos Allegre and Anna Zucchetti introduced me to a special water supply in 1994. That was the well to the village of Quebrada Verde where, with their help and that of all the villagers, there is now a safe water supply. Thank you Marcos and Anna.

The following books have also been invaluable:

Waterborne Disease: Epidemiology and Ecology by Paul Hunter, published by John Wiley and Sons.
Toxic Cyanobacteria in Water: A Guide to Their Public Health Consequences, Monitoring and Management edited by Ingrid Chorus and Jamie Bartram, published by E&FN Spon.

Finding Water (Second Edition) by Rick Brassington published by John Wiley and Sons Ltd.

The Manual on Treatment for Small Water Supply Systems, by WRc-NSF Ltd and the Drinking Water Inspectorate.

Michael Price's Introducing Groundwater, published by George Allen and Unwin.

N.F. Gray's Drinking Water Quality: Problems and Solutions published by John Wiley and Sons.

King Cholera: The Biography of a Disease by Norman Longmate published in 1966 by Hamish Hamilton.

The figures have been adopted from the following sources to whom grateful acknowledgement is made. Figure 1: M. Price, *Introducing Groundwater*, George Allen and Unwin (1985), Figures 2–4, 6, 7 and 9: R. Brassington *Finding Water* (Second Edition), John Wiley and Sons Ltd (1995), Figure 5: *The Manual on Treatment for Small Water Supply Systems*, WRc-NSF Ltd (2002) and Figure 8: John Murphy (2003) Marshall's Pumps Limited, Oldham.

This work is a summation of many years of looking at small and private water supplies, the help of all the above-mentioned people and many others is gratefully acknowledged. Any mistakes, of course, I claim for myself.

Finally, I would like to gratefully thank my wonderful wife Valerie who has been thoroughly supportive throughout and who has selflessly looked after me and our three children – Catherine 16, Joseph 12 and Beatrice 7, while I spent hour after hour in front of a computer. They have also been remarkably patient with their Dad, but then they are wonderful as well. It is to all four of them that this book is gratefully dedicated.

David Clapham
July 2004

1 Introduction

The importance of water

Water (H_2O) is a fascinating substance and has many properties that make it unique. It is, for example, the only chemical on earth that is naturally found as a solid, a liquid and a gas. It is the only material that expands when it is cooled and contracts on melting. As its temperature lowers, ice starts to form and at about 4°C, its volume starts to increase. It is 9 per cent bigger by the time it becomes solid ice. This increase is due to the molecules rearranging themselves into rigid structures, more widely spaced than before. If this did not happen, ice would be denser than water and sink to the bottom of the oceans. This would cause northern lakes and seas to freeze solid. Without this property all the creatures living in the water would die.

Water is a result of explosive hydrogen reacting with corrosive oxygen. It continually polymerizes at room temperatures forming long molecular chains. The chains are kept together by hydrogen bonding, which is about 90 per cent weaker than normal covalent bonding. Hydrogen bonding is a type of electrostatic interaction between atoms in one molecule and hydrogen atoms in another. The hydrogen atom has no inner shells of electrons to shield the nucleus and there is an electrostatic interaction between it and electrons on an oxygen atom in a neighbouring water molecule. So they stick together but are relatively easily separated. If these bonds were slightly stronger, water would remain as ice up to 100°C.

Water also has a much lower boiling point than would be expected. Counter-intuitively, the amount of energy needed to turn ice into water and water into steam is much higher than for other substances. For example, the amount of energy needed to heat water by 1°C is ten times that needed for the same mass of iron (Matthews, 1997). It has also been suggested that hot water freezes more quickly than cold water. Although not everyone is convinced, this strange property has been acknowledged since the time of Aristotle in the fourth century BC. It is called the Mpemba effect. Erasto Mpemba was a Tanzanian schoolboy who found that he could make ice cream more quickly if he warmed the mixture first. Trust a schoolboy to find that out.

'Out of water We have created all things alive' – The Qur'an. Water is vital to the survival of every creature on the planet, whether they are a plant, an animal or microorganism. Plants use water to maintain structure and rigidity; animals use water to move nutrients and waste products in, out of and around their bodies. It is also home to many of them; living organisms have been found in clouds, ice, water-bearing rock and boiling geysers. Many creatures have therefore had to adapt to the amount of water in their environment, whether they live in the desert and have to survive with very little or in the ocean where there is an abundance. Water also forms the basis for a great deal of speculation about life on other planets, because where there is water there is life. A recent *New Scientist* article (Anon, 2002a) reported that scientists have confirmed that there are vast reserves of water ice under the surface of Mars. This was considered vital news for the search for life on Mars. The temperature of the ice on Mars is around $-50°C$, but even this would allow some form of microbe to survive. The NASA and European Mars space probes launched in 2003 aimed to land in areas of the planet that looked like they were fashioned by flowing water (Mullins, 2003).

Life is thought to have begun in the oceans around the hot thermal vents, bubbling up from fractures in the earth's mantle. Eventually, aeons later, creatures crawled out of the ocean on to dry land, but they never went far from a source of water. Where human settlements have sprung up, water is the vital component that decides where they are sited. All the major towns and cities in the world have a source of water close by, usually a river flowing to the sea. Man can live for many days without food, but if deprived of water, survival is short. A decrease of 10 per cent is dangerous; in children it can be fatal. Only oxygen is more important for the continuance of life. Different books give different figures for the amount of water in the human body, the amount ranging from 50 to 70 per cent water. This range must depend on whose body it is, its shape and size. I suspect that around 62 per cent is the usual amount (Luhr, 2003). Not only does water have to be constantly available for life but we also need a sufficient supply to remain alert and healthy.

The total amount of water on the Earth is about 1,400 million cubic kilometres (Price, 1985). Of this more than 97 per cent is seawater covering about 70 per cent of the Earth's surface. Another 2.2 per cent is held in the polar ice caps. Most of the remaining 0.8 per cent (about 8 million cubic kilometres) is found in the ground. Half of this is greater than 800 m below the surface. The small amount of fresh water easily available to mankind – that which is found in lakes, rivers and streams – is less than one-fiftieth of 1 per cent of the total amount. It is a precious resource and should be looked after.

If I have failed in my attempt to persuade you that water is very important, consider the details of a news item in the *Cape Cod Times* (Anon, 2001c). The headline reads – 'Turkish Women Boycott Sex'. The article goes on to say, 'Angry wives exile husbands from the bedroom until the village water system is fixed.' Apparently the women of the village of Sirt in southern Turkey began their protest as a joke about a month before, but it then became serious. The village has about six hundred inhabitants and a twenty-seven year old water system that often breaks down. For several months before the protest began, the women had to line up in front of a trickling village fountain and carry their water home in large cans, sometimes over a distance of several kilometres. Apparently they got the idea for the boycott from a popular 1983 Turkish film, where women used a similar ploy to get their men to work. In fact, the idea was first used in a play by the Greek author Aristophanes, where the women of Athens withdrew their favours in an attempt to force their men to make peace with Sparta. Unsurprisingly, this direct action by the women spurred the men of the village into sorting the water system out. Blood, sweat and tears all contain water.

Not only is water a prerequisite for life itself, but for many creatures, particularly human beings, that water has to be clean. If it is contaminated with pathogens (i.e. organisms that will cause illness) or it contains poisonous substances, health will suffer. Unfortunately for us, water is easily contaminated and will absorb chemicals and microorganisms as it moves through the hydrological cycle. It is therefore vital that we protect drinking-water sources and, if the water becomes contaminated, we have to make it safe. This book is designed to help do that.

Water in the developing world

Most people in the developed world, whether their water is from a large company or a small utility, are virtually guaranteed to receive safe drinking-water all the time. There is a small group of people however, who drink water from small or private water supplies whose water is not guaranteed to be safe or on tap all the time. This book has therefore been written as a reference for anyone who either has a small water supply or is responsible for ensuring their safety. In addition, many people who live in the developing world have to obtain their water from small supplies that are woefully substandard and dangerous. Hopefully this book will help to improve the quality of these water supplies as well. After all, a spring or a well has the same problems, whether it is in the wilds of West Yorkshire or the Peruvian Andes.

A definition of small water supplies

This book is about those water supplies that are fed by springs, wells, boreholes and watercourses, providing water to individual houses, farms and businesses, small settlements and villages. Supplies that are provided independently of large authorities or corporations. In some countries, they are called private supplies and in others community supplies or even non-community supplies. It does not cover locating sources of water, medium to large-scale water collection and storage, distribution networks or small municipal supplies with traditional water treatment in place. There are other books that deal with those types of problem perfectly adequately. This book attempts to pull together and give an easily understood explanation of all the various aspects of small or private supplies in one volume: safety, protection, risk assessment, sampling, sources of pollution, microbiological and chemical considerations, geological aspects, outbreaks associated with small supplies, water treatment and testing.

Throughout this book, I have used the hyphenated term 'drinking-water' for the substance and drinking water for the act of consuming it. This is an attempt to avoid confusion, but I apologize if it doesn't. Drinking-water is the noun used by WHO in the title of their guidebooks on quality, and if it is good enough for them it is good enough for me.

The hydrological cycle

There are several mechanisms that have an effect on the quality of water, and it is important to be aware of them when planning a new water supply, renovating an old one or trying to interpret sample results. The term 'hydrological cycle' describes the movement of water around the planet (Figure 1). We are aware of parts of this cycle as we can see water in the rivers and the sea and as it falls as rain, hail or snow. The hydrological cycle is a natural process and begins (if a cycle can technically 'begin') in the

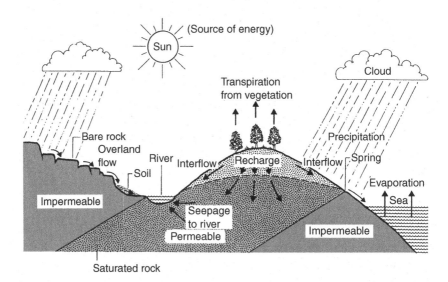

Figure 1 The hydrogeological cycle (from Price, 1985).

oceans and other large bodies of water. The energy that is absorbed from the sun's rays causes some of the water to evaporate. As it rises in the air, it forms clouds, which are then blown onto the land. The water vapour usually remains suspended in the clouds until the temperature falls and the water condenses. The water molecules are attracted to small particulate material in the air that carries an electrostatic charge – this process is called nucleation. When the droplets become big enough, gravity pulls them earthward, where they fall as rain, unless the temperature of the cloud is low enough and snow crystal are formed. If the rain passes through very cold air on the way down, the droplets become hailstones. These various forms of water falling through the air are collectively known as precipitation. A typical thundercloud contains more than 50,000 tonnes of water (Price, 1985). If all the water in the atmosphere were to fall equally over the earth it would produce about 27 mm of rain. The annual average rainfall globally is nearer 900 mm. This means that the water held in the atmosphere is recycled about thirty-three times per year and at any one time is only 0.03 per cent of the world's fresh water supply.

A lot of the rain falls back to the oceans and short-circuits the hydrological cycle. The water that falls on solid ground can lie on the surface and be re-evaporated by the heat of the sun. It can also move downward through the soil and be taken up by the roots of trees and plants. These then move the water up to the leaves where it re-enters the atmosphere in gaseous form. This process is called transpiration. The joint processes of water leaving the land via evaporation and from plants are called evapotranspiration. The amount of evapotranspiration has to be carefully calculated by water engineers when trying to work out how much water is available for subsequent collection and usage for large-scale drinking-water purposes. Water that does not return to the air via evapotranspiration, either runs over the land until it meets a river or a lake or it moves down through the soil and rocks where it becomes groundwater. Groundwater can move into and out of a river depending on the amount of moisture in the land surrounding the river. This process is known as seepage. The groundwater slowly moves downwards until it eventually meets an aquifer, an underground river, it emerges as a spring or seeps into a river. All these will eventually lead it back to the sea, where the process begins again.

An important point to bear in mind is that some of these processes take a short time and others take thousands of years. Ice at the North and South Pole has been there for thousands of years and glaciers may contain water that fell as rain or snow hundreds of years ago. Underground water can also be very old. The waters drunk at the spa in Bath in the Southwest of England for example have been underground for over ten thousand years. Enterprising business people in several parts of the world are bottling some of this ancient water and are selling it at premium prices. It is sold as being 'pure', as it fell as rain before humans could contaminate it with their industrial pollution. Unfortunately, things are not that simple. As the water leaves the surface of the ocean it brings with it minute quantities of various salts. As it forms rain-drops, the water picks up natural and man-made contamination in the form of particulate matter. As it falls from the sky, gases and pollutants dissolve in it. The gases – CO_2 and SO_2 in particular – turn the water into a dilute acid. Once on the ground, the water picks up microbiological and chemical contamination. The chemical contamination increases as the water passes through the rock structure and the chemicals within it dissolve in the dilute acid. Once the water re-emerges from the ground, it can be further contami-nated by animals or by a variety of man's activities as it flows back to the sea. This is obviously a very simplified version of the hydrological cycle and what happens to water; there are many complex activities going on within the process outlined here. However, it is this movement around the earth that brings water into contact with the various substances that contaminate it and cause illness and disease to millions of people every year.

Geology

The geology of an area has a marked effect on the quality of the water that passes through it or is stored in it. Rock is primarily classified into three basic types, these are igneous, sedimentary and metamorphic. Igneous rock is formed directly by the crystallization of cooling molten magma from deep in the earth or from the larva of volcanoes and includes granite and basalt. Sedimentary rocks are those that have formed from sediment, usually under pressure. The sediment can be from pre-existing rock that has been broken down or weathered, for example sandstone or from organic material such as plant material – coal and small sea creatures – limestone. Metamorphic rock is formed when pre-existing rock is subjected to either chemical or physical change by heat, pressure or chemical activity. Marble and slate are both metamorphic.

Groundwater is not normally found in large underground lakes or reser-voirs; it is contained in the pores or voids of the rock structure. These water-containing rocks are known as aquifers. The amount of water that reaches these water storage aquifers is called the recharge. Some rocks make better aquifers than others. It is the storage capacity of the rock and its ability to give up its water easily that makes a good aquifer. This property is called its porosity – the more pores it has – and thus the bigger storage capacity and the ease with which water will move through it, the more porous it will be. Sandstone and limestone are examples of good aquifers. Many rocks however have little or no porosity. They do not hold water and do not allow water to move through them. These are called impervious rocks. Granite, shale and slate are examples of impervious rock. Because of the pressure on them, all rocks deeper than about 10 km below the surface of the earth are impervious.

The upper layer of rock and soil that contains some water, but where not all the pore spaces are full, is called the unsaturated zone. Underneath this layer, where the pores are full of water, is (predictably) the saturated zone.

Water in the saturated zone is known as groundwater. The line between these two zones is called the water table. To find where this surface is, engineers normally dig down to a level where the hole begins to fill with water. Because water gradually moves down towards the sea, this surface is not absolutely flat and its position will depend on many factors, including the type of aquifer, the surrounding topography and whether there are impervious rocks below it.

Small aquifers can be found above the general water table in hills and mountains on independent areas of impervious rock. These are called perched aquifers and are often found in areas where springs are located (Figure 2). They can also be used to supply boreholes and wells because they are easier to reach than deeper aquifers. Above a perched aquifer is a small, domed 'perched water table'. The water moves downwards and outwards over the edge of the supporting impermeable layer of rock. Because the aquifer is near the surface the water quality can vary, for example due to the effects of the weather, and its microbiological quality can be worse than in a deep aquifer where the microorganisms from the soil have had time to attenuate. Sometimes this water is of such variable quality or so insufficient for the needs of the supply that a borehole is drilled through it into a more useful aquifer below. Care must always be taken when doing this to ensure that the water from the perched aquifer does not contaminate the one underneath.

Some aquifers, known as confined aquifers, are sandwiched between layers of impervious rock and are fed by water falling on hills and mountains from some distance away. This results in pressure building up in the aquifer and if a well or borehole is drilled through the overlying rock, the water gushes upwards, without the need for a pump. This type of well is known

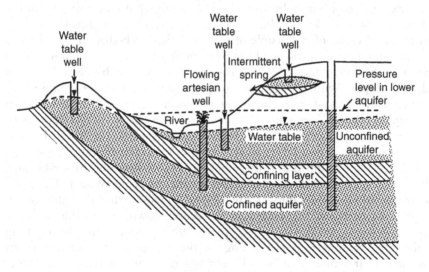

Figure 2 Confined, unconfined and perched aquifers (from Brassington, 1995).

as an artesian well. The name is from the Latin – *Artesium* and is named after the Artois region of northern France. A useful fact to discuss over your next bottle of Stella! The level of the water table is different to that which it would be if the aquifer were unconfined. This unconfined water level is known as the 'potentiometric surface' and it is this level that the plume of water rises to in an artesian well.

The areas of land in which artesian wells can be found are usually basin-shaped; hence they are sometimes called artesian basins. The London area of the UK is sometimes said to be an artesian basin. Other people are not convinced of this and say that the wells of London may never have risen to the surface by themselves. The cause of this confusion is that some time ago, due to the huge number of wells and underground workings, so much water was pumped out of the aquifer underneath London that the water level fell quite considerably. This was by as much as 90 m in some places. Recently this pumping has been reduced, and the water level is rising again. This may cause problems in the future if London Underground's tunnels begin to fill up with water.

Groundwater moves downward through the soil and porous rock under the influence of gravity. Sometimes it will reach the sea or move into and out of a river or a stream. Otherwise, it will continue downwards until it reaches a layer of impervious rock. When it reaches this level the water will travel along the top of the rock until the seam runs out and the water continues its downward journey through more porous rock and along more impervious rock or until the surface of the impervious rock meets the open air. The water will then emerge, either as a spring, a seep hole or at the mouth of a cave. Springs can often be located by looking at geological maps for faults in the rock or places where impervious strata underlie a band of porous rock.

When a well or borehole is drilled into an aquifer, it is important that the depth of the well is more than sufficient to provide enough water for the year round needs of the people using it. To check whether the borehole is deep enough, water is pumped out of it for several hours. If it cannot keep up with the demand, or the water is of poor quality, the borehole has to be drilled deeper. Again it is important that where aquifers are separated by impervious layers of rock, the drilling engineers make sure the water from the upper aquifer does not contaminate the one below it. The amount of water entering the borehole is dependent on the type of rock that the aquifer is made of and how quickly it allows the water to move through it. As the water is pumped out of the borehole the water table around it falls in a curving gradient downwards. This is called the cone of depression. The new level of the water table is known as the drawn-down water table. Many textbooks represent this cone and the drawn-down water table as being perfectly symmetrical. Of course, in real life the cone would be anything other than symmetrical, depending on the nature of the different rocks around it. It would also be different upstream and downstream of the gradient of the water flowing through the aquifer.

If the well or borehole is designed to abstract a lot of water, small test boreholes may be drilled around it to check the effect the main borehole will have on the water table. If the water table is severely affected and is lowered by several metres, this may have a deleterious effect on the surrounding land damaging its ability to grow crops. Ideally the amount of water drawn out of an aquifer by a well or borehole should be in balance with the amount of recharge. When too much water is extracted, this is sometimes known as water-mining. Other than the effect on crops, existing wells and boreholes surrounding the over-pumped borehole will eventually stop being deep enough to obtain water themselves and may have to be abandoned or re-drilled. Finally, if the water table is lowered to below sea level, the emptied pore spaces become infiltrated by seawater. The water from the borehole then becomes salty and undrinkable. This process is known as saline intrusion. This over-abstraction is happening in many places around the world where populations are growing and the demand for water rising. One such place is Lima in Peru. Lima is situated in a coastal desert region, where the water table is going down by about a metre a year. Unless something is done to reduce the amount of water abstracted, saline intrusion from the Pacific Ocean is expected to begin in the year 2005 (Alegre, 1999).

The structure of the rock and the time the water spends underground will also affect its micro-biological content. The rock structure of sandstone or fine gravel, for instance, will have a physical filtering effect and thus tends to sift out and remove microorganisms. The longer a body of water is underground the more likely it is that the microorganisms in it will die off. This is due to the lack of nutrients and the natural ageing of the microorganisms. Aquifers are not normally a suitable environment to encourage microbiological growth, particularly by those organisms most associated with causing disease in human beings.

As it moves underground, the rainwater starts to dissolve chemicals in the rock. As well as the acidity, the amount it dissolves will be partially dependent on four other factors. These are

- the size of any particulate material it comes into contact with;
- the type of rock matrices it passes through;
- the depth the water travels to; and
- the speed of the water.

The first two are important because they determine contact area. Very small particles have a comparatively large surface area. Thus, soil and fine particulate material like clay will present a much larger surface than igneous or metamorphic rock and more chemicals ions can be leached from them into the water. The more rock that the water passes through and the slower it travels, the more contact time and the greater opportunity there will be for chemicals to be dissolved.

The study of water and its relationship with geology is known as hydrogeology. Knowledge of the basics of this subject is useful in understanding the reasons for some of the problems associated with small water systems, both their quality and quantity. Where public water supplies are taken from large aquifers, ones that provide a lot of water on a constant basis, a great deal of research is carried out to investigate the nature and extent of the aquifer and the quality of its water. Trials take place to ensure that the water supply is adequate, whether microbiological organisms are present and what the range of its physical and chemical contaminants are.

However, private water supplies are not normally drawn from the large, well-known aquifers that feature in large-scale hydrogeological maps, they usually get their water from small, seldom studied aquifers. According to Morris (2001) less than 1 per cent of private water supplies in the UK have had their catchments investigated. Throughout the world, there are hundreds of different aquifers tapped into by small water systems and the land above them is used in many different ways. The type of aquifer, the surrounding geology and the use of the land above it cause wide variations in the quality of the water.

The exact nature of the water in the vast majority of aquifers feeding individual small water sources is likely to remain poorly understood. There is not enough interest or resources available to find out about the quality of the supplies, either by the responsible authorities or the people who get their drinking-water from them. To become aware of the potential problems of the water in small systems, simpler and cheaper methods than hydrogeological studies are called for. Water sampling is the main method used at present, but as Chapter 4 shows, relying on sampling alone gives a very false impression of the potential for contamination of small water supplies. It is therefore important to know about the different types of small water supplies, how easily they can be contaminated and what needs to be done to prevent the contamination entering the water in the first place. Each type has specific problems that can be anticipated and thus dealt with.

Of equal significance is the protection of the individual supply, the nature and physical attributes of the land surrounding the source and potential sources of contamination. These can significantly affect the water as it flows over the surface, passes through the overlying soil and rock and into the supply. These investigations are usually called 'sanitary surveys' or risk assessments. The chapter on sanitary surveys will explain how they can be used to assess the area around a small water supply, its effect on water quality and what needs to be done to ensure the water is kept safe to drink. Several people, including Morris (2001) have attempted to classify the risks to small water systems from the catchment geology and provide an appraisal of groundwater source vulnerability. There are several sophisticated sets of assessment forms for springs, wells, surface waters and boreholes, but for smaller supplies the main problem is much more one of contamination from pollution sources in the vicinity, rather than the underlying geology.

For people wishing to make themselves more aware of the geological nature of the areas where private water supplies are found, there are several useful sources of information, particularly universities. Most countries have a geological institute or government department. This organization will have surveyed the country and produced maps or books that will indicate the nature of the rocks in various localities. In the UK, the British Geological Survey produces 1:50,000 scale geological maps. There are 359 of these covering England and Wales. Each area has two maps, split into 'solid' editions, that detail the older pre-Quaternary rocks and 'drift' editions that detail Quaternary and later superficial deposits. Sometimes these maps are combined into 'solid and drift' editions. The Quaternary period is basically the last 1.6 million years and is characterized by ice ages. The maps give a good general picture of an area, its underlying rock structure and details of any significant faulting. Larger-scale maps of 1:10,560 are available in the libraries of the British Geological Survey based in Keyworth, Nottinghamshire and Edinburgh, Scotland. Many of these maps are old editions, completed when the land was surveyed in the late nineteenth and early twentieth centuries, but of course, not much changes in geological terms in a couple of centuries. Each set of maps also has a companion book known as the Sheet Memoir. These fascinating volumes were written to support the maps and give much more detail of the nature and structure of the ground with photographs, pictures of local fossils and other useful information. As well as the geological maps there are hydrogeological maps covering larger areas of the country. These tend to be of less use, as the detail is insufficient to get a good picture of the small aquifers supplying private water supplies. The British Geological Survey has also produced a series of shorter books under the heading of the British Regional Geology series. They cover twenty major geological areas of the UK and are written in a modern, reader-friendly fashion. These slim volumes are a good introduction for anyone wishing to finding out about the geology of an area, without having to learn a lot of geological terminology.

2   Small water supplies

Public supplies, those provided by large utilities, water undertakers, municipalities or private companies usually provide drinking-water that is of uniformly good quality and which is getting better all the time. Small or private water supplies, in contrast, are often of dubious quality and will probably remain so for a considerable while. Even within this category, domestic supplies are five times more frequently contaminated than commercial water sources and the smaller the supply is, the more likely it is to be contaminated (Rutter *et al.*, 2000). For a variety of reasons, water from small sources can periodically contain microbiological contamination, dissolved metals, various organic chemicals and small particulate matter. This may have been washed off the surface of the land or dissolved from the soil or the rock the water passes through. The water will also have physical properties such as acidity and hardness that depend on the surrounding geology. Most urban dweller can rely on their tap water being cheap and safe to drink. For those adventurous souls who live in or visit remote rural areas, a simple glass of water may contain organisms that can make them ill, or in the case of *E. coli* O157 may kill them. This is not to say that you will become ill as soon as you swallow that refreshing glass of water or that farmers are dying in droves from their well water. The human body will build up immunity over the years to provide a measure of protection and the quality of private water supplies varies a great deal according to the weather, so you may be lucky.

It is a common misconception that water coming out of a tap is always of the same quality. This may be because with mains water it usually is. Taps in the city and taps in the country look the same and it easy to assume that their water quality is similar. However, with small supplies, particularly springs and wells, quality can change hourly. When a sample of water is sent to a laboratory, the results are only a picture of that particular slug of water, a 'snapshot' of it at the exact moment it was taken. The quality of small water supplies varies according to a number of factors. These are the recent weather patterns, the geology of the land surrounding the source, how long the water has been in the ground, whether there are any sources of contamination in the catchment (both fixed and walking), what measures have been put into place to protect the supply and what, if any, treatment is installed. If you are responsible for ensuring the safety of private water supplies or you have your own small supply, a thorough understanding of these processes will enable you to evaluate the health risk from the water, organize better ways of assessing water quality and help you realistically interpret the results of any sampling.

In many countries around the world, environmental health staff or sanitarians have a responsibility for ensuring the safety of drinking-water from small and private supplies. In the UK, they are required to be aware of the quality of the water in their area, to catalogue and categorize the supplies and to test the water on a regular basis. They also have to inform the consumers of the results of any sampling. This is often a thankless task. Some consumers of water from small supplies pay scant heed to its quality and regard the person from the local authority as a meddler in what is essentially their own affair. When they are asked to contribute to the cost of this sampling, this only adds insult to injury. One example may help prove the point. For many years, a farmer in the Oldham region of Lancashire had a private water supply that was so bad, so full of the waste produced by his sheep, on such a regular basis, that he was eventually persuaded, threatened and cajoled into installing a water treatment system. The system included filters, pH balancing and twin collection tanks. A small metal tube, about 5 cm in diameter separated the twin tanks. After a short while, the farmer rang the installer of the treatment system in a highly agitated state. Not only had he been forced to put the system in against his better judgement and not only was it far too expensive but now the water had stopped coming out of the tap. When the installer investigated, he found that a small rodent had entered one of the tanks through the overflow pipe and swum to the connecting tube. Unable to swim back to where it had come from, the unfortunate creature had given up the will to live, died, rotted and expanded to fill and thus block, the connecting tube. When the farmer was informed of this, he insisted that the connecting pipe be replaced with a larger one, so that next time something died in it, the water would not stop coming out of his tap.

This may be an extreme example (or it may not, if other tales that I have heard are anything to go by) but it illustrates the problem. Private water supplies are often old and their construction and the location of the source forgotten. Misunderstandings about them abound and technical if not legal, responsibility for them often lies with no one in particular. When asked about the source of a supply, the owner or user will often vaguely wave an arm in the general direction of where they were told the source was when they first moved in. This is usually up-hill, but not always. Sometimes, if they are bored, they will go along with you to help find it, at other times they will not. In order to find the supply, you need to develop a Zen-like talent to read a field, a glen and a moor. An enthusiasm for lifting up large flat stones and an ability to spot likely signs of water sources, such as broken fencing, deep muddy pools and strategically placed abandoned milk crates or pallets is also a bonus. Once you have located the source, you are often rewarded with a visual explanation as to why the last set of sampling results was so bad. One report of an investigation into an outbreak of poisoning from a private water supply could not decide whether the contamination had come from the sewage that had been sprayed over and around the source or from the two dead sheep in the collection tank (Duke *et al.*, 1996). In order to improve water quality it is strongly recommended that a visit to the source be made; to see how the supply is protected from the elements and the attention and droppings of animals both large and small.

In some quarters, it is considered that users of private supplies normally choose this type of water as a conscious act, that small water supply users are fully aware of the health risks involved and have decided to accept these risks because they prefer the taste or do not want to buy water from mono-lithic corporations. Whilst this is a legitimate idea, I wonder about its accuracy. It is true that there is a movement that sees spring water as a natural and safe alternative to big water companies forcing chemical-filled water onto innocent consumers. There is a marvellous quote in a book called the Green Pages, which was a bit like Yellow Pages for environmentalists. The author, who lived in rural Scotland, stated that his water came from the hill at the back of his house 'via a fairly primitive filter system' (Button, 1988). He mentioned that it occasionally got blocked by a dead rabbit and some-times came out brown and peaty after heavy rain. He said however that, 'I do trust it when it comes to drinking it straight from the tap. After rural Scotland I find some of the water in the southeast and East Anglia pretty revolting'. He expresses surprise that people in those areas do not complain more about their water. I am sure they would, if it came out of the tap coloured brown after it rained and had the occasional dead animal in it. If water is subject to that level of contamination, there is no reason why *Campylobacter, Cryptosporidium, E. coli* O157 or other pathogens could not also be present; but I doubt whether any amount of information about the relative risks of drinking from this supply would have brought about a change in viewpoint. Nevertheless, I would have thought that the decision to move to a rural property would not normally be influenced by the type of water source. You may prefer the country to the urban landscape for a variety of reasons, but would the water supply be a major one? If being on a private supply were so popular, you would expect a number of people to reject their public supply and drill a private borehole. I am not aware of this happening very often and I suspect that a small water supply usually providing free water is a much stronger draw.

There is a wide range of small water systems; in fact no two will be exactly alike. They range from large works, similar to a mains water sys-tem, supplying treated, piped drinking-water to hundreds of people, down to small, untreated springs used by one individual. Supplies are usually cat-egorized according to their originating source, that is, a spring, a borehole, a well, a surface supply (stream or lake) or a rainwater collection system. The type of supply will affect the quality of the water and knowing about how they work will help decide the most appropriate way to protect water from contamination.

Nowadays, when a new water supply is necessary, most people will have a borehole drilled. Sometimes however, a spring supply may be available locally that can be tapped into. When a new water source is provided, the construction should be sound, capable of withstanding physical damage and designed to remain wind- and weather-tight for many years to come. During installation it is also important that the water supply is adequately protected from any source of contamination and that the various protection mechanisms are regularly maintained afterwards. Small water sources are not always looked after however and although the present occupier may be interested in ensuring a good supply of safe water, the next one might not appreciate what needs to be done. A sound construction will help protect the water supply even in these circumstances.

For new supplies, it is recommended that an experienced contractor is used, one who can professionally assess the potential problems of the source. The contractor should test the water several times to establish the quality and quantity that can be expected. They should also take into account how these may change over the year. The water will often have certain characteristics that should be adjusted before it can be drunk. The water may be acidic or hard or may contain high levels of metals such as iron and manganese. Although these particular metals are not a health-related problem, the water will tend to discolour clothes when they are washed and stain sanitary fittings. If present in large enough quantities they will also make the water taste bitter. Only when the water has been tested for its physico-chemical qualities and microbiological contaminants can a treatment system be planned. Even with water from a new borehole that is consistently free from pathogens, disinfection must be seriously considered in order to ensure safety and in case of unforeseen accidents. A minimum of chlorination, reverse osmosis or ultraviolet disinfection is recommended. Most small water systems however, have been installed for some time and many leave much to be desired.

The various protection mechanisms that are important for each type of supply are described here but a watertight system from source to tap is essential, as is a safety protection zone around the source. This should normally include a strong fence to keep out any animals. There has been some discussion as to how big this zone should be to protect the water and the health of the people who drink it. For groundwater supplies, it is often recommended that the protection zone has a radius of 50 m from the source. This is based on the theory that any water travelling to the supply from outside the safety zone will have been underground long enough for all microorganisms to have died. Morris (2001) has recently pointed out that these calculations are primarily based on the survival of bacteria. Protozoa such as *Cryptosporidium* and viruses have been found to survive far longer than bacteria, so the 50 m radius is technically too short. It may even be a dangerous idea to rely on a generalized safety zone, because it will give people a false sense of security. Many small water systems do not have the extra protection that chlorination brings and fences will not keep out disease-carrying small animals and birds. It must be said therefore, that although a 50 or 100 m exclusion zone around a public water supply borehole may be ideal, for private supplies it is not normally an option and if suggested, might be greeted with derision. Of course, many suggestions to owners to improve private water supplies are greeted with derision, but the listener should at least be given some logical options. Animals must obviously be kept at a reasonable distance from water supplies, particularly wells and springs. Nevertheless, sources are often located in fields owned by someone else and it would be difficult to justify to a farmer that a fence, 100–200 m in diameter, must surround the spring in the middle of the cow pasture; it would interfere with the amount of grazing for the animals and thus the profits. The farmer would also find it particularly galling if he or she provided the water to the householder free of charge. Wells and boreholes are also often found close to farm buildings and it may be even more irritating for the farmer to be informed that no animals are allowed within 50 m of the farmyard. What should the recommended distance therefore be? Whenever a health official is giving advice, they need to get the maximum, sensible distance possible. Whatever the size of the exclusion zone, it will never be a complete replacement for a final, end-on disinfection system, but it will reduce some of the pollution and as with all multiple-barrier systems, every little step helps.

Springs

Spring supplies are usually found in hilly and mountainous areas. They are naturally occurring and often easily located because they are the beginning of a small stream or pool (Figure 3). Some springs emerge from deep underground where the water has travelled a great distance through porous rock, sand, etc. (Plate I). The effect of this is that the water will have been microbiologically purified to some extent and, if it can be adequately protected from contamination where it surfaces, will be reasonably safe to drink. The spring however may not be coming from deep underground. Many springs are regularly contaminated by faecal material and it may be that the layer of impermeable rock the water is travelling over is just under the soil's surface and there has been no filtering involved whatsoever. Of course, the water may also be a mixture of types, from several separate catchment areas, coming together at one geological fault or impermeable rock stratum. A deep supply, fed by rainwater from many kilometres away may become mixed with easily contaminated water from a shallow spring system, just prior to where it emerges. Thus, a spring supply that normally consists of good quality water may sometimes become contaminated in times of

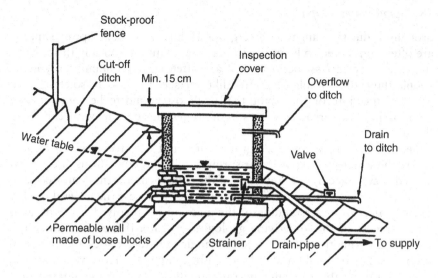

Figure 3 A spring supply (from Brassington, 1995).

Plate I Spring supply from Machu Pichu, Peru. Dating *c.*1500. (See also Colour Plate I.)

drought or due to changes in underground water movement. Spring supplies are often considered to be groundwater sources and technically this may be the case, nevertheless, because they are often easily contaminated, many people think they should be treated like surface supplies. To solely rely on source protection for something that is almost bound to be contaminated with pathogenic organisms at several times during the year can be a dangerous idea.

Because it is travelling underground, it is not easy to work out exactly where the water in any particular spring supply has come from. It could be argued that an accurate knowledge of the supply catchment is unnecessary, as well as technically impossible but, in order to assess the general quality of the water, an understanding of the mechanisms involved is useful, as is an awareness of the underlying geological picture. If these are known, a rough approximation of the water's changing characteristics and potential problems can be made and subsequently dealt with. However, it is also important to be able to eliminate contamination from the area immediately surrounding the spring. Source protection is vital and will shield the water from high loads of rapidly changing contamination.

The problem with a spring is essentially one of getting water from a natural hole in the ground into a pipe, without allowing any contamination by soil, animals or birds. There are two main methods. One is to either insert the pipe up the hole or to dig into the rock for a while and then position the pipe. The other method is to build some form of collection chamber at the point where the spring emerges from the ground and have an outlet from the chamber attached to the pipe leading to the tap. Collection chambers (or spring boxes) are made of a variety of materials and use an assortment of construction techniques. A well-built spring supply should ensure that whatever method is chosen, the water is protected. Collection chambers should be made of impervious material such as engineering brick or concrete, with a watertight, substantial lid, locked in position. This will prevent unwanted attention from the public and help them resist the temptation to put something nasty in the chamber. The top of the chamber should be at least 30 cm above the surrounding ground level to prevent floodwaters entering. The collection chamber will also need an overflow pipe. The overflow pipe should have fine metal mesh around the end to prevent access for insects, leaves, litter, small animals, etc.

If the 'shoved-up-pipe method' (I would appreciate suggestions for improved terminology) is chosen, the builder should ensure the pipe is pushed as far into the spring as possible. It should be securely fixed in place and covered to prevent mechanical damage with a large amount of concrete sloping towards ground level. The concrete must be skimmed to provide as smooth and waterproof a surface as possible. It is important that there are no flat areas to allow for ponding. Where the hillside has been dug into to find a more solid and better-protected starting point for the pipe, similar methods are used. The concrete should extend as far as possible away from the spring. I would recommend that this is at least 2 m, but in practice it is often much less. It is virtually impossible to know what the size of this apron should be; very little science has been done to establish the diameter against contamination ratio. If someone asks for your advice, you should go for the maximum amount that you can talk the builder or owner into – the important thing is to protect the drinker's health and the size of the apron cannot be accurately gauged unless it is put in and the water tested over an extended time period. Once it has been fitted however, it is unlikely that the builder will return to extend it, even if the water is less than satisfactory. Where the spring is connected directly into the pipe there will have to be a pressure release point or overflow, further down the system.

Suitable access points and rodding eyes should also be incorporated into the system, otherwise maintenance and clearing blocked pipework will be very difficult. The source of one village supply that I investigated was located at the bottom of a cliff. Unfortunately this became completely lost due to rockslides. As far as source protection was concerned this was quite good – it was safe from any form of contamination not involving dynamite but, if any maintenance is ever needed, it will take weeks of digging before the end of the pipe will be found.

All springs must be surrounded with a stock-proof fence at least 4 or 5 m away, the further the better. This it to keep animals out and prevent them leaving their droppings all around the collection point. There should also be a small ditch around the spring, just inside the fencing. This is to collect any rainwater running downhill from above the source and channel it around the spring and away. It is placed inside the fence to stop it attracting animals as a watering hole. Some people consider that the fence itself attracts sheep and cattle, as it gives them something to scratch against. I consider that a fence is much better than nothing, but this is one reason why it must be located a good distance away from the source.

Inevitably, some spring supplies will be constructed in the summer and the source connected to the property during an extended dry period (I realize of course, that this is a rare scenario for the British Isles). However, when the autumn rains start, unexpected problems can occur. The pretty little vale where the spring is located suddenly becomes the bottom half of a rushing mountain stream and the whole affair is a couple of feet underwater. The brickwork is being damaged by the debris brought down by the torrent, the lid has been washed away and the little lamb that watched the building of the collection chamber with such wide-eyed interest, has drowned upstream and become lodged next to the overflow pipe. This will explain the subsequent outbreak of cryptosporidiosis in the village below but also illustrates the importance of thinking about all the possible things that can go wrong with a supply and making sure you have put something in place to stop them happening. This situation is not just a fictional illustration, it actually happens. In North Yorkshire, there is a large old house on a private water supply. The pipes to the house are about 4 km long. The cost of digging the trench up to the spring, buying the pipes, building the pressure breaks and connecting them all must have been tremendous.

The cost of planning the spring collection point must have been zero. The builders used the shoved-up-pipe method (but without any concrete protection) and there is a wooden pallet on the top of the pipe to keep it in place. Every autumn the pallet is washed away as the area turns into a mountain stream. To cope with the constantly polluted water, a complex treatment system had to be installed. This makes no logical sense whatsoever.

A type of spring supply that is sometimes encountered but is not what it seems is the one that is really an old land drain. Land drains are usually butt-jointed clay pipes or plastic pipes with holes cut in them. Older ones can be made with stone slabs or flat rocks protecting the sides and top of the trench. They are laid in a suitable pattern, usually a 'herringbone' formation, at a small distance underneath the surface of badly drained fields. They collect the rainwater and channel it to the bottom of the field where it runs downhill until it meets a stream or a more permeable surface layer and disappears underground. Land drains therefore dry out the field and allow it to become more productive. Unfortunately, one generation's land drain outlet can become the next generation's spring supply. After a few years memories fade and what many people fondly believe to be a pure mountain spring is, in reality, the sump to a field of cows. Because land drains collect water from just under the surface, as soon as there is a downpour the droppings of any animals grazing in the field quickly contaminate the water supply.

Once you are aware of what a well-constructed spring collection point should look like, it enables you to improve the water quality of the normal types you meet. Those that are of any age will be built of local materials but later ones are usually of brick or concrete. Many lids that were once close fitting and sound will now be misshaped or rusted. Other lids will be completely missing. Sometimes they will have been replaced by a rough approximation, such as a large stone slab, a milk crate, a wooden pallet or a few planks. The sides may need pointing if they are brick or cracked or crumbling if made of concrete and no longer watertight. Some springs have open collection chambers or run out into little storage ponds either natural or artificial, with the outlet in the bottom of the pond or chamber. Others will form the head of a watercourse with the pipe collecting the water laid in the bed of the stream. Fencing is invariably broken or missing and most of the fence posts will have been knocked over after acting as scratching posts and being unable to withstand the attention of a particularly itchy cow. If you develop an interest in private water supplies, you will soon enjoy finding new variations on their design and come to marvel at the inventiveness and artistry, if not hygiene awareness, of their creators. My latest favourite is a spring that is covered by the whole side of a shed, complete with door. You expect the door to open up at anytime and a zombie-like being to rise out of the waters.

It should be pointed out that it is always best to stop pollution from getting into the water, rather than trying to remove it once it is there. Where the water is consistently poor, an alternative source should be identified if possible. The main aim of any collection system, whether it is a spring, a well or a borehole, should be to take the water from the ground without exposing it to any adulteration. A treatment system is a back up and should only be used to remove any microbiological contamination or excessive chemical pollution the water has picked up underground. Any safety system should have a multiple-barrier approach, so that if one fails, the people who drink the water are still protected by the others.

One possible problem with an existing spring supply is actually finding the source. When a sanitary survey is being carried out or follow-up testing used to identify the cause of a sampling failure, there is sometimes confusion as to where the supply originates. There can be several sources that may or may not feed into the supply. To help with this, harmless, fluorescent tracer dye can be put into the possible sources and the water at the tap run until the dye starts to come through. The usual dye is called fluorescein and is a very bright green but other colours are available to enable several potential sources to be tested at the same time. Where the dye gets so diluted that it cannot be seen by the naked eye, it can sometimes be found using an ultraviolet lamp. This will cause the water to glow faintly in a darkened room. However, these dyes adhere to clay particles in the soil and are often completely removed by the time the water gets to the tap. In such cases, a more accurate tracer test is available using bacteriophages. These are viruses that attack bacteria and are used for a variety of reasons, including ease of identification. Another important one is that they are harmless to man – it is never a good idea to put pathogenic material into a private water supply! Bacteriophages are produced in microbiology laboratories in large numbers, that is 10^{15} per litre. They can be identified in water at a level of about fifty per litre, so they can be isolated even where there are large dilution factors. Where there is a need to use them, contact should be made with the local public health laboratory or a university with a microbiological or public health department. Cross-checking with the authority's medical advisors before using them is also recommended.

Wells

The word 'well' sometimes causes confusion. In many parts of the world, it is used to describe two types of supply, the mechanically drilled borehole, which I will discuss later and the hand-dug well. For the purpose of this book, I will refer to them separately and except where the meaning is obvious, will use the word 'well' when I mean the traditional hand-dug form.

Wells are one of the best-known and more traditional forms of small water supply and everyone has been brought up with nursery rhymes and fairy tales about them (Plates II and III). The oldest well ever discovered is 7,300 years old and was discovered in 1991 by archaeologists in the town

Plate II Well supply, the village of Quebrada Verde, Lima, Peru. (See also Colour Plate II.)

of Erkelenz, Germany. It was 15 m deep and the sides were supported by split oak timbers (Seegers, 1992). In some castles (such as Chateau Le Rochepot in Burgundy, France), the well was also a form of escape in times of siege – the well would have a secret tunnel (adit) dug just above the water level leading to the outside (Plate IV). Most wells are about a metre in diameter, normally over 100 years old, although they can be much more and they tend to produce water of poor quality. Constructed by hand, a well was usually dug where no spring or other easily obtained source of potable water was available. The digging would proceed downwards until a water-bearing layer of rock, gravel or shale was found that appeared to produce an adequate supply of clean water. The flow would be measured by filling buckets or pumping the water out and waiting until the surface level rose again. If this took too long, the diggers would carry on downwards until there was enough water produced to satisfy the demand. The interesting question to me has always been how far would you go downwards before giving up and trying somewhere else. Where to dig would always have been the first problem, but

Plate III Well supply, Santa Catalina Monastery, Arequipa, Peru. (See also Colour Plate III.)

it has been said that if you dig down far enough almost anywhere in the British Isles you will eventually come to water (Brassington, 1995). Some people have relied upon water diviners, who profess to use naturally given gifts to locate the place where the well diggers should start to shovel. The results are sometimes very impressive and those who claim to have the gift often appear genuine and as nonplussed as everyone else about their abilities. I however remain unconvinced.

As the well is being dug, it has to be supported by a wall to prevent it collapsing on top of the diggers. This wall should be impervious to water until well below the level of the water to be used for drinking. Filling the space

Plate IV Well supply and emergency exit, Rochepot Chateau, Burgundy, France. (See also Colour Plate IV.)

around the outside of the wall (the 'annular' space) with an impervious substance such as puddled clay can do this. In practice, this often does not happen or it is not sufficiently watertight to guarantee there will be no infiltration of water through the upper walls. Water from the whole depth of the well may therefore percolate through and run down to the water supply at the bottom (Figure 4).

In developing countries wells are still dug, although this is usually by mechanical means, with sealed pre-formed concrete rings being used as the walls. This system is a bit more reliable for keeping extraneous, potentially contaminated water out of the well. The wall is normally raised above ground level to prevent rainwater, flooding and spillages running into the well. Often the well is covered with a small housing unit that allows a rope and bucket to be lowered into the water. This does not always happen and in practice, as with springs, the tops of wells can be an interesting assortment of constructions, mainly in need of repair. Nowadays the rope and bucket system has often been replaced by some sort of pumping device and the provision of a more suitable cover to prevent dust and rubbish getting in (as well as spoiling the traditional game of dropping poor pussy in). Over the years, electric pumps have also gradually replaced traditional hand pumps.

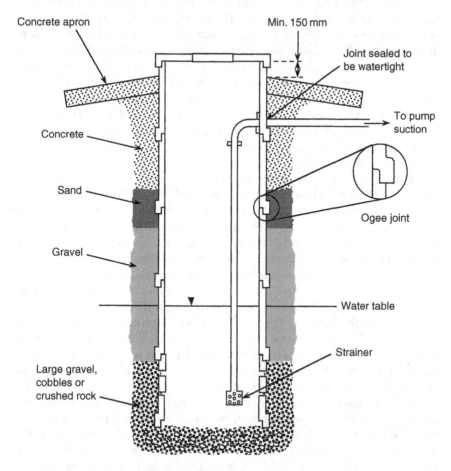

Concrete apron

Min. 150 mm

Joint sealed to be watertight

To pump suction

Concrete

Sand

Ogee joint

Gravel

Water table

Strainer

Large gravel, cobbles or crushed rock

Figure 4 A well (from Brassington, 1995).

It is rare for a well to be regularly maintained, so if a portion of the wall collapses, it will probably not have been repaired and the area where the collapse occurred may allow polluted water in. It will also cause soil and debris to be washed into the well. Added to this mixture can be rainwater or spilled water from the surface. Wells are usually left to look after themselves until a particularly major problem occurs, such as running out of water, so it is unlikely that they will ever have been cleaned or the bottom layer of silt and detritus (and any dead animals) periodically removed. If wells are ever renovated it is interesting to watch the faces of the people who drink the water when animal skeletons, rotting toys and items of discarded rubbish are hauled out.

An unprotected well can sometimes be a mixture of different waters with different characteristics if it has crossed through several layers of permeable and impermeable rock when it was being dug. This would be necessary if the higher permeable layers had insufficient amounts of water for the well's needs. Occasionally, during drought conditions, the water in one or more of the layers will dry up or reduce to a trickle; this may have a major effect of the quality of the water. In some instances, this effect may be so severe that the treatment system will no longer be able to cope. Wells, like boreholes are considered by some to be groundwater sources, with cones of depression forming around them and the water being of a reasonably uniform quality and quantity. It is more practical, as with springs, to think of them as surface water supplies – substandard, subject to variable water quality and easily contaminated.

Another problem with certain wells is that iron bacteria build up at the bottom of the well where they grow by metabolizing the naturally occurring iron in the water. They include the following types of bacteria: *Sphaerotilus, Leptothrix, Crenothrix, Clonothrix, Phragmidothrix, Liekeella, Sphaerotilus* and *Haliscomenobacteria*. The number of bacteria can build up to become quite substantial and when they start to die off, they decompose and impart a sulphurous 'rotten eggs' taste to the water. Although it is not toxic, it is very unpleasant to drink or bathe in. The usual treatment for this problem is to periodically put sufficient chlorine into the well to kill all the bacteria. This is called shock chlorination. It is sometimes called super-chlorination, but that term is more correctly applied to mains water treatment where excess chlorine is added part way through to reduce organic contaminants and then lowered to normal levels at the end of the process. After the chlorine has killed the bacteria, some method of flushing the well or running the contents to waste has to be employed. Another method is to introduce heating elements or steam pipes into the water and boiling the bacteria to death. Treatment has to be repeated periodically as the numbers tend to build up regularly. Treatment systems can be installed to remove taste and odour problems from private and small water supplies but iron bacteria will often test the capabilities of the most efficient treatment systems.

Boreholes

Of all the different types of small water supply, boreholes are usually the most consistent, taking groundwater from deep underground (Figure 5). The borehole is a deep, narrow hole drilled by a specialist company and the company's experience and skill usually ensures that the water is protected from contamination. For private water supplies or small systems, a borehole is typically 100 or 150 mm across. As well as domestic private supplies, boreholes are also drilled to supply public water undertakers or factories. In these situations, where more water is needed, the borehole will obviously be a lot bigger. The quality of the water from boreholes, particularly bacteriologically, makes them a good source of water. It has been filtered by the

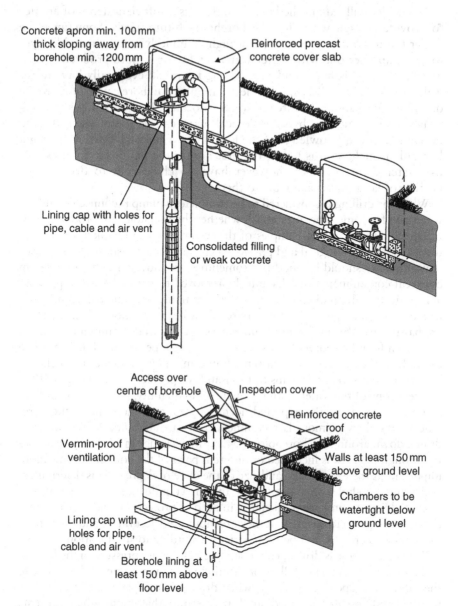

Concrete apron min. 100 mm thick sloping away from borehole min. 1200 mm

Reinforced precast concrete cover slab

Lining cap with holes for pipe, cable and air vent

Consolidated filling or weak concrete

Access over centre of borehole

Inspection cover

Vermin-proof ventilation

Reinforced concrete roof

Walls at least 150 mm above ground level

Chambers to be watertight below ground level

Lining cap with holes for pipe, cable and air vent

Borehole lining at least 150 mm above floor level

Figure 5 A borehole (from Brassington, 1995).

rock structure in the aquifer, subterranean nutrient levels are low and the intake should be so deep that any bacteria in the water have died before they reach it. Water from rivers and lakes on the other hand, even in pristine condition, can easily be contaminated and subject to sudden changes in quality, dependent on rainfall or snowmelt.

The driller will extend the bore until there is a sufficient depth of aquifer to provide enough water for the borehole's future needs. Sometimes the driller has to go past one aquifer through its underlying impermeable layer to reach another one with better water quality or a more productive rock matrix. If water is not found, the drilling team will eventually have to try drilling somewhere else. Sometimes a number of bores will have to be drilled before a suitable source of water is located. Rather than just picking somewhere at random, hydrogeological maps can be useful when drilling boreholes, as can knowledge of the local area. The initial cost of drilling a borehole is high, so it is best if the first bore is the only one. Some people use water diviners, others however have no choice but to drill on the portion of land that is available to them.

When the drilling is considered to be complete, a pump is connected and the water run for some time to establish whether the borehole will produce a sufficient quantity of water. Samples of the water will also be taken to identify the water's bacteriological and physico-chemical characteristics. The bacteriological quality should be good but sometimes groundwater is high in certain chemical contaminants that have to be removed to make the water potable. These are usually aesthetic, such as the commonly encountered aluminium, iron or manganese, but sometimes borehole water will contain chemicals that are hazardous. Various harmful substances such as lead, uranium and radon have been found in boreholes (as well as in other types of supply). The prime example at the moment is the arsenic found in many Bangladeshi boreholes. This is causing severe problems to thousands of people (Smith *et al.*, 2000).

The testing of the water, both for quality and quantity, should go on as long as possible to mimic the expected normal operating conditions of the borehole. Three days or more is recommended as a minimum. As the water is drawn down around the borehole the water may change in quality as its velocity through the rock is increased or different parts of the aquifer become more important as the exact shape and size of the cone of depression is determined. Wells and boreholes that are 'under the direct influence of surface water' (a United States term) are particularly in need of protection, and special attention must be given to ensure that they are monitored accurately and their treatment systems are always operating strictly according to specification.

When a borehole is drilled, either slurry is kept under pressure in the bore to make sure the sides do not fall in, or sheathing is put in to support them. Where the sides are supported by slurry, when drilling is complete, the sheathing is installed and the slurry pumped out. It is important that the sheathing prevents water from higher levels getting into the drinking-water. This is particularly important where a higher, more easily contaminated aquifer is bypassed to reach a lower, purer water source. In the annular space, a suitable watertight packing material should be used to protect the water supply further. In the bottom portion of the borehole, where the aquifer provides the water, the sheathing has to support the surrounding aquifer (unless it is self-supporting) but has to be sufficiently permeable to allow enough water in.

A permanent pump has to be connected when the borehole is finished. This is either located in a purpose-built structure above the borehole or is a submersible-type pump that is carefully lowered down to the bottom of the hole. With submersible pumps, the head works will be smaller and consist of an access point, the sheathed electrical wires entering the bore and the water pipe leaving it. As with all water sources, it is important to protect the head works of the borehole from risk of contamination and physical damage. It should for example, be sited in a place away from traffic, so that the head works and pipework do not get run into or run over. This may seem blindingly obvious but a recent outbreak of waterborne disease from a UK private water supply was probably caused by damage from heavy traffic driving over waterpipes under a car park (Reacher *et al.*, 1999). The bore-hole head works should also be raised to prevent accidental ingress by flood-water. Animals should be kept away and the housing and covers kept locked so that access is restricted. Even with care, boreholes can become blocked or gradually filled in over the years. In the borehole of a UK food factory, recently re-drilled because water volume had decreased, three pumps, a set of keys and a toolbox were removed (Murphy, 2003).

The growth of iron bacteria was discussed in the section on wells and can also become a serious problem in boreholes, causing strong odour and taste problems. These problems can be quite spectacular with showering becoming almost impossible due to the smells being given off. Treatment for this problem in boreholes is the same as with wells. It is again possible to install specific treatment to remove the taste and odour but the treatment often has to be site specific with trials of different media to find the best combination. Of course, it is very important to ensure the iron bacteria do not enter the borehole as contamination on the bit used for drilling. All borehole contractors should therefore sterilize their equipment with a strong chlorine solution prior to any drilling operations and when any training of drilling operators is taking place, the importance of this procedure should be emphasized.

If a borehole is abandoned it is important to ensure that it is properly capped to prevent any future contamination of the aquifer. Many problems can be caused where abandoned and forgotten boreholes have not been sufficiently sealed or filled-in. It then becomes an easy route for any contamination from the surface to pollute the aquifers below. Of course, when an aquifer becomes polluted, particularly from fuel oil or organic chemical spillage, it is virtually impossible to ever entirely clean it up. Pouring concrete down the abandoned hole or using impervious clay or other suitable local materials can prevent this contamination occurring.

Surface water

Surface supplies use water from lakes, streams and rivers (Figure 6). Fortunately, this type of source for small supplies is not often encountered

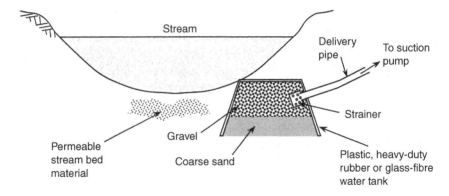

Figure 6 A surface water source (from Brassington, 1995).

and should only be used as a last resort. When public water utilities use water from surface supplies there will be a robust treatment system installed, but this is far less likely with private supplies. The water quality from surface supplies will always be very poor as any open body of water is subject to pollution from birds on the water, creatures in the water and run-off from fields along the banks. Not only is the quality poor, but also it will often be subject to rapid quality changes depending on the weather, if a cow defecates close to the inlet or a sheep falls into the water and dies. It should be borne in mind that, if this is a possibility, there will always be a cow or sheep prepared to sacrifice itself, just to prove the point. Dilution, settlement and natural die-off of bacteria provide some diminution of pollution loading in surface waters but the rapid movement of water in streams or rivers or caused by short-circuiting across lakes does not always allow for bacterial die-off or settlement of small particulate matter. Because of their poor quality, surface supplies should only be used where the householders have unsuccessfully attempted to drill boreholes or dig a well and there are no alternative spring supplies nearby. Of course, another reason for having to use a surface supply is that the householders are too poor to have any other form of water provision. Unfortunately, in many parts of the world this is the only option for millions of people.

Sometimes when a borehole is not producing sufficient water, a surface supply has to be used to augment or replace it. Where this happens the owner may forget and believe that the water continues to be of the same microbiological quality as the original. I have visited several properties where this has happened and it reinforces the need for a sanitary survey to assess water quality rather than merely relying on periodical sampling. I once tested the water used by a guesthouse in the Yorkshire Dales, which I was informed had a well-protected spring supply with concrete storage tanks and sealed lockable covers. When I inspected the tanks, reality was somewhat different. Although they were of a substantial concrete construction with suitable lids, the larger of the tanks had a blue pipe coming out of it with what used to be an airtight cast-iron lid keeping it in place. The pipe then fed into the adjacent babbling brook and followed it upstream about 150 m where a large stone sat on the end of the pipe to keep it on the bottom. According to the map, this watercourse was about 4 km long with sheep grazing on both sides of it. This arrangement not only used an inherently dangerous water supply but also allowed all sorts of small to medium-sized creatures to investigate the tank, fall in and drown, eventually settling on the bottom and decomposing. The guesthouse that used this supply had a water treatment system of course. However, 12 per cent of supplies with treatment are found not to be working sufficiently to kill faecal bacteria (Nichols, 2001). Even if the treatment system did work, it would probably not remove the *Cryptosporidium* and *Giardia* that would inevitably be present in the water from time to time or any pesticides the farmer or the Forestry Commission used on the slopes above the stream.

For surface supplies, a pipe is run from the lake or stream nearest the property to the house or a storage tank. Rather than just laying the pipe on the bed of the lake or river, it should be buried and connected to a particle filter to protect it from unwanted attention and coarser pollution. The stream or riverbed can be dug out to a sufficient depth to allow an encased box of sand and gravel to act as a filter around the pipe. Ideally, it should also have a fine metal mesh filter attached. This provides further filtering action as well as stopping the sand and gravel entering the supply. Not only does the filter box remove particulate matter from water but also bacteria that often adhere to these small particles, so some of the microbiological load is also reduced. When this type of supply is installed, the water should be allowed to run to waste until it becomes clear. Because it is clear however, does not mean it is safe to drink. That should never be assumed.

A slightly better quality of water can be obtained by a variation on surface water provision. This involves digging a hole in the land next to the lake or stream, close enough for the water to filter through the soil and rock matrix into the receiving pipe. The hole should be big enough to allow a sufficient rate of water flow to the pipe. This is called 'bankside filtration' and can be reasonably successful, although it must never be relied on to produce water that can be drunk without further treatment. Once again, a fine particulate filter of sand and gravel should surround the pipe and a metal mesh fixed on the end of it when the hole is back-filled.

There is then the question of how to get the water from the stream or lake to the properties it is intended to serve. Gravity is not usually any help, for the obvious reason that most houses are built above the surface level of an adjacent lake or a river. Normally therefore, an electric pump is used to transfer the water to a storage tank above the property. The water is then fed under gravity to the house. Another type of pump that is occasionally encountered is known as the 'hydraulic ram'. This is famous nowadays mainly for the bizarre way its name can be translated into and out of other languages by automatic computer programmes (e.g. the 'water sheep' device). The hydraulic ram uses the power of the mass of the water moving in a stream or river to raise a much smaller amount up a pipe. It sits at the side of the river in a housing about 1.5 m cubed. It is not necessary to describe the exact mechanism but the hydraulic ram can be very useful in areas where electricity is not available or is too expensive to be a realistic option. If employed, they are easily found as they emit a regular, slowly repeating thud from what appears to be the middle of nowhere. As with the other types of supply employing pumps, the water from a surface supply will need to be kept in a storage tank or pressure vessel. This is so that there is not only a sufficient volume of readily available water in the system when loads are high but also allows the pump to run for a set amount of time to fill the tank.

Rainwater collection

Because of the inherent dangers of microbiological contamination, this type of supply is only considered where there is absolutely no alternative. The set-up is very simple as the roof is used as the collecting area and the rainwater is channelled via the gutters and downpipes to a storage unit. These storage tank or rainwater jars are only filled during heavy rain. When a major rainstorm begins, the first flush of water goes to waste taking the leaves, bird droppings and other debris with it. After a time the downpipe is moved so that it runs directly into the storage tank. I suspect that there has been little work devoted to the time and ferocity of downpour needed to remove say, a seagull's dropping, from a tiled or slate roof (and even less on disseminating the information to the poor people who have to use this system). Seagull droppings are notoriously high in pathogenic bacteria as they scavenge from a variety of places including sewage works and waste disposal tips. One recent study found that that such dropping had upwards of 2.6 million faecal coliforms per wet gram and a substantial number of *Campylobacter* species (Jones, 2003). The water from rainwater collection systems is therefore often of very poor quality and must be accompanied by a good disinfection regime.

In the UK and other developed countries, roof collection systems are rare but not unknown. They are more common in tropical regions where there are no suitable alternatives. In Thailand, for example, many northern villages store their water in concrete rainwater jars about 1.8 m tall and 1 m in diameter. They were mass-produced for the government and are installed with a close fitting concrete lid. Because of the high volume of rainwater during tropical storms, the roofs are scoured and the rainwater entering the tank appears to be reasonably good compared with a lot of shallow groundwater supplies. In addition, because the water is stored for long periods in the jars, the bacteria naturally die off. Testing for faecal coliforms has shown that the water is of reasonable quality in most cases (Pinfold, 1993). If the jar is fitted with a tap to obtain the water, rather than opening the lid and using a scoop, the water will remain potable for some time.

Desalination

Where the only possible source of sufficient water to meet the needs of the population is seawater, desalination plants are sometimes installed. These are usually distillation units where the seawater is heated and the evaporated steam is condensed onto a cooled collection tube and then piped to a storage vessel. Some of these plants can be quite large and produce sufficient water for a small town. They normally produce very good quality water, but the energy needs of such a system are expensive. They are therefore normally found in the Middle East, in some of the rich Arab states, and on small tropical islands, providing water to tourists and the better-off local inhabitants. They are also found in areas of the world where people buy desalinated bottled water because the local public supply is either very irregular or so contaminated that it is not fit to drink. To make this a profitable business the cost of the water will be high, disadvantaging the poorest in society. Most desalination plants are purpose-built package plants. Another form of desalination plant uses reverse osmosis treatment. As with all treated water, a problem can occur if the water is contaminated after filtering. It therefore needs to be carefully protected, particularly in smaller bottling plants.

Fog collection

Fog collection or fog farming is a system found in a few mountainous desert areas with particular meteorological conditions, such as the mountain slopes of the Atacama Desert along the western edge of South America. Although the geographical regions where it is being tried have virtually no rainfall (the Atacama is the most arid desert in the world), a very thick fog often covers the land as clouds blown in from the ocean rise up the mountain slopes. The technique consists of fixing long lengths of nylon mesh at right angles to the prevailing wind. The heavily fog-laden air blows over the mesh and the water in the fog condenses on it. When sufficiently large droplets are formed they run down the netting and along collecting vessels into pipes that lead to the village below. If the nets are kept clean and maintained, the water from this system is considered to be of reasonable quality and much better than dying of thirst.

There is a problem however, in areas where the nets are sited immediately downwind of large urban conurbations. This was found when they were installed on some of the hillsides around the Peruvian capital of Lima to provide drinking-water to some of its shanty villages. The lead in the air from traffic fumes was so excessive that the amount that dissolved into the water made it unfit to drink.

Other associated components

Whenever a sanitary survey is carried out or initial sampling takes place, an inspection of the whole of the water supply system is important. The source, the tanks and any exposed pipework should be carefully examined. If a major problem is found it should be put right before any sampling takes place. There is no point in sampling the water to highlight a problem you already know about or to sample the water to find a problem that you could have identified at no further cost (apart from your time) by just walking around the water system.

Pipework

The pipework from a small water source should have several characteristics. It should be watertight; allowing neither potential drinking-water out or more importantly, contaminated water in. It should also be made of materials that do not contaminate or taint the water it carries. It must be capable of withstanding the pressure of the water in it, particularly if it is sourced some way up-hill from the premises. It should be adequately connected with watertight seals to the source or pump at one end and the storage/treatment system at the other. Hopefully the pipework will be blue, so you know that there is water inside it (rather than electricity or gas) and buried at a suitable depth to prevent it freezing in the winter, normally about 30 cm. However, rather than burying it properly, at least one water supply owner has solved this problem by running the water continuously in the winter to prevent it freezing. Pipes should be protected from the gnawing attention of rodents and be designed so that they do not disintegrate if the water in them is either acidic or contains a great deal of scouring particulate matter. Finally, there should be a plan showing where it runs so that if there is a problem it can be located. Modern or renovated supplies will usually have polyvinyl chloride (PVC) or alkathene pipework, which is specifically designed to convey water safely.

As would be expected, the pipes in many small water supplies will fail to meet these requirements. Older supplies may have iron pipes, which can become corroded, particularly where the water is acidic or aggressive. The corrosion may lead to the ingress of contaminated water or leakage with a consequent loss of pressure. A few of the larger private water supplies may have iron pipes with pitch fibre linings. These linings have been found to leach polynuclear aromatic hydrocarbons into the water. WHO and others consider some of these to be potentially harmful (WHO, 1993a). Pipework with pitch fibre linings on mains supplies have now largely been removed by the water utilities, but they remain a possible problem in private supplies.

Long lengths of garden hosepipe are occasionally used to connect a water source with the premises using the water. These are not suitable as they have not been designed to withstand physical damage, they may leach unacceptable amounts of chemicals into the water and the connections are not suitable for long-term drinking-water supply. Another form of 'pipework' sometimes encountered consists of brick channels or small culverts made from large flat stones found in the field where the water system runs. These are usually associated with very old supplies, installed before the availability of cheap pipework or an appreciation of the need to keep drinking-water free from contamination. These are obviously not watertight. As well as encouraging small mammals, insects, etc. along their length, they also have a tendency to cause the ground above them to collapse as the water running through it surcharges from time to time and gradually wears away the soil. These channels are often in out-of-the-way places and as part of your sanitary survey, you might be the first person to walk above them for some time. It is somewhat embarrassing to explain to a householder about the need for safe drinking-water as you fall through a hidden hole in a field and contaminate their water supply with your wellingtons.

Water storage tanks

A storage tank is capable of performing several useful functions at the same time. It can extend the life of any pumping system because the pump operates for short periods to keep the tank topped up, rather than continually running or being used every time the tap is turned on. A large tank also allows some grace if there is a fault in the system. If suitably constructed, it will also give a constant water pressure at the tap or central heating boiler. This prevents problems with their operation due to fluctuating water pressure, as well as stopping the particularly distressing problem of the water suddenly getting really hot or freezing cold (I'm never quite sure which is worse) in the middle of taking a shower.

Another benefit of a storage tank is that because it should be large enough to slow the water to a virtual standstill, it will allow several biological and physical processes to take place. Where water flows above a certain velocity it will keep particulate material in suspension, once it is slowed to below that speed, such as in a storage tank, particles will begin to settle out. The heavier the particle, the sooner it will fall to the bottom as the flow rate deceases. This process will reduce the cloudiness (turbidity) of the water and thus improve its acceptability for drinking. This action also improves the biological quality of the water, this is because microorganisms carry a slight negative electrostatic charge and will tend to be attracted to particles with the opposite charge and sink to the bottom with them. This property is used in larger water utility water treatment systems where suitably charged chemicals are added to the water to speed up the process. In water tanks however, the process will often happen naturally, albeit to a lesser degree. Settlement cannot however, be relied upon to remove all pathogenic organisms – very small *Cryptosporidium* oocysts for example would take months to settle out (Ives, 1993). Adequately sized storage tanks also tend to smooth out fluctuations in the quality of the raw water, thus ensuring a more consistent loading for the filters and preventing contaminant breakthrough from excessive pollution events.

The other natural process involved is a reduction in the number of microbiota that occurs whenever bodies of water are left for some time. This property is also utilized in water storage reservoirs attached to water treatment works, surface water supplies from large lakes and in a sewage treatment process known as waste stabilization ponds. The earliest water treatment systems used by water companies in the 1800s, collected water from rivers (often seriously polluted) and kept it in reservoirs for a while before pumping it to the public. The purifying processes involved the natural ageing and dying-off of microorganisms from the effects of low nutrients and natural predation. Most bacteria causing waterborne illness in man are happiest when they are kept at the temperature of mammalian bodies and in a gut environment with a plentiful supply of nutrients. This is also why, when we look for them in the laboratory, they are often put on agar broth and incubated at 37°C. They are at somewhat of a disadvantage therefore, when excreted into a cold world and washed into the comparatively nutrient-starved environment of a naturally occurring body of water or piped into a storage tank. To make matters worse they may be surrounded by other microorganisms, which are used to these conditions and are looking for something to eat. Of course, this process will only reduce the numbers of microorganisms and will never lower them to levels where water treatment is considered unnecessary. Pathogenic organisms have survived for millions of years by regularly being excreted into water and then re-ingested, reproducing and causing illness in another individual. *Cryptosporidium* for example, will last for over six months in virtually any environment encountered in the UK, due to its strong outer coating. However, any water purification scheme should employ a multiple-barrier system and this is where the storage tank plays a useful role.

Outbreaks of waterborne disease and contamination of small water systems often follow periods of heavy rain. During and immediately afterwards, highly polluted water can often be washed into springs and wells under the influence of surface water. If a system can be devised whereby the first flush of contamination is run to waste, the quality of the water entering the tank would be improved, reducing the amount of treatment necessary. There is no reason why this could not be used as long as the process was automated and did not rely on the owner of the system having to be present every time it rained. This can be solved if the inlet pipe has a number of small holes along the bottom, where it feeds the tank. This would ensure that during normal flows all the water would enter the collection tank but during heavy rainfall there would be so much water that most of it would overflow the holes and run to waste (Horan, 1994).

Storage tanks have their own dynamics. Where they are used as a form of initial water treatment, the design of the tank and the position of the inlet and outlet pipes are important, as they will affect the final water quality. If the tank is too small, when it is filled or emptied there may be mixing of the water with settled contaminants. The inlet pipe should be designed to enter below the surface of the water and baffles need to be installed to reduce turbulence and facilitate settlement. If the outlet pipe is positioned too high in the tank, it may remove floating detritus from the water and if it is too low, it may pick up settled sludge from the bottom of the tank. It is not a good idea to rely on the initial storage tank as the sole treatment system; nevertheless, as part of a multiple-barrier approach, it is worth considering. I am not aware however, of any serious work that has compared the usefulness of a large storage tank before treatment with a post-treatment tank of the same size and quality of design.

Storage tanks should be constructed so that they hold over a day's worth of water (the size will therefore depend on the amount needed by the particular premises or group of buildings). They should be constructed of sound, robust, watertight materials – neither allowing leakage out nor contamination in. They should be fitted with an overflow pipe that allows excess water out but has a fine-meshed outer covering to stop birds, insects, rodents and small animals getting in. A dead mouse or vole that has fallen into a water tank, thrashed around a bit and drowned is not only difficult to get out, but the particular flavour they impart to the water is not to everyone's taste.

The tank should also be provided with a watertight lid that is large enough to allow cleaning (including vole and mouse removal). Where they are sited outdoors, it is also necessary for the lid to be lockable or have a strong iron bar bolted and padlocked over it. This stops the unwanted attention and unauthorized lifting of the lid by passers-by and prevents the local little boys association jumping in for a joke (or worse). There should also be a small ditch around the tank to collect rainwater running past, diverting it safely away. As well as being situated away from possible flood plains, the top of the tank and the lid should be well above ground level so that rainwater and mud will not pond around and over it. Strictly speaking, this should not matter with a watertight tank and lid, but the tank will be in place for many years and, if insufficiently maintained, may not always have that watertight lid. In fact, after a few years it may not have a lid of any sort. Like the water source itself, a stock-proof fence or strong wall should surround any tank to stop animals getting anywhere near it. Once again, this should not strictly matter if the tank is suitably constructed, but it is always vital to have several barriers to prevent contamination and you should not merely rely on the tank's watertightness to safeguard water quality (Plate V).

The storage tank should be well maintained and regularly cleaned out, shock-chlorinated and flushed through. The lid should be checked and replaced if damaged, the overflow pipe should be looked at to ensure the 'fine mesh screen' over the end has not been torn or removed and the

Plate V Protective building to a private water supply storage tank, West Yorkshire. (See also Colour Plate V.)

fencing should be checked to make sure it is still keeping animals out. This will often never happen. Once a tank is installed, it is a traditional rural custom to completely forget about it. In some cases, not only to forget to maintain the tank, but also to forget its whereabouts or even its very existence. When, as part of a sanitary survey, you ask about the storage tank, you may have to find it yourself. You may also be told that there is no tank, you will then have to go and find it anyway.

Once located, tanks come in a wide variety of types and conditions. Those that I have come across include holes in the ground without any particular walls and others with a leaky stone-sided construction. Some have missing lids, others have assorted lids of large pieces of stone, milk crates or wooden pallets. Others have reasonable lids but they have been wedged open – in one case to let a pipe in from a cow-trough two fields away. Alternatively, the lid will be firmly locked into place and the key will have been missing for years. These are invariably the tanks that you need to inspect or take a sample of the water from. In one supply in the north of England, the lid consisted of three very large (i.e. 2 m long and 0.5 m wide) blocks of concrete cemented together. As it was necessary to inspect the inside of the tank, it was a major operation to gain access. The householder was a trifle put out a week later, having replaced and re-cemented the lid, to be asked to open the tank up again. Other storage tanks have brick walls

Plate VI Inside view of storage tank to a private water supply, West Yorkshire. (See also Colour Plate VI.)

in need of pointing. Another tank that I particularly remember, served a caravan site and had a glass roof, like a conservatory. Unfortunately there were several panes missing allowing easy access for birds. It also had a particularly fine collection of ferns growing around the walls. They obviously appreciated the greenhouse-like atmosphere in the tank, as no doubt, did the microbes in the water (Plate VI).

Pressure breaks

These are mainly employed in upland spring supplies where the source is located some way uphill from the premises and the water is conveyed downhill in sealed pipes. The pressure in the lower end of the pipe would tend to burst the pipe or damage the plumbing system, so a device is needed that relieves this pressure and prevents the possibility of such damage; this is called the pressure break. The normal type of pressure break is similar to a small storage tank. Water comes in at one end to fill it, there is an outlet pipe situated a little way below the water level running to the premises (or another pressure break further down) with an overflow for excess water to run to waste (Figure 7).

Pressure break need the same protection measures as storage tanks. They should be watertight, have a sealed, locked lid that is large enough to allow

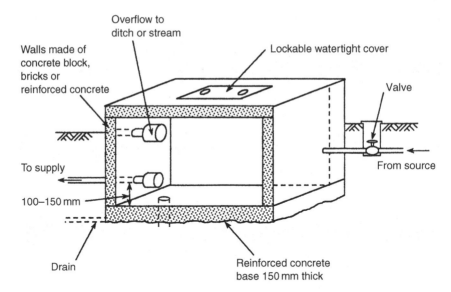

Figure 7 A pressure break tank (from Brassington, 1995).

for cleaning and maintenance. The overflow should have a mesh cover to prevent the entry of unwanted dirt or small creatures. The top of the tank should be above ground level and there should be a stock-proof fence and rainwater divergence ditch. It should be regularly inspected and maintained to ensure the integrity of the structure and to check that the overflow has not become completely blocked, thus taking away the point of having a pressure break in the first place. As with many features of private water supplies, reality often provides a different picture to the ideal one described here. The pressure break is likely to have been neglected and in need of maintenance. The overflow may allow a variety of creatures into the water supply and the lid may be damaged or have been replaced by the British Standard borrowed pallet. The tank will probably need cleaning and have a build-up of settled debris at the bottom.

A water supply should be protected from contamination during its whole length and pressure breaks should therefore be actively looked for when carrying out a sanitary survey. As with storage tanks, the pressure break may have been forgotten about by the owner or not mentioned when the property last changed hands. It is as well therefore to try and follow the line of the pipes from the source to the property. I appreciate that trying to follow the line of a pipe that may be a few feet underground is easier said than done, nevertheless there are occasional clues. Pipes are usually straight, so a line-of-sight can indicate the possible route, although this is not guaranteed. Sometimes the line of the pipe can be seen where there is a dip in the soil level caused by the settling of the back-fill after installation. Occasionally, where the pipe is leaking or the conduit is made of stones slabs, the line can be identified by a difference in the colour of the grass above it. This can be particularly obvious during a drought. A similar phenomenon occurs in leaking drains and sewers. In that case, the point where the drain is leaking coincides with a splendid growth of nettles. This plant likes a nitrogen-rich soil, so it grows particularly well where there is a reasonable amount of sewage and water.

Although the creative ability of the owners of private water supplies have been somewhat tested by the pressure break, there is one example of the homespun approach, that has not yet been covered in other textbooks. During one inspection of a spring supply, an environmental health student and I were following the line of a pipe from the spring, back to the house. Halfway across an otherwise boulder-free field, we saw a large slab of mill-stone grit. When we lifted the stone we found a medium-sized hole, in the centre of which was a carefully whittled piece of wood sticking vertically out of the ground, the shape and size of a small truncheon. The student being typically enthusiastic immediately picked the piece of wood up to examine it. This had the effect of covering her with water as the wood had been tightly wedged into a hole in the water pipe. The idea appears to have been that if the water pressure became too much for the integrity of the pipe, the wood would be forced out of the hole, thus releasing the pressure. The resultant loss of flow would then alert the householder to the problem. The student had to stand over the fountain, jabbing the wood randomly into the soil hoping to find the hole in the pipe. Despite my shouting words of encouragement from a suitable distance, this took longer than would be expected and it was a particularly sad and sodden trainee that dripped all the way back to the office.

3   Physical properties of water

Water is often referred to as the universal solvent and it has the ability to dissolve virtually everything it comes into contact with. A glass of water for example, will contain molecules of the glass in the water. On its way from the ocean to falling as rain on the plains in Spain, water will always pick up chemical and physical contaminants. On reaching the surface of the earth, it may go through a bird dropping or cowpat as it passes on its way into the soil. Within this faecal material will be microbiological organisms that can transfer into the water. Some of these may reproduce and grow. Others will lie dormant waiting to be ingested; they may be pathogenic or benign. As the water passes vertically downwards, it comes into close contact with soil particles. Soil is a complex mixture of organic material from decomposed plants and animals, ground-down rocks and microorganisms, most of which will have passed through the gut of worms. Because the particles are small, they have a comparatively large surface area for the water to dissolve chemicals from. The water will also pick up small solid particles that it incorporates as a suspension. After the soil, the water will pass through subsurface rock and once again can dissolve chemicals from the geological structure.

The atoms of many chemicals bond together through simple electrostatic attraction. These bonds are weakened by the polar nature of water. This allows some of the atoms to escape and become surrounded by water molecules. They then continue as charged ions in the water. This solvent capacity is partly the reason for some of the health problems associated with contaminants in water. A glass of water that has 0.005 µg/l of a chemical of molecular weight of about 100 in it, will contain a billion molecules of that substance (Whitehead, 1992). Because of this solvent ability, it is necessary to set limits for substances in water rather than trying to eliminate them completely. When scientists voice this fact, it can lead to worries. The public wants 'pure' water but unfortunately this is impossible.

Tastes and odours

UK maximum concentration allowed: 3 dilutions at 25°C.
WHO Guideline Value: None set.

Most water quality regulations include standards for taste and odour as water can easily take up unpleasant chemical flavours. These may be apparent at the tap or when the water is heated. Tastes or odours that are too strong make the water unfit to drink. This is whether or not the substances causing the problem are toxic or whether they are acting individually or in concert. The standards are not measured in the same way as for other parameters (in mg/l or μg/l), because an individual substance is not being looked for. The test is carried out at a standard temperature and is measured in dilutions. The water should no longer taste or smell of the contamination after it has been diluted three times with distilled water. In the Private Water Supplies Regulations 1991 (HMSO, 1991b), the legislation says that when sampled, the water should be smelt or tasted unless there may be a problem with it. I cannot imagine that anyone in their right mind, who samples drinking-water from private supplies for a living, would ever risk actually tasting it, either on site or in a laboratory.

Conductivity

UK maximum concentration allowed: 1,500 μS/cm at 20°C.
WHO Guideline Value: None set for conductivity but 1,000 mg/l for total dissolved solids.

Conductivity is a measure of the ease with which electricity will pass through a substance and is used to give an idea of the amount of dissolved chemicals in water. It is well known that water will conduct an electric current. This is why light switches have to be fitted away from showers and why it is considered so dangerous to throw live electric wires into baths. Completely pure water will carry very little electric current however but, as we have already seen, absolutely pure water will never be found outside of a laboratory. In fact, it is the dissolved chemicals in the water that transfer the charge. The more dissolved chemical in the water, the easier it will be for the current to pass through it. The constituents of dissolved chemicals (the ions) have an electrical charge. Common salt, sodium chloride (NaCl) for example, when dissolved forms two charged ions, Na^+ and Cl^-. If an electrical current is applied to the water, the positive and negative ions will move through the water to the opposite poles; the more ions in the water, the more current will pass. A simple battery powered meter, with a two-pronged probe can easily be constructed to demonstrate this. When the prongs are put in the water, a reading can be taken that measures the amount of current and thus the level of dissolved chemicals. In Europe, conductivity is measured in micro-siemen per centimetre (μS/cm) at 20°C. The two parts of the probe are therefore usually one centimetre apart. The reason for a standard temperature being set is that it is easier for the electricity to pass through the water at higher temperatures. When taking readings in the field, the difference in conductivity at different temperatures can largely be ignored. The temperature of the water should be noted, however, to ensure consistency.

WHO guidelines for drinking-water quality do not mention conductivity as such, but instead have a guideline for total dissolved solids. They both measure roughly the same thing, so a standard for either is quite satisfactory – the higher the amount of dissolved solids in water, the greater will be the conductivity. Conductivity is not a problem in itself and just because it is above a certain level does not mean that the water will cause ill health. The European Union has therefore set a conductivity guideline (but not a legally binding maximum) of 400 μS/cm.

The reason why conductivity can be a useful indicator is that it simply and easily indicates when the water contains a high level of dissolved chemicals and although some may be benign, others may not. If the conductivity is low it means that it may not be necessary to carry out expensive laboratory diagnoses to find out exactly what these chemicals are, particularly if resources are limited. This is not to be generally recommended, as individual waters may have low conductivity but the actual chemicals they contain can be hazardous. Another reason why conductivity can be a useful indicator is when the conductivity of the water suddenly goes up. This may indicate that a pollution incident of some sort has happened and, as with turbidity, suggests other contamination will now be in the water, particularly pathogenic organisms. This often follows a sudden bout of heavy rain, but may also indicate that a chemical spillage has occurred somewhere in the supply's catchment area. Conductivity is a cheap method of testing water but should only be relied on where the sampling officer has a good knowledge of the local geology and has carried out a series of more detailed studies to establish the usual contents of the water in the area.

Hardness and softness

UK maximum concentration allowed – no standard set in legislation.
WHO Guideline Value: None set.

Hardness is the property of water that makes it difficult to get a lather when using soap. Softness, if you are used to washing in a hard-water area, is the property where you can never get rid of the lather and are unable to rinse yourself clean when having a bath. The hardness or softness of water is related to the amount of certain polyvalent cations (positive ions having more than a single electrical charge) dissolved in the water. These are mainly calcium ions, particularly from calcium carbonate but can also be magnesium ions. In fact, other divalent metal cations can help cause hardness, such as iron, manganese, barium, zinc and strontium, but they are found in such small amounts they can be virtually discounted. Monovalent cations (those with only a single charge) do not cause hardness. All natural water sources contain some of these dissolved salts, so that all water has some hardness, but the more of them there are, the greater will be the amount of hardness. Water that has less than 60 mg/l of calcium carbonate is usually considered to be soft water. Naturally hard water is usually found in concentrations above 100 mg/l. Hardness above 200 mg/l may cause scaling problems in water fixtures. People can taste the hardness when concentrations are between 100 and 300 mg/l and above 500 mg/l the water becomes unpleasant to drink and is usually unacceptable to consumers. In some communities amounts above this are drunk but only where there is no alternative source of water.

There are two types of water hardness, temporary and permanent. Temporary hardness is removed when the water is boiled; this is the process that leaves deposits of calcium carbonate on water heaters and kettles. It is caused by calcium and magnesium bicarbonates – $Ca(HCO_3)_2$ and $Mg(HCO_3)_2$. They are formed as water containing dissolved carbon dioxide (CO_2), passes over sedimentary rock containing calcium and magnesium carbonate. When heated, the bicarbonate looses the water and CO_2 and the carbonate is precipitated out. Permanent hardness is formed as the cations pass over rocks containing sulphate ions. This produces calcium and magnesium sulphate. In water, this dissolves to form Mg^{2+} and Ca^{2+} ions. Permanent hardness is not removed by boiling and the water needs chemical treatment to reduce it.

Hardness is dependent on the type of rock the water has passed through, for example, if the local rock is limestone or chalk, the water will be hard, if the local rock is millstone grit or granite, or the water has not spent long underground, it will be soft. In England, most water southeast of a line between Scarborough and Lyme Regis is classed as hard to very hard. The rest is a mixture of slightly hard – the Midlands and Lancashire, and soft water – Devon, Cornwall, West Wales, North Yorkshire and Cumbria. In Scotland and Wales, water is usually soft. Soft water areas also naturally coincide with mountainous regions where the geology is mainly hard granitic rock.

Why is hardness and softness a problem in private water supplies? The 'furring' caused by boiling water with temporary hardness will coat the heating elements. This insulates the element, thus raising water heating costs and eventually blocking heating pipes. In addition, some foods cooked in hard water loose colour and flavour. Soft water tends to be more acidic and more aggressive, thus dissolving relatively more chemicals (e.g. iron, aluminium and manganese) from the rocks as it passes through them on the way to the collection point. Some of these chemicals may have health implications, others are aesthetic parameters that may nevertheless cause staining or impart an astringent taste. Soft water will also dissolve copper and lead from plumbing, causing further possible health problems. Eventually the water will damage plumbing installations by causing leaks where the copper or iron has been dissolved. It is considered therefore that slight hardness can give a protective layer of calcium carbonate to metal water pipes.

Hardness is measured in mg/l but this is based on the equivalent amount of calcium carbonate that would cause the same amount of hardness. There are also US measurements of parts per million or a German scale of 'degrees of hardness' (dH). Apparently one dH equals 17 ppm. To add to the confusion there are definitions of different types of hardness. Total hardness is a measure of both temporary and permanent hardness. Carbonate hardness is another term for and measure of temporary hardness. Similarly, non-carbonate hardness is another name for and measure of permanent hardness. Calcium hardness and magnesium hardness are, I hope, self-explanatory. All these terms however are still measured in equivalent amounts of calcium carbonate.

Because hardness is a linear quality there is no distinct separation between levels, but an approximate scale is shown in Table 1. WHO has not set a guideline value for hardness, as drinking hard water does not cause health problems. Indeed, calcium and magnesium are vital trace elements and children for example, should be encouraged to have a diet high in calcium for correct tooth and bone development. One problem that has been reported however is that eczema is sometimes made worse in hard water areas. This is probably because of the additional irritation from the physical act of washing the skin with more soap and less lather than for other, more complicated reasons.

Rather than being a health problem, in various studies dating from 1957 (Zoetman and Brinkman, 1976), hardness has been found to have a protective health effect. This produces a reduction in the number of deaths from cardiovascular disease, with mortality rates falling as hardness increases (Lacey, 1981; Powell *et al.*, 1982). Studies around the world have confirmed this in various towns where both types of water have been supplied and differences in stroke and heart disease mortality have been associated with hardness levels in many UK studies. Gray (1994) points out that there is a general decrease in deaths by 0.8 per cent for every additional 10 mg/l of calcium carbonate, up to 170 mg/l. The reasons for this have not been finally concluded. The calcium may be the reason, but it is found in a great many foods and the amount provided in water is comparatively low. On the other hand, it has been suggested that the amount of magnesium provided in food is only just sufficient for normal needs, so the amount found in water may make a comparatively large contribution. In Ontario,

Table 1 Levels of hardness of water

Water type	Equivalent $CaCO_3$ (mg/l)
Soft	<60–75
Slightly hard	75–150
Hard	150–300
Very hard	>300–350

for example, the water has 40 mg/l of magnesium but the average dietary amount is only 200 mg. Magnesium is important for myocardial (heart muscle) function and health recovery after a heart attack. It may be, therefore, that it is the magnesium in the hardness that is responsible. A recent study in Sweden, however, suggested it was both the magnesium and the calcium that were important protective factors (Rubenowitz *et al.*, 1999).

Conversely, it may be the absence of hardness, that is the softness of the water, which increases the death rate from heart disease. Because softer water tends to contain additional lead, mercury and other trace metals, it may be that it is the presence of these chemicals that causes the problem. One study found that in two areas of the UK (Glasgow and London) with different levels of water hardness, heart attack rates were similar but, after taking other factors into account, the death rates were significantly different (Crawford and Crawford, 1967). It appears therefore, that rather than causing heart attacks, one of the metals found in soft water may reduce the ability of the body to recover from them.

Water softeners are ion exchange units and they often replace the calcium and magnesium ions in hard water with sodium ions. Sodium is not recommended for several medical conditions and people are often put on a low salt diet to counteract these problems. It would be very unwise therefore to re-introduce sodium through drinking-water treatment. Equally important, badly maintained water softeners can build up microorganisms within the medium.

Some people will nevertheless want to soften their water for a variety of reasons. These include reducing damage to heating elements, saving children with eczema from unnecessary skin problems, reducing soap usage and preventing the scum forming properties of hard water on their sanitary ware. It is import that if this is done, the water for drinking and cooking should not be softened.

Hydrogen ion or pH

UK maximum concentration allowed – 10 pH value.
UK minimum concentration allowed – 6.5 pH value.
WHO Guideline Value: None set, although skin problems occur at levels greater than pH 11.

The pH scale is a measure of a water's acidity. It runs from 1 to 14 and the lower the pH, the more acid the water is. Water with a pH of 3 or less is highly acidic and with a pH of 10 or above, very basic (you may know this property as alkalinity, i.e. that the water is alkaline, however these are not strictly interchangeable terms and alkalinity or buffering is explained in more detail later). 'Pure' water has a pH of 7; this is classed as being neutral. Rainwater is usually slightly acidic with a pH of between 5.2 and 5.5. This is because the carbon dioxide that has dissolved in it from the atmosphere has formed carbonic acid. Pure water in equilibrium with atmospheric carbon dioxide has a pH of about 5.6.

If the pH of water from a small supply is above 7 this will be due to the geological formations it has passed through, probably limestone or chalk. It also indicates that it will probably be hard water and therefore will contain calcium or magnesium. If the pH is low, this will tell you that the water may not have passed through much subsurface structure at all or has travelled through rock strata such as millstone grit or sandstone. It will be soft water and therefore may be aggressive to plumbing installations. As it passed through the geology of the catchment, acidic water will have leached metal ions from the rock and soil. It will also tend to contain more potentially dangerous chemical salts than neutral or basic water. Although pH in the ranges normally encountered does not, in itself, mean that the water is dangerous, it can affect health in several ways. For all these reasons, knowing its pH can tell you a lot about the water.

Where does the pH scale come from? It is based on the negative logarithm of the number of hydrogen ions in the water. In water, H_2O molecules are unstable and dissociate or separate into hydrogen ions (H^+) and hydroxide ions (OH^-). In actual fact, the H^+ ions are hydronium (H_3O^{2+}) ions, but for simplicity, they are represented as H^+ ions. In any sample of water, the water molecules are constantly separating (H_2O becoming H^+ and OH^-) while pairs of hydrogen/hydroxide ions are always combining (H^+ and OH^- returning to H_2O). In 'neutral' solutions of water, the relative rates of these two reactions is equal and the concentration of hydrogen ions and hydroxide ions is equal. These ions represent a very small proportion of the total amount of water – approximately 1.008 g of hydrogen ions and 17.008 g of hydroxide ions per litre. The concentration of these two ions is therefore about 1×10^{-7} moles per litre each (a mole is a standard chemical measure and is defined as the amount of a substance that contains as many units as there are atoms in 0.012 kg of carbon). It is this $^{-7}$ that is the negative logarithm that becomes the pH of 7. It is therefore the number of hydrogen ions in the water that decides the pH. As the water gets more acidic, the number of H^+ ions increases from 10^{-7} to 10^{-6} and onwards to 10^{-1}, that is pH 7 to pH 6 down to pH 1. It is technically possible to alter the balance between the two ions so that the pH is less than one but this is a very acidic substance and will not be met outside very specialized laboratory conditions. Just for good measure, there is also a pOH scale. This is for hydroxide ions and as the pH goes up, the pOH goes down. The temperature of the water will also affect the pH/pOH equilibrium, so standard measurements are taken at 25°C.

The more H^+ ions in the water, the more reactive it is. This is because metal ions are in competition with the hydrogen for binding sites on the molecules of various substances the water comes into contact with; as the H^+ concentration goes up, the hydrogen will take these binding sites and force the metal ions into the water. A similar effect takes place as the water gets more basic and the OH ions start to become more reactive. Bleach is very basic for instance and will destroy biological tissue as well as reacting with chemicals such as colour dyes. Because the pH scale is logarithmic, a reduction of one pH means that there are ten times more hydrogen ions in the water; a reduction of two will mean a hundred-fold increase. The pH of water changes depending on the substances dissolved in it, SO_2 and CO_2, for example, are easily dissolved in water and will therefore upset the H^+ and OH^- balance. CO_2 is nearly 30 times as soluble in water as oxygen, so although there is only a relatively small amount in the atmosphere it will have a proportionally larger effect.

Table 2 outlines the approximate pH of some common liquids. If water is very acidic or basic it will damage eyes, skin, mucous membranes and other soft tissue. Problems have generally been found to occur above pH 10 and below pH 4. Below pH 2.5 skin damage is severe. Another potential health problem is that pH will alter the ability of chlorine to disinfect water. As pH also affects flocculation and many other processes, it is carefully controlled throughout large-scale water treatment. WHO has previously set guidelines for pH based on aesthetic quality; in 1984 it recommended acceptable levels of between 6.5 and 8.5 (WHO, 1984). The latest guidelines however set no absolute figures as WHO is moving away from setting suggested levels for non-health related parameters. The guidelines merely state that the optimum range is between pH 6.5 and 9.5 (WHO, 1996a). The UK limit is presently between pH 6.5 and 10.

A concept that sometimes causes difficulties is 'buffering'. This is the property of water or soil to resist changes in acidity. Soils that are rich and loamy have a buffering component, as do some waters, depending on the dissolved chemicals in them. Buffering basically means the ability to keep the pH stable as acid is added. In water, this property is also called alkalinity. There will be a variety of substances in water that affect it in such a way that adding set amounts of acid to the water will not result in a steady change in its pH. If the water has sufficient buffering capacity, it can absorb and neutralize the added acid without significantly changing the pH.

Table 2 pH of various common liquids

Liquid	pH
Battery acid	1
Lemon juice	2
Coca cola	2.5
Vinegar	3
Wine	3.5
Milk	6.5
Distilled water	7
Sea water	8
Bleach	13

Imagine that a buffer is a bit like a sponge. As more acid is added, the 'sponge' absorbs it. Its capacity is limited however and once it has been used up, the pH changes rapidly when more acid is added.

Alkalinity is measured in terms of an equivalent amount of $CaCO_3$ buffering capacity. This is because there are many different chemicals dissolved in the water that affect alkalinity and consequently the neutralization of the hydrogen ions, but they all have the same effect. The two most important bases contributing to alkalinity are the carbonate ion (CO_3^{2-}) and the bicarbonate ion (HCO_3^-). These are also the ones normally associated with calcium in hard water and this is why hard water is usually alkaline. As with pH and hardness, the geology of a catchment and its soil usually determines alkalinity. Limestone or chalk areas tend to produce alkaline (as well as basic) waters. Areas where the rock is igneous tend to have water with little buffering capacity. Those with well-leached soils, such as high moorland areas, also have waters with little capacity to resist pH changes. Water with high alkalinity (greater than 100 mg-equivalent $CaCO_3$/l) tends to have pH values above 8, while low alkalinity (less than 50 mg equivalent $CaCO_3$/l) waters are usually acidic.

Turbidity

UK maximum concentration allowed: 1 NTU.
WHO Guideline Value: None set – complaints will be expected at 5 NTU.

Turbidity is simply the cloudiness of water and consists of suspended, un-dissolved organic and inorganic particulate matter. The turbidity of the water is dependent on source water quality and the effectiveness of any treatment or storage. It can consist of silt, dead organic matter from animals and plants or living organisms such as algae. In drinking-water from groundwater sources, the turbidity is mainly caused by inorganic silt particles. These are also known as suspended solids. If the water is allowed to stand for a while, the turbidity will often settle out and become sediment. Turbidity is measured in nephelometric turbidity units (NTUs). Despite the long name, 'nephelo' is just the scientific word for cloudiness, so nephelometric merely means the measure of cloudiness. The term FTU (formazin turbidity units) is also used as a measure of turbidity and equates to NTUs. Laboratories can also use particle counters to give a more sophisticated, scientific approach to this basic measure.

Turbidity has been recognized as an important measure of water quality for many years. In 1855, the distinguished scientist Michael Faraday invented a simple measure that used a piece of white card and a measuring stick. When the card was held underwater, the shorter the distance under the surface before it disappeared from view, the dirtier the water was. He famously used this to show how contaminated the River Thames was. An equally simple measure of turbidity is sometimes still used today. All that is needed is a thin black-cross painted on the bottom of a tall measuring jar, which is then slowly filled with water. The sooner the cross disappears from sight, the more turbid the water is. Initially, if waters of known turbidity are poured into the measuring vessel, marks can be made on the outside, at the depth at which the cross disappears. Using this scale, readings can be taken for water where the turbidity is unknown.

WHO has not set a health-based guideline for turbidity, so is it important? The answer is yes and the reason is fairly obvious – the cloudier the water, the more likely it is to contain unwanted chemical and microbiological contaminants and if water is normally turbid it is probably easily contaminated. Particulate matter will also protect microbiological contaminants during disinfection processes; they will for instance, shade bacteria from ultraviolet light. Chlorine is also less effective in cloudy water and there are two reasons for this. The particles use up some of the chlorine as they react with it themselves and their rough surfaces give some physical protection to microorganisms. Water is therefore normally required to have less than 5 NTU for disinfection to be effective. WHO has not set a health-based limit because turbidity, like pH, is not necessarily harmful itself. It has however recommended that filtered water should have an average of 1 NTU with a maximum of 5 NTU in any one sample. This is usually the level at which turbidity will also be noticed in a glass of water. The European Union requires member states to have regulations limiting turbidity to a maximum of 4 NTU.

The other problem with turbidity is aesthetic. People do not like drinking cloudy water, even if it is safe to drink. It looks unpleasant and taste can be adversely affected. Because of these problems, where an official drinking source contains safe but turbid water, people will naturally be very suspicious of it and often start to use a clearer but less safe supply. Perhaps the most famous example of this was the pump in Broad Street, in the Soho area of London, during the cholera epidemic of 1853. Some people, who subsequently became ill, took water from the contaminated Broad Street well because the water looked cleaner and tasted better than other wells nearer their homes (Snow, 1854). Unless there is very little choice, people will not use small water systems with turbid water. Even if the supply has been used for a long time before becoming turbid, it may quickly be abandoned. Of course, in many developing countries they do not have this choice.

A final problem, particularly for river water, is that very turbid water will quickly block physical filters such as reverse osmosis membranes or slow sand filters. At times the Yellow River in China, for example, is so loaded with silt that it cannot be used as drinking-water at all. To counter this, river-water often has to be stored in reservoirs by the side of the river (bankside storage) to give it time to settle out and allow normal treatment processes to take place.

As with many other problems in an inadequately protected small water supply, turbidity is more likely to be present following heavy rain or snowmelt. Thus, if turbidity suddenly goes up, this will indicate that unwanted contaminants may be present and the water should not be drunk without being boiled. As is well known however, just because water looks pure, it is no guarantee of its safety.

Insufficiency

As well as problems with physical or chemical contamination, during the summer months private supplies can occasionally run dry or have their flow severely restricted. This is particularly so when spring supplies do not tap into large aquifers. When new properties are built in rural hamlets they often connect to existing supplies, thus lowering the amount available to everyone else. In some instances, this can cause problems between neighbours, particularly when newcomers are used to unlimited amounts of tap water for watering the garden, cleaning the car, etc. When supplies run dry there are several alternatives available, short-term measures, such as bowsers or bottled water can be provided, voluntary restrictions can conserve supplies or an alternative supply located and used. If the existing supply is a borehole, it may be possible to enlarge or deepen it. Some rural properties have two water supplies, the main one where the water is of good quality but unreliable and a second one where the water does not taste so good but does not run out in the summer. Where problems are regular or severe, enforcing authorities should consider whether to take formal action to improve the supply. These powers are available in the UK under section 80 of the Water Industries Act 1991 (HMSO, 1991a).

4   Risk assessment

Some things seem to become a fashionable worry, creating hectares of newsprint, whilst others receive virtually no attention. Compare the interest in genetically modified food or mobile phone masts with that of lead in drinking-water. There has never been a recorded case of human illness with the first two while lead has been poisoning people for over two thousand years. Unfortunately, the potential for a substance to cause illness is not directly proportional to the attention it attracts from the public, the media or politicians. For various reasons, certain contaminants in drinking-water attract more than their fair share of interest and if you are involved in drinking-water safety and have to explain health-related information to people, you should have an understanding of the issues that motivate them and how this affects their perception of potential problems. It will also shape your approach if you get involved in encouraging communities to improve their drinking-water supply. Whether or not something is a risk to their health, people will have perceptions regarding it that affect what, if anything, they will do about it. These perceptions will also determine the attention it is given by legislators and influence the amount of money spent on quantifying the 'problem' and reducing any risk.

All activities and all substances carry some risk. The box on page 54 is an example whereby an ordinary substance like water can be made to seem potentially much more dangerous than it is by overstating the risks and praying on people's mistrust of all things chemical and scientific, particularly if it is associated with big faceless corporations.

Many things are considered to be hazardous to health. Although it is possible to lessen it, this risk can never be reduced to zero (Morgan, 1992). Paracelsus, a very perceptive fifteenth century physician and writer is quoted as stating that 'everything is poison, nothing is poison, it is the dose that makes the poison'. It is therefore important to assess the amount of risk associated with a particular substance and compare it with other hazards, both self-imposed and unavoidable, faced by human beings and the environment. One obvious consideration is the danger involved in not using the substance. It has been argued for example, that the use of certain fungicides is safer than non-application, due to the subsequent risks associated with

DiHydrogen monoxide is colourless, odourless, tasteless, and kills uncounted thousands of people every year. Most of these deaths are caused by accidental inhalation of DHMO, but the dangers of DiHydrogen monoxide do not end there. Prolonged exposure to its solid form causes severe tissue damage. Symptoms of DHMO ingestion can include excessive sweating and urination and possibly a bloated feeling, nausea, vomiting, and body electrolyte imbalance. For those who have become dependent, DHMO withdrawal means certain death.

DiHydrogen monoxide

- is also known as Hydric acid, and is a major component of acid rain;
- contributes to the greenhouse effect;
- may cause severe burns;
- contributes to the erosion of our natural landscape;
- accelerates corrosion of many metals;
- may cause electrical failures, and decreased effectiveness of automobile braking systems;
- has been found in excised tumours from terminal cancer patients.

Contamination is reaching epidemic proportions

Quantities of DiHydrogen monoxide have been found in almost every stream, lake, river and reservoir in Europe today. But the pollution appears to be global; traces of the contaminant have even been found in Antarctic ice. On the east coast of the UK alone, DHMO has caused millions of pounds of property damage.

Despite the evident dangers, DiHydrogen monoxide is often used

- as an industrial solvent and coolant;
- in nuclear power plants;
- in the production of polystyrene and other packaging materials;
- as a fire retardant;
- in agriculture and research into domestic animals;
- in the distribution of pesticides. Even after washing, produce remains contaminated by this chemical;
- as an additive in certain 'junk foods' and other food products;
- excessive amounts of this substance are used in the brewing and distilling industries, both in manufacture and distribution.

Companies dump waste DHMO into rivers and the oceans, and nothing can be done to stop them, as this practice is still legal. The impact on wildlife is extreme and cannot be ignored any longer.

The horror must be stopped!

increased mould growth on stored grain. This is because some of the moulds can produce carcinogenic (cancer-causing) substances known as mycotoxins. Grain is not produced by farmers in small quantities and used 'fresh'; it is harvested once or twice a year and must be stored. The risk to health, assessed from the number of stomach cancers arising from the growth of natural mycotoxins during storage, is much higher than the probable risk to health from the minute amount of fungicide residue in the stored grain (Lee, 1992).

Perception of potential risks has been the subject of much research. A study in Oregon has shown that while people generally have a good idea about which causes of death are more numerous than others, there is a tendency to over-estimate rare causes of death and under-estimate common ones. The highest over-estimates are found in the most dramatic events (Lichtenstein *et al.*, 1978). Many contaminants of water will fall into this low-risk, over-estimated category. To illustrate this, there are far fewer deaths due to pesticide usage in the UK than are caused by bulls (Lee, 1992) and the number of non-fatal accidents involving hospital admittance are less for pesticides than those involving flowerpots (Morgan, 1992). Nevertheless, the public's risk perceptions produce much more vociferous campaigns to reduce the use of pesticides than those for fewer bulls or flowerpots.

To the average human being, perception of risk is dependent on several separate considerations. These include whether the risk is a new one, rather than something that people have become used to. The perceived risk to health from nuclear rather than conventional power generation is an example. Both carry risks to the health of the workers involved in the industry and to the public. These dangers include nuclear accidents and carcinogenic chemicals produced by the combustion of fossil fuels. Public opinion is that nuclear power generation is the more dangerous in both the short and long term. This is particularly untrue in the short term with deaths from coal-powered stations occurring at over ten times the rate of those in nuclear establishments (Morgan, 1990). Over a longer term, the risks appear to be very similar. The public attitude however, may well significantly alter future policy on power generation. Judicious questioning can often upset widely held beliefs. Many people would claim that they want a 'natural environment' whereas what they may really want is carefully controlled countryside as seen in the UK. Many parts of Europe were rife with 'natural' malaria-bearing mosquitoes and impenetrable forests hundreds of years ago but not many people would want a return to that. Although an abundance of wildlife is seen as being a good thing, surveys have revealed that a majority of people want bees but not wasps, butterflies but not moths, ladybirds but not flies and buttercups but not nettles (Whitehead, 1992).

Another preconception involves the difference between inevitable risks and those that are accepted voluntarily. An early study showed that people were prepared to accept risks from a voluntary activity such as smoking that were roughly one thousand times those that they would agree to when the risk was unavoidable (Starr, 1969; Morgan, 1990). An illustration of this concerns substances called polynuclear aromatic hydrocarbons that are occasional contaminants of drinking-water. Some have been found to cause cancer in laboratory animals and WHO has produced guideline values for several of them. They are included in European water quality legislation as parameters considered injurious to health, with strict controls on the amounts allowed to be drunk. Polynuclear aromatic hydrocarbons, however, are found in many foods, especially those that have been burnt during the cooking process. They are also found in fresh vegetables, vegetable oils and at relatively high concentrations in smoked meat. The call for these products to be removed from the food market, where the risk is entered into voluntarily, is much smaller than the pressure to ensure that they are not present in drinking-water, where there is no choice over whether they are consumed or not.

The following chemicals are all naturally found in food and have produced various ill effects on laboratory animals when they were fed them in suitably large amounts. We have been eating them for centuries but because they are natural there is no call for any of these foods to be banned on health grounds.

- Acetaldehyde (apples, bread, coffee, tomatoes) – mutagen and potent rodent carcinogen
- Aflatoxin (nuts) – mutagen and potent rodent carcinogen; also a human carcinogen
- Aniline (carrots) – rodent carcinogen
- Benzaldehyde (apples, coffee, tomatoes) – rodent carcinogen
- Benzene (butter, coffee, roast beef) – rodent carcinogen
- Benzo(a)pyrene (bread, coffee, pumpkin pie, rolls, tea) – mutagen and rodent carcinogen
- Benzofuran (coffee) – rodent carcinogen
- Caffeic acid (apples, carrots, celery, cherry tomatoes, coffee, grapes, lettuce, mangoes, pears, potatoes) – rodent carcinogen
- Catechol (coffee) – rodent carcinogen
- Estragole (apples, basil) – rodent carcinogen
- Ethyl alcohol (bread, red wine, rolls) – rodent and human carcinogen
- Ethyl benzene (coffee) – rodent carcinogen
- Ethyl carbamate (bread, rolls, red wine) – mutagen and rodent carcinogen
- Furan and furan derivatives (bread, onions, celery, mushrooms, sweet potatoes, rolls, cranberry sauce, coffee) – many are mutagens

- Heterocyclic amines (roast beef, turkey) – mutagens and rodent carcinogens
- Hydrazines (mushrooms) – mutagens and rodent carcinogens
- Hydrogen peroxide (coffee, tomatoes) – mutagen and rodent carcinogen
- Hydroquinone (coffee) – rodent carcinogen
- Methyl eugenol (basil, cinnamon and nutmeg in apple and pumpkin pies) – rodent carcinogen
- Quercetin glycosides (apples, onions, tea, tomatoes) – mutagens and rodent carcinogens

No human diet can be free of naturally occurring chemicals that are rodent carcinogens. Of the chemicals that people eat, 99.99 per cent are natural.

(Bruce Ames, PhD and Lois Swirsky Gold, PhD, University of California)

Thank-you to the American Council on Science and Health for the information (ACSH, 2002).

Other factors that affect the perception of risk are as given here.

- The extent to which the risk is seen as likely to cause a catastrophe;
- The extent to which the risk seems under control;
- Whether the risk is viewed as a potential for reduction in gains or an increase in loss. When some doctors were asked in an experiment if they would prefer treatment that gave a 10 per cent mortality rate or a 90 per cent survival rate, the majority preference was for the latter (Morgan, 1992). Both statements, of course, mean the same thing; and
- The set of mental processes that people go through to make sense of complicated issues. These are known as 'heuristics'. They are automatic problem-solving procedures, using simple but necessary thought processes. One heuristic is to weigh risk based on the 'availability' of information. This causes easily recalled risks to be perceived as more dangerous.

All these influences weigh against many of the 'contaminants' of water, even where the actual risk is low. This will cause attention, political pressure and money to be directed towards their control and away from other, more deserving cases. For many years, water experts and chemical manufacturers have railed against these perceptions and the attentions of 'crusading' environmental organizations, claiming that they are unfair and unscientific.

They are now coming to the realization that they have to live with this concern and must alter the way they inform the public about risk. It is pointless explaining in great detail to a group of homesteaders about *Cryptosporidium* and *Escherichia coli* O157: H7, if their eyes glaze over and you are only being seen as a bureaucrat trying to baffle them – or even worse, getting them to spend money. You need to say that they are drinking water with sheep faeces in. In fact, you may be even better off using a simpler, Anglo-Saxon word for faeces. If the problem is one of getting a group of middle-class environmentalists to install chlorination, they may be more worried about chlorination by-products. A simple explanation of the benefits of chlorination, the safety factors applied and the comparative risks of other everyday chemicals, may be better than threatening them with formal action. You could also tell them about the sheep faeces in their water as well, just in case.

It has to be said that no scientific study can ever prove something is safe. The example often used is sometimes known as the 'black swan argument'. Scientists can study swans and note that they are always white, year after year. They can never prove that all swans are white however, because at some point in the future a black swan may swim past and all previous conclusions become void. This is not an argument for or against banning certain chemicals (or swans for that matter) but to put in perspective demands for proofs of safety by politicians and the media, when scientists are reporting the results of health studies. The recent controversy over MMR (measles, mumps and rubella) vaccinations is a case in point. No matter how often research shows that the vaccination is the safest option, because no one will say it is absolutely safe (as that is impossible), other people will claim that it means there is still doubt. The result is a rise in the incidence of the dangerous diseases which the vaccinations are designed to prevent.

The MMR debate also highlights another interesting aspect of human perception. The question in people's minds is whether there is a connection between autism and the vaccination. Symptoms of autism often start to develop at about the same time the jab is given. A distraught parent will therefore associate the inoculation with the initial signs of illness. It is understandable for the parent of a child who becomes seriously ill to make these connections. To them, the two events will always be connected, no matter how much evidence to the contrary is presented. They may be totally unconnected however; if I drink a single cup of tea on a particular morning instead of the usual two and then have a bad day at work, it does not automatically mean that a reduction in tea consumption led to the high stress levels.

When reporting on controversies where the majority of informed opinion considers that something is reasonably safe, the press will search out people who believe it is dangerous, so that they will appear to be putting forward a balanced report. Although it may seem that this style of reporting is even-handed, it is disingenuous. Because equal weight is given to both sides, no matter what differences there are in the amount of scientific support, it legitimizes the more tenuous viewpoint.

A final reason why public opinion appears to be being manipulated to consider certain substances in water to be of greater concern than strict scientific opinion would expect is the probability of data being published in scientific journals. Data that show something is dangerous are more likely to be published than data showing that there is nothing to worry about, with a series of negative results to prove it. Consider an experiment designed to test whether football boots are carcinogenic when fed to rats. If the results were positive, it is fairly likely that the paper will be written up and the shocking news splashed across the front pages of the tabloid press. If the experiment showed no connection between football boots and cancer would the researchers write the experiment up and forward it to scientific journals for publication? Most respectable scientific journals have many authors wanting to be published and editors may not wish to fill their journals with a series of papers explaining that a wide variety of substances are of no health significance whatsoever. Bad news sells newspapers and scientific journals alike.

Although private water supplies appear to be an inherently greater risk to the health of the individual consumer, than mains water there does not appear to be the equivalent amount of attention paid to them by the government, pressure groups or the media. When the first European Drinking-water Directive (EEC, 1980) was passed, it was eleven years before regulations were brought in for private water supplies. Pressure groups are continually expressing the view that public drinking-water supplies in the UK are unsafe and becoming more dangerous. In 'Britain's Poisoned Water' (Craig and Craig, 1989), the authors state that their biggest problem in writing the book had been the potential for the over-use of words like disgraceful, scandalous, dangerous and squalid. It would appear therefore that they could not possibly find the language to hope to describe the problems of private supplies. However, there is not a single comparable publication addressing the issue of private supplies. One of the reasons may be that those responsible for the control of private supplies are usually the same people who drink the water (Glicker, 1992). There may be an additional mental risk-reduction if expensive improvements are required.

Another reason for the difference between the uproar over scares about the contamination of mains water and the silence over private supplies is the unknown nature of the risks from public supplies. The problems attracting attention at the moment include potential carcinogens such as pesticides and disinfection-by-products. Epidemiological and medical proof concerning risks from these substances is scarce. It appears much easier to attract attention to these unknown, but potentially widespread lethal effects, than the well-documented, but less attractive aspects of gastrointestinal diseases. As Packham (1990) has wisely pointed out, 'From a global point of view our priorities may seem rather strange, and a visiting Martian could be forgiven for asking why it is that on one part of the planet so much effort is devoted to the elimination of an unproven link between water quality and cancer, while elsewhere tens of thousands die daily from eradicable disease.'

Certainly, some research is pointing towards problems in the developed world caused by essential trace elements being lost from our diet due to an obsession with purity (Anderson, 1992). Illnesses amongst some children in Australia have been found to be related to inadequate levels of iodine, iron and chromium in their diet. These trace elements had previously been provided by minor contamination of their food and water.

Rainfall and weather

When you study the outbreaks of waterborne disease from small or private supplies, in the overwhelming majority of cases they were preceded by heavy rain. Of course there still needs to be a source of contamination and someone has to drink the water, but it is the rain (or snowmelt) that is perhaps the single-most common factor affecting the quality of the water. It is for this reason that it is vital to take the weather into consideration when sampling from small water supplies and when interpreting the results of that sampling. The effect of the weather is so important that unless the sampling of springs and wells is done immediately following heavy rain, it will normally be pointless.

A recent study in the US (Curriero *et al.*, 2001) found that over half of all waterborne disease outbreaks occurred after the heaviest 10 per cent of precipitation and 68 per cent followed the heaviest 20 per cent. Outbreaks associated with surface supplies were also statistically linked to heavy rainfall during the previous month. Interestingly, water from groundwater supplies was also linked to heavy rainfall but there was a time lag of two months between the precipitation and the subsequent illness. This study indicated that not only are the failure rates of microbiological testing of water supplies raised after rain, but they are sporadically followed by illness. This linking of poor quality sampling results with actual illness is very rare. Although the study looked at all types of supply, it should be noted that smaller systems are much more likely to have insufficient protective systems to enable them to cope with the effects of heavy rain.

The weather affects the quality of water in small supplies in several different but interdependent ways. One is the simple effect of heavy rain washing animal droppings, soil and detritus into a badly designed or insufficiently protected spring, well or borehole. Another is that heavy rain will flush microorganisms through the soil and into the water supply more quickly. The faster the velocity of the water through the subsurface infrastructure, the less time there will be for microbiological die-off. Where ponding occurs on the top of any spring box or storage chamber, it will find its way in through any hole, no matter how small and wash whatever was also on the top of the chamber in with it (such as bird or animal droppings).

Lighter rain will often only wet the surface of the soil and thus have a much lesser effect.

When the rain falls is also important. In the summer, after a prolonged dry spell, there will be many more animal droppings lying about in the fields ready to be washed into a water supply. As the autumn progresses, the majority of droppings that could have some of their contents washed into the supply will have; so only fresh cow pats will take part in this process. In the winter, farmers tend to keep animals inside, so there is less excrement on the land to contaminate the water. As the spring arrives, animals will return to the fields and moors, increasing the amount once again. In addition, in the spring, because there are many younger animals about, the *Cryptosporidium* content of the dropping will go up. This is because *Cryptosporidium* 'favours' young creatures over older ones, both farm-animals and humans. The figures on *Cryptosporidium* incidence show a rise in the spring.

Another factor is the physical condition of the soil. Following a prolonged dry spell, the soil has a low moisture content and will, to some extent, be hydrophobic – that is, it repels water. This effect can be seen when you try to water a very dry plant. The water tends to stand on the surface and only soaks in slowly. Once it is moist however, much more water can be poured into the pot and be absorbed by the soil. This is essentially what happens with rainwater. At first only a small portion is absorbed by the soil and, if there is a heavy downpour, most of it will tend to run over the surface, downhill to a stream or easily polluted water supply, washing any detritus with it. After a further period of rain the soil becomes damp, water is absorbed much more easily and it will be held within the soil structure. If there is further rain, it will start to move vertically downwards, until it reaches an aquifer.

As more rain falls, the water table will rise. As the saturated zone nears the surface, the water can no longer just flow vertically downwards. It starts to run at a more horizontal angle, following the surface contours of the land. The effect of this is that rainwater can enter a water supply from further away, rather from vertically above it and because the water has travelled for longer in the top portion of the earth, it will have travelled for a greater time in the soil layer and thus been in contact with the small soil particles for longer. Consequently it will have a larger amount of dissolved chemicals in it. I am obviously simplifying what actually happens to explain this point and in reality the way water travels through the ground is a much more complex affair, but the general principle can be appreciated.

If the rain continues to fall, the whole soil and rock structure will be in the saturated zone and the fresh rainwater will merely run over the surface of the ground. This increase in overland flow may cause a greater scouring effect on surface detritus and excreta. In conditions like these, many sampling officers do not consider it appropriate to sample private water supplies, it being far too wet to travel to remote farms and tramp about sodden fields.

Other meteorological conditions will have much less of an effect than rain. Very dry weather will reduce contamination, particularly by microorganisms, because there is no water to transport them to the water supply. A similar effect will be noted in the winter if the ground freezes or there are significant amounts of snow. Once again the water will not be moving. As with rainfall following a dry period, once the snow melts and the ground thaws, polluting material can be transferred into the supply, with a consequential reduction in water quality.

The effects of the weather on water quality will typically be most immediate and severe on those supplies that use surface water or are close to the surface, are fed via faults in rocks or have inadequate protection mechanisms. Other supplies, particularly correctly drilled deep boreholes, will either not be affected by the weather or any changes to water quality will be reduced by the filtering effect of the aquifer. It is obviously much better to have one of these well-protected supplies, where the water is reasonably consistent, than one where quality is compromised every time it rains. But, if the supply being investigated is one that is adversely affected by rainfall or snowmelt, what can be done? One option is to dispense with it and have the property connected to the public supply. Unfortunately this is not often possible, as those supplies that can be easily connected to the mains usually are, and the remaining ones are located too far away to make a connection economically viable. This is not always the case however and it is possible to come across properties with their own supply, living next door to those on a public supply. This situation has usually arisen due to personal preference or an inability to appreciate the potential benefits of connection. Another option is to carry out a thorough sanitary survey to identify the reasons why the supply is being contaminated and if possible, put them right. Alternatively it may be worth considering drilling a new borehole. Although this has an initially expensive outlay, once it has been safely installed, the water will normally be of an acceptable and consistent quality. In addition, there are no regular expenses from spare parts (except disinfection equipment) and unlike many treatment systems, if the pumping mechanism of a borehole fails, it does not fail to danger. If all these options do not appeal, a competent contractor who can guarantee that it will consistently provide safe water should install a water treatment system.

Sampling programmes

There are two things wrong with most private or small water supply sampling programmes. The first is that they do not always tell you what you want to know; the second is that what they do tell you is often misinterpreted. So, what is it that you want to know if you sample water from a small supply? The answer is – will the water damage human health, either in the short or long term? Over and above ascertaining whether a water supply complies with national legislation, the primary purpose of sampling is to quantify risk by measuring the levels of various contaminants in the water. However, in order to find out whether the water will be detrimental to human health, it should be sampled when the supply is likely to be at its worst and for the parameters that are most likely to indicate the presence of pathogens or dangerous chemicals. It is pointless for instance, knowing that water from a spring supply sampled in August was pathogen-free, if the next time it rains it will be awash with *Cryptosporidium* and *Campylobacter*. Sampling programmes should therefore always take into account the known effects of rainfall.

Unfortunately, authorities in most of the UK (and many other countries) ignore this and the sampling of private water supplies remains essentially random. This approach produces a general microbiological failure rate of about 30 per cent in the UK (Rutter *et al.*, 2000). Correctly organized sampling programmes, ones that take account of the effect of precipitation on water quality, are more likely to have failure rates of 70 or 80 per cent (Humphrey and Cruickshank, 1985; Barraclough *et al.*, 1988; Barrell, 1989). This research shows that about 50 per cent of the negative results in a normal sampling programme (that is the 'satisfactory' ones where no coliforms have been found in 100 ml of water) are likely to be unreliable (false negatives). The real situation is even worse because where a sample of water fails, at least one more additional sample will be taken. The one-third of samples that are presently found to fail the standard will therefore contain many repeat tests so the actual number of *supplies* providing the false negatives will be more than 30 per cent of the total.

Therefore, it seems that at least half of the results that say the supply is satisfactory would, if the sample were taken at the correct time, prove to be unsatisfactory. For *Cryptosporidium* this may be even more of a problem, it has been calculated that nine out of every ten 'spot samples' will underestimate the oocyst density and thus the risk of infection (Gale, 2001). Because private water supply monitoring programmes normally require long periods between samples, a supply could be officially classed as satisfactory for several years whilst in reality it is frequently contaminated. Some may argue that if there is little better than a fifty–fifty chance of the sampling programme finding that a water supply is safe, perhaps you may as well not bother and toss a coin instead.

The problem with the interpretation of sampling data, particularly microbiological results, is due to the varying nature of the water from a supply. There is a belief occasionally encountered (thankfully rarely nowadays) that water quality in private supplies is a uniform entity, that is, the 100 ml that has been sampled is somehow representative of the whole water supply, not only throughout the system but also throughout the year. Unfortunately it is not, it is representative of that 100 ml of water, nothing else. If you train people in simple microbiological laboratory techniques, you occasionally observe their amazement that one water sample, if separated into several different aliquots, can give a variety of results. Each separate sub-container of water has different amounts of indicator bacteria in them and sometimes these differences are quite substantial. If you amplify this concept to consider small water systems as a whole, the problem becomes much worse. The quality of water in the vast majority of private water supplies change from hour to hour, day to day. Different results will be obtained when sampling at the source, the inlet, the outlet, at the tap and within a storage tank. Results will even differ if the sample is taken from assorted points within the storage tank.

If the laboratory result says that a random sample from a private water supply contains zero faecal coliforms per 100 ml, does that mean the water is safe to drink? Of course not, it may mean that you merely missed the particles with the bacteria on when you sampled the water and if you had taken another sample, somewhere else in the system or a few minutes later, it would have failed the standard.

If this information is so obvious why don't all authorities sample within a day or two of heavy rain and incorporate this into their sampling policies? The reason is that most laboratories are not solely dependent on one authority or small water supply sampling programme for their income. Consequently, they give a set day or number of days to each authority for its water-sampling programme. For authorities with large numbers of private water supplies, this will be every day and these authorities will normally only appoint sufficient staff to carry out this daily programme, with all the sampling evenly spread across the year. They will not have a flexible workforce, easily moved on to sampling when the weather is bad or several members of staff waiting around for periods of heavy rain. In smaller authorities, there may only be sufficient small water supplies to provide enough samples to fill one or two days a week. The laboratory can then arrange other specific sampling days with the other authorities in its area so that the laboratory's technicians are working at a constant rate. This is a sensible arrangement for the laboratory manager, but not for a health official who wants to know whether a water supply is likely to contain pathogenic bacteria at some time during the year.

Laboratories prefer water samples to be delivered as quickly as possible to reflect the condition of the water as it was at the source. They also like them delivered early, so that all the necessary work can be done on the sample before the laboratory technicians go home for the day. In the US some utilities are allowed to post their samples in, but in the UK, sampling is usually carried out by an environmental health officer or technician going out early in the morning, travelling some distance to remotely sited small water systems and returning to the office via the laboratory in the early afternoon on the same day – week in, week out. This is not designed to identify poor quality water supplies. Luckily, a few of the more enlightened laboratories and local authorities are beginning to organize more flexible sampling arrangements. However, there are many more authorities that are not flexible, and the time, effort and cost of sampling to the authority (and the person whose water has been sampled) is wasted in a lot of cases.

A week or so later the results are received from the laboratory and checked to see if any of the parameters that have been found in the water fail any of the standards laid down in the legislation. If there is a treatment system on the supply, a failure will at least show that it needs repair or replacement. In a few cases, re-sampling takes place without any investigation of potential problems. If the re-sample is satisfactory, it is sometimes assumed that the problem must have sorted itself out and the water supply is thus declared fit to drink, perhaps for the next five to ten years. A local authority once contacted me to ask how many times a supply must fail the water quality requirements before formal action should be taken. I pointed out that if the water had been found to contain faecal organisms, even once, then it had contained the contents of something or someone's bottom – and that was enough for me to expect some action to be taken to improve the supply, informally at first maybe, but eventually formally.

If there is a bacteriological failure of a water supply and a re-visit is made to re-sample the water before improvements are made, what are the possible results? There are two obvious alternatives; one is that the second test shows that the water is still unsatisfactory. Does this help? There is an argument that, if you intend to take formal enforcement action, such as serving a statutory notice to improve a private water supply, you should be certain of your facts and two or more positive samples for faecal coliforms would help prove that a notice is justified. If it were appealed against, the Secretary of State (who is the person to whom appeals must be submitted in the UK) for example, would probably want to see the results of a series of samples before agreeing to confirm the notice. A second failure therefore will only be useful if formal action is seriously being considered. There is of course, another argument that goes along the lines of 'if you have to use formal action to obtain improvements to a water supply containing faecal material, you are dealing with lunatics'. Who would need to be coerced by a local authority to remove sheep droppings from a drinking-water supply? If a single sample has shown the presence of faecal indicators, then it has shown that there is potential for further contamination of that supply at other times throughout the year.

The other alternative is that when you re-sample, the laboratory report indicates that the water is now satisfactory. This happens quite often. Unless the supply is fairly poor and frequently contaminated, the initial sample was reasonably lucky to find a problem in the first place – an accident that probably occurred because the sampling followed a significant precipitation event a day or two earlier. If the second sample does not similarly follow rain, it may quite easily produce a satisfactory result. However, what does this second result tell you? That the original sample did not contain animal droppings – of course not. That the water is now satisfactory and will remain so for half a decade – again, of course not. That the people responsible for the supply have had it checked and all the problems have been corrected – more than likely not, if my experience is anything to go by or finally, that the laboratory made a mistake the first time round – once again, of course not. Mind you, this supposition has some popularity amongst a few local authority staff members, mainly the older ones, and even more particularly amongst those who have never put a foot inside a laboratory other than to drop a sample off. So, what does the negative result of a second sample tell you? Unless treatment has been fitted or the supply protected from further contamination, it tells you absolutely nothing.

Re-sampling should normally only occur as part of an examination of the whole system, with samples being taken in a logical sequence from the tap back through each stage of the supply up to sampling directly from the source. The point where the contamination starts to show will be the place to start looking for the cause of the problem. It may not, however, be the only problem in the system. These investigations of the different parts of a small water system will be subject to the normal sampling problems – the pollution may not be there at the time of sampling or the 'wrong' slug of water may be tested. The probable causes of contamination at each stage should be identified and put right and once all the obvious problems are remedied, the system should then be re-sampled at a time when a failure would be likely. Ideally the actual cause of the contamination will be found and put right, but some supplies will remain subject to some contamination no matter what protective measures are put in place.

Many local authorities only look at private water supply sampling results as discrete entities. A problem is identified, the supply is improved and the sampling programme moves on. It is rare that results from the whole sampling programme are investigated to see if there are patterns evolving or specific problems are associated with particular areas or times of the year. One authority in Yorkshire sampled its private water supplies for both copper and nitrates for years without ever considering what the results were telling them. Year after year, they never had a failure for copper and for many years they never had a failure for nitrates. Of course, what the results were telling them was that they were wasting their money. Because the situation was never considered as a whole, these conclusions were never drawn and the sampling programme continued unabated, wasting money every time a sample was dropped off at the laboratory. Copper was not a problem because, despite the water being fairly acidic, runs of copper piping were not long enough for the problem to become sufficiently serious. Nitrates were never going to be a problem as there was no arable farming in the area nor any general applications of fertilizer to land. This would have been known by anyone with a general knowledge of the area. If the local authority wanted to be more scientific about it, a few trial samples could have been taken at an appropriate time (after muck spreading or sewage sludge application), from the few systems close to where this type of activity took place.

There was, in fact, one supply where a sample eventually showed a significant failure of the nitrate standard – so did this justify the inclusion of nitrates in the sampling programme? Not really, the sample was also full of total and faecal coliforms and the investigation of the problem would have been carried out anyway. This identified the cause as leakage from the farmyard area into the water supply. Once that was put right, the nitrates and the coliforms levels went back to normal.

When sampling private water supplies, where should the sample be taken? How can any of the sampling points be representative of the supply itself? The answer is that they cannot. The kitchen tap is often the sampling point, but sampling from the tap doesn't tell you anything about the source. It only shows what that particular piece of water is like, after it has travelled through the ground, tanks, pipes and the tap itself, picking up various contaminants on the way. The UK theory is that if the sample from the tap is satisfactory, the source water and the supply system must be satisfactory. If you sample from the source, as happens in the US, and it is satisfactory, this does not tell you whether the system itself contaminates the water after it has left the borehole or spring. Sampling from the kitchen tap appears to be a more sensible way of organizing a sampling regime, but only if it is considered that one sample every year or so will tell you something important about a water supply.

If a sample is taken from a private supply in the summer or during a dry spell, the result may well come back as nil faecal coliforms per 100 ml of water. This could mean that the water is satisfactory. It could however easily mean the sample was taken at the wrong time. This type of sample is known as a 'high risk zero'. It is a term used to describe a negative sample where there is a good likelihood that another one taken just before or after it or taken from a different part of the substance being sampled would have a different result. This is statistically highly likely to occur where small volume samples are taken from water that has not been treated to a very high standard (i.e. small water systems). Because pathogens are not evenly distributed throughout the water, spot sampling will not always choose the portion of water with the 'average' number of organisms. This particular feature is known as 'variance of pathogen densities'. Statistically there will be an under-estimate of the risk from random sampling. For instance, more than half of one hundred spot samples in one test under-estimated the arithmetic mean *Cryptosporidium* oocyst density by a factor of over ten times (Gale, 2001). Spot sampling also over-estimates the net pathogen removal efficiency of any treatment system (Gale and Stanfield, 2000).

Another possibility for this negative sample is that it contains pathogenic organisms. Due to differences in longevity, some pathogenic organisms can be present in a water supply long after the usual indicator organisms have died off. *Cryptosporidium*, for example, can be found in water several months after it has ceased to show signs of the presence of faecal coliforms. Local authorities do not routinely look for pathogens such as *Giardia*,

Enterococcus or *Clostridium perfrigens*. All these survive longer than faecal coliforms and interestingly, there have been no studies that show the number of faecal coliforms in a private water supply correlate to the chances of becoming ill.

Nil results will often mean that the owner of the supply is told that the water is satisfactory and needs no treatment or disinfection installing. With the possible exception of borehole supplies, water quality in over 70 per cent of small water systems is likely to contain faecal material at some time during the year. Whatever the results of a sampling programme, the owner of a supply should be informed of this fact. At the present time, it is the responsibility of the sampling officer to show that the supply is unsafe and, as can be seen, that is not an easy thing to do. It has been suggested that this arrangement needs turning on its head; should it not be the responsibility of the owner to prove that the water is safe, particularly in commercial operations? If the water is not proved to be consistently safe to drink, some form of disinfection should be applied. Ideally the disinfection system should also have some residual effect in order to continue killing microorganisms up to and including the tap.

Despite the present drawback of relying on sampling as the only method of risk assessment, most authorities that have responsibility for water quality will have some form of organized programme for the supplies in their area. In Europe and some counties in the US, this will be a statutory requirement. European countries are covered by the 1998 Drinking-water Directive 98/83/EC (European Union, 1998) but the Private Water Supply Regulations of 1991 covering England and Wales and 1992 for Scotland (HMSO, 1992) were the result of the previous Directive (EEC, 1980). They require a certain frequency of sampling depending on the population using the source or the volume of water used. New regulations are due soon, but the important features of the present legislation are discussed in further detail later in Chapter 19.

If sampling is to take place, how should a sampling authority organize a monitoring programme? Laboratories need to be contacted and sampling arrangements re-drawn to allow monitoring to be allied to rainfall events. This may result in fewer samples being taken because there is only a limited number of days when it rains (you may, of course, disagree with this statement if you live in the north-west of England). They should however give you a much better picture of water quality. Borehole sampling can fairly safely take place in the drier summer months with surface water, wells and spring supplies being tested in the autumn, spring and during milder winter periods. In order to identify supplies likely to contain physico-chemical contaminants, results from whole catchments should be studied to identify any patterns. The geology of an area or its local farming practices may alert sampling authorities to potential problems and when they are most likely to occur. It has even been suggested that for the very smallest of private water supplies sampling officers could test the water on-site for basic physico-chemical properties. The results of this would determine whether there was any need for a full suite of chemical sampling to take place. This would not replace microbiological sampling however, as these parameters do not correlate with faecal contamination.

A better way of establishing water quality is a sanitary survey and that will be dealt with in the next section. However, both sanitary surveys and water sampling should compliment one another. As part of a robust approach, judiciously sampling the water from a private supply can help to reinforce the predictions of the survey. It can also help to hone the quality of the sanitary survey methodology by confirming or denying whether its assumptions are correct.

Another use of sampling is to confirm to the people drinking the water that there is something wrong with it, as predicted by the sanitary survey or the opinion of the health official. It may be a surprise to some readers of this book (though I suspect not to most environmental health people) that members of the public occasionally find it difficult to believe officials from the local authority when they tell them that something needs doing to protect their health. A laboratory report can therefore help to persuade an owner that there actually is something wrong with their water supply. Of course, this can be counter-productive when a supply that is truly appalling and obviously open to all sorts of contamination, provides a sample of water that is devoid of any contamination whatsoever. Sampling to confirm the results of the sanitary survey should therefore be taken, as with all samples, when the supply is most likely to fail.

Taking the sample

Despite the drawbacks of sampling programmes, it is important that every time a water sample is taken, the process is the same. Sampling for chemical parameters is reasonably simple and just requires the tap to be run for about two minutes, the appropriate bottle from the laboratory being filled (carefully to prevent contamination entering other than in the water) and the top put back on tightly. Microbiological sampling is more complicated and the following is a standard method used by public water supply operatives and those who sample private water supplies. I would like to thank Dr Gary O'Neill, Oswyn Parry and Yorkshire Water Ltd for allowing me to share their methodology with you. It would be advantageous if all authorities used this system, so that there is uniformity in the process.

When sampling private water supplies, it is usual to take water from a tap that is directly connected to the water supply and not one that is fed from a storage tank in the house. Normally this means the cold tap in the kitchen. This is because sampling officers want to find out about the water that is supplied to the property and not water that may have been contaminated by assorted fixtures within the premises. Taps come in a variety of shapes and material. They can be single taps or mixer taps and made of plastic or brass. Brass taps can be chromium-plated, gold-plated or plastic coated. The hand-wheel can also be made of clear plastic, onyx or a similar stone material. Within the tap, there may be an insert that is either copper or plastic. This can be fixed in with a screw or removable by pulling. A dental mirror can be used to see whether any inserts are present and how they are fixed.

When taking a sample, the tap should be disinfected by one of two methods. It is important to ensure that the tap itself is not contaminating the water and these methods have been proven to reduce bacteriological failures from dirty or leaking taps. The traditional method of 'sterilizing' taps is by flaming. This is done with a propane or butane burner (a blowtorch). A methylated spirit burner is not suitable, as the flame does not get the tap hot enough. Obviously flaming should not be used with plastic taps, gold-plated taps, plastic taps that look like metal taps, plastic-coated metal taps and taps with plastic inserts that cannot be removed. If a leaking tap has to be used, a protective layer of clean tissue paper should be wound round the tap to soak up the leak and stop the drips running into the sampling bottle.

After the inserts have been removed, protect any plastic or stone hand-wheels by tightly surrounding them with several layers of aluminium foil. The outside of the spout of the tap should then be cleaned with tissue paper moistened with a chlorine solution (prepared as per manufacturer's instructions) or a disinfectant cleaning-wipe. Another tissue or cleaning-wipe should be used to wipe inside the spout to see whether there is any slime present. If any slime is found, the whole of the inside of the tap should be cleaned with a chlorine solution-soaked cotton wool swab on a wire probe or similar device. The tap should then be run to ensure that all debris is washed away. Any flammable material should be removed from the area (many environmental health officers claim to know somebody who has set fire to kitchen curtains or melted a tap during their early career but I would suggest that these tales are mainly apocryphal). Then the whole of the spout of the tap should be heated evenly until steam starts to appear. If the tap is one where the water tends to dribble out when it is turned off, do not wait for steam to appear – you will ruin the tap. However, the tap should still be heated so that when it is turned on again, steam is produced. The flaming should not overheat the tap but the point of the exercise is to boil all internal and external contamination.

If flaming is not an option then disinfection using a chlorine solution is necessary. Again, protect the water sample from any contamination by leaks and remove any inserts. Also clean the inside and outside of the tap. Then swab the whole of the inside of the tap's spout with a chlorine solution-soaked cotton wool swab on a wire probe or similar device. A good swab should fit the inside of the tap reasonably well and produce a slight amount of suction when removed. Using a plastic pipette, squirt chlorine solution up the spout as far into the tap as possible. Repeat this process several times to ensure thorough mixing and contact. Leave the tap to disinfect in this condition for at least two minutes. Turn the tap on for a little while to ensure that all debris is washed away and leave running for five minutes. Ensure the tap is running slowly to stop splashes from the sink re-contaminating the tap. Not everyone who gets a glass of drinking-water is going to go through this procedure and the water will not therefore be strictly representative of what is drunk in the household. Neither is it exactly representative of the water being supplied to the house, because some contamination from the tap can still be left even after thorough disinfecting. However, this method will be a more than robust way of removing as much contamination as is possible prior to the sampling process.

When taking any sample, it is important that there is no contamination caused by bad practice on the part of the sampler. This is doubly important where microbiological sampling takes place. A water supply should not be condemned as unfit because a careless sampler touched the inside of the cap of the sampling jar with his/her fingers. First, therefore, make sure that the bottle being used is reasonably fresh from the laboratory and still sealed. This is to ensure that it is sterile. Where the water is from a chlorinated supply, the bottles should contain a small amount of sodium thiosulphate. This uses up any remaining chlorine left in the water and stops any further disinfection taking place. This makes certain that the sample represents the exact quality of the water as it leaves the tap. If the supply is unchlorinated, this chemical is not necessary. After sterilizing the tap, run the water from it at a steady rate. Make sure that you do not do anything to re-contaminate the tap. After the tap has run for a little while, cautiously remove the top of the sampling bottle so that you do not touch any part of the top of the bottle. Carefully hold the outside of the cap and make sure it cannot be splashed or otherwise contaminated. Do not put the cap down. Fill the bottle until it starts to overflow and then replace the cap. If anything happens to contaminate the sampling bottle before the sampling has finished, start again.

When a sampler wants to test for lead in a water supply, it is sometimes necessary to take two samples – the first (non-flushed sample) is to show the amount of lead that has leached overnight from the pipes in the house, the second (flushed) sample shows the normal amount of lead in the water as it enters the property. However, a lot of houses with lead waterpipes have a run of copper piping before the kitchen sink. What happens in practice therefore is that the first drawing of water from the tap has actually been standing in copper piping. If possible therefore, check the waterpipes and slowly run sufficient water to clear any that has stood in copper. Domestic copper piping will hold about 750 ml of water per metre of pipe. Measure the amount of water and this will tell you how far it has travelled. When the water that has stood in the lead pipes is starting to come out from the tap, the non-flushed sample can be taken. For the flushed sample, the usual advice is to allow the water to run for a minimum of two minutes to ensure that the sample of water is fully representative of the water from the supply. Even so, many properties have lengths of piping that take longer than two minutes to run water from the outside of the property to the tap. So where possible, further measurements should be made to ensure that sufficient water has been run. A recent study found that rather than getting the base level, flushing the water for two minutes actually raises the amount of lead in the water (Lloyd, 1999). After the two minutes, the water that has been standing in the lead pipes started to come through. After several more minutes, the water that was outside the premises arrived and the lead content of the water went down again. The normal advice for taking flushed samples (i.e. run the water for two minutes) can therefore result in findings that are not actually representative of the water supplied to the house and perhaps five minutes would be more sensible.

When sampling from a tank or directly from the source, the sampling process presents a little more difficulty. Although water quality within a tank or spring-box will vary at different heights and depend on where you sample in relation to the inlet and outlet, it is important that if you only take one sample, you try to obtain a representative one. If water is taken from the surface you will potentially allow some floating contamination into the bottle and if the sampling bottle goes too near the bottom, settled mud or debris may enter. Also try to avoid the sides, in case they have biofilm attached. It is possible to rig up a device that will enable a laboratory bottle to be dropped into the water using string and Brassington (1995) details a suitable design. The bottle is lowered quickly so that the majority of water that fills it is from the desired depth; but take care to ensure that the bottle does not slip from its ties. Another alternative is to use a sterilized bucket and then pour the water carefully into the bottle. Only as a very last resort should the sample be taken by hand. Ideally the sample should be taken by the person whose water supply it is, so that they are aware of what is happening. In such cases, it is important that hands and forearms are scrubbed clean and wiped with a disinfecting wipe. The cover of the tank, if there is one, should be carefully removed and placed upside down so that it cannot be contaminated. The bottle should be held so that any water going into it does not come into contact with the hand. Remove the cap as explained previously and keeping the bottle vertical, quickly place it about 10 cm below the surface of the water and allow it to fill. Pull the full bottle out of the water and put the top back on quickly. Finally carefully replace the cover. Remember that there is always the potential for contaminating somebody's drinking-water supply and the greatest of care must be taken. However, all small water supplies should have a disinfection unit as part of the system and this should ensure that any contamination from the sampling process is eliminated.

After collecting the sample, the bottles should be kept in an insulated cool box containing frozen ice packs. The ice packs should be put into the cool box just before leaving the office or laboratory. All bacteriological samples in the UK should be delivered to the laboratory within 6 hours. Nevertheless, the samples should always be returned to the laboratory as soon as possible. The sample bottles also need labelling and the appropriate laboratory forms filled in. Experience has shown that it is important to complete all the paper work as soon as the sample is taken to avoid possible confusion later on. Another piece of advice is to ensure you have a surfeit of labels, at least two ballpoint pens and a pencil close by. If you do not have a spare, any ballpoint pen will stop working when needed – I don't know why, but it just will. It is difficult to correctly label twelve different bottles back in the laboratory, when the only identification marks are scratches you have made with your fingernail on the paper label. In the UK, analysis has to start within 12 hours of the sample being taken (30 hours in the US), although recent research has indicated no loss of accuracy if this is extended to 24 hours. A final piece of advice is to try and get to know the people who work in the laboratory. They are experts in what they do and can give very useful advice on sampling techniques in strange places or where you think a private water supply may have a particular problem.

Sanitary surveys and risk assessment

If a regular sampling programme is not the best way to identify potential problems in small water systems, then what is? The answer is a risk assessment, where the whole supply can be examined, from the area of land surrounding the source up to the tap. This enables potential problems to be identified, assessed as to their severity and for measures to then be put in place to prevent or reduce the possibility of contamination getting into the water. This is also known as a sanitary survey, a sanitary assessment or an environmental survey. Pickford (1991) explained that, 'sanitary inspections are among the essential elements of an effective drinking-water quality surveillance and control programme'.

It is not always necessary to take bacteriological samples to be reasonably certain that pathogenic material will be in a water supply. A thorough inspection of the spring collection point or the top of the well, the surrounding area and the piping system including tanks, pressure vessels and any treatment system, will often identify potential pathways for contamination, even if they are not all operating on the day of sampling. Because the weather has such a major impact on water quality, a sanitary survey can be carried out during a dry spell, when sampling would be pointless. In fact sanitary surveying when the weather is warm and dry is a reasonably pleasant way to spend a day. Conversely, wandering about a field, looking for the hidden source of a small water supply in the rain tends to lose its charm when you are cold, wet through and have just fallen over in the mud.

When visiting a private supply, especially for the first time, it is important to carry out a sanitary survey in conjunction with microbiological and physico-chemical testing. Lloyd and Helmer (1991) state that, 'The complementary nature of a sanitary inspection cannot be over-emphasized and the sanitary inspection should at the very least reveal the most obvious points of contamination risk and, can provide a robust and conservatively safe method of risk identification.' These authors also confirmed that the sanitary survey generally 'reveals more of the chronic risks of contamination than can be revealed by a single and costly bacteriological examination. This is not an argument for dispensing with bacteriological testing but for an economical and intelligent approach where funding is limited'.

Environmental health officers and sanitarians have traditionally used a form of sanitary survey when assessing risk in food premises; the otiose nature of random sampling during food hygiene inspections has long been accepted. For food production, the emphasis is on ensuring that the people responsible for the safety of the food examine the way it is produced and work out how the practices and processes involved can adversely affect it. They ensure the safe production of food by removing or reducing to a minimum risks like cross-contamination hazards and putting in place protective measures such as temperature control and monitoring. It has long been accepted that by having staff and management that understand hygiene concepts, following good practice and allying this to the satisfactory condition of premises and utensils, safe food will be provided. Unlike water, sampling during a food inspection has never been the only method of determining safety. The food industry itself is moving further away from reliance on 'end product testing' towards stricter process control. Inspectors of food premises expect proprietors to use a formal process of risk assessment to ensure food safety. The system generally used is Hazard Analysis, Critical Control Points (HACCP). The Pilsbury Corporation devised HACCP for the American space programme in the 1960s. It is a quality system that relies on identifying those places in the production process where there is a potential food safety risk. These points are then assessed to see which ones are important (critical) to the safety of the food. The next stage identifies what needs to be done to prevent these critical control points causing a problem, such as altering the system or ensuring strict control of temperatures. Finally, the system requires scrupulous monitoring of the factors affecting the control points to ensure the food is always kept safe. A simple version of this method can easily be used for identifying and controlling potential problems in private water supplies, and by applying a systematic approach consistently safer drinking-water can be obtained.

Once the source of the supply has been located (which, it must be said, is often easier said than done), the sanitary survey should look for problems and possible causes of contamination all the way back to the tap. Faecal contamination does not usually arise below the soil surface except for anomalies such as leaking septic tanks and drains close to the water supply. Where this risk can be discounted, the main point of contamination is the area where the supply meets the open air. The survey should identify potential sources of contamination such as manure heaps or streams and note the absence of simple measures involving fencing to keep animals away and drainage ditches to divert surface water around the collection area. Similarly, if there is insufficient concrete protection around the point of issue, no close fitting, lockable lid or for spring supplies, the collection pipe is not securely fitted as far as possible into the source aquifer, the details should be logged and the possible risk assessed. If there is a collection reservoir, this will also need careful examination. Storage tanks can often cause more of a hazard than the collection point and should therefore receive similar attention. Close inspections of wellhead construction and the area surrounding it will reveal potential points where contamination could enter the water supply. Although boreholes are usually less likely to need attention, a sanitary survey should be carried out to ensure that the borehole is not one that is easily contaminated (i.e. under the influence of surface water) and that problems cannot be caused either by surface water running down the outside of the tubing or via defects in the piping or water storage system.

Any competent person can draw up suitable checklists to help identify all the possible pollution sources and ensure that the survey is carried out in a logical fashion. The person inspecting the site should also be suitably trained, so that they are able to identify the most serious problems and alert the owner to the need to put them right. These simple and obvious safety measures should be an integral part of an assessment of every small water supply system, unfortunately they rarely are.

Recently, several more formal versions of sanitary survey risk assessments have been produced. These attempt to quantify the level of risk. Kay (2003) has produced a good one, as has Lamb *et al.* (1998). A third has been designed for the National Trust (1996) by Cranfield University. They are all being looked at by the Scottish Executive as part of their consideration of new regulations for private water supplies following the 1998 European Union Drinking-water Directive. The one by Lamb *et al.* can also be found in the appendices of the *Manual on Treatment for Small Water Supplies* (WRc-NSF, 2002). Yet another version is proposed for the website of the Chartered Institute of Environmental Health, produced by Morris of the British Geological Survey. Unfortunately, all these risk assessment methods suffer from not having been subject to extensive testing in the field by local authorities around the country. When they are, comparisons can be made with the actual results obtained from water sampling and the scoring systems refined to enable them to be more accurately predictive. During one workshop, every example private water supply assessed was classed as being a high risk (Kay, 2003). It is important therefore, that before one of these methods or a combination of them, becomes the universally accepted standard, it is proved to be a reliable predictor of water quality across a wide spectrum of supplies.

Where formal action may be taken following a survey by an enforcing authority, a well-considered and openly documented procedure should be followed. In that way any such action can be consistently applied to every supply that may be at risk. Attention should be given to identifying which problems will initiate advice and which will require formal action. The stages in the process should be clearly identified and followed. A scoring system for the survey, allied to any sampling results should make it clear to everyone involved as to what will happen and what measures will be taken if improvements are not made. It is often advantageous to have these measures approved by the appropriate political body responsible for public health. This process should include some form of consultation with stakeholders. It is important that consultation is carried out not only with the people who may have to spend money to put things right where failures occur but also medical authorities and those members of the public who may be expected to drink the water.

In conjunction with sanitary surveys, it is usual to take a sample of the water. Because someone is already at the property, this is relatively easy and cheap. The sample may confirm the suspected composition and quality of the water in the supply as identified by the on-site survey and interpretation of local geological information. It may also reveal problems not unearthed during the risk assessment process, but as discussed previously, this is not guaranteed.

Sanitary surveys do not tend to identify potential chemical contamination with as much accuracy as they do microbiological problems, but they can prove useful, particularly in conjunction with geological knowledge of the area. The presence of lead in the water may be suspected if the water is soft or aggressive and there is lead piping in the house. Nitrate contamination will be a consideration in areas where arable farming takes place and chemical contamination may occasionally be an issue if a private supply is near to an industrial area, an isolated factory or a solid waste disposal site.

It should not be assumed that once a sanitary survey has been carried out, it is the end of the matter. Subsequent visits to the supply need to include an examination of protective measures such as fencing, overflow-pipe insect proofing and treatment systems to ensure they are still satisfactory. Although many sources appear to have been unchanged since the Middle Ages, there should also be a check to see whether there have been any recent alterations or new potential causes of contamination. Whenever the opportunity presents itself, it is sensible to make sure that all the individual parts that make up a private water supply are all still operating safely.

5   Water and health

The spread of major epidemic diseases such as cholera and typhoid has been associated with drinking-water since the mid-nineteenth century. It was Dr John Snow, a pioneer epidemiologist, who had reasoned that because of its effects on the alimentary tract, the cholera 'poison' must be introduced into the body via the mouth. In families, the spread could be explained by contact with soiled bedding, but across areas or along streets he concluded that the explanation must lie with the water supply. Snow got the chance to test this theory during the 1853 cholera epidemic. He found that of the two competing companies supplying water to London, one supplied water from the Thames taken from the centre of the city, while the other had recently moved its intake upstream to a comparatively clean part of the river. At that time, the supply of water to houses in London was reasonably randomly allocated between these two companies. Each company would have its own feed pipe running down a street, with different houses connecting into it. Snow was therefore able to rule out any confounding reasons for people getting cholera, such as occupation or class. He visited many houses to establish the link between the different degrees of disease from the two water supplies and found that the water from the polluted part of the river caused much more illness than the upstream supply. He called this his 'Grand Experiment'.

It was during this investigation that the famous incident of the Broad Street Pump occurred. Snow was asked to investigate the cause of a vicious outbreak of the disease in the Golden Square area of London. Because he already suspected water as the cause, he looked at the five pumps (private water supplies) in the area. Four of them were visually contaminated and therefore probably not very popular but in the fifth, the Broad Street pump, the water was clear. A local resident however, told him that it had smelled offensive just the day before. Subsequent investigation by Snow showed that of the people affected by the disease, the vast majority had drunk water from this pump. In his own words:

> On proceeding to the spot, I found that nearly all the deaths had taken
> place within a short distance of the pump. There were only ten deaths

in houses situated decidedly nearer to another street-pump. In five of these cases the families of the deceased persons informed me that they always sent to the pump in Broad Street, as they preferred the water to that of the pumps which were nearer. In three other cases, the deceased were children who went to school near the pump in Broad Street...

With regard to the deaths occurring in the locality belonging to the pump, there were 61 instances in which I was informed that the deceased persons used to drink the pump water from Broad Street, either constantly or occasionally...

The result of the inquiry, then, is, that there has been no particular outbreak or prevalence of cholera in this part of London except among the persons who were in the habit of drinking the water of the above-mentioned pump well. I had an interview with the Board of Guardians of St James's parish, on the evening of the 7th inst [September 7, 1854], and represented the above circumstances to them. In consequence of what I said, the handle of the pump was removed on the following day.

(Snow, 1854)

By getting the handle of the pump removed, Dr John Snow elegantly brought to an end a cholera outbreak that had caused over 500 deaths.

It was left to another member of the Parish Inquiry Committee, the Reverend Henry Whitehead, to discover the most likely cause of the outbreak. This was from the washed soiled nappies of a child that had died of cholera at 40 Broad Street on 2 September 1854. The house drains were later shown to have decayed brickwork that allowed sewage water to pass from the drain into the well.

Further examples of the connection between water and disease can be found in the work of William Budd. Budd demonstrated the contagious nature of typhoid fever and how the infection is spread from faeces via water. Proof that treatment can affect the ability of water to spread disease was first shown in 1892. The contaminated River Elbe in Germany supplied two towns. The town of Altonia escaped the cholera because its supply was filtered whereas the adjacent town of Hamburg did not (Dadswell, 1990).

Since then, water has continued to be associated with disease. In the UK, as in other developed countries, the major waterborne diseases of cholera and typhoid have diminished as water treatment and disinfection have improved and an ever-increasing number of people have been connected to public water supplies. A study of the major recent outbreaks of waterborne disease shows that the majority have been associated with unchlorinated or defectively chlorinated supplies. Only two outbreaks of cryptosporidiosis and one incident of chemical gastroenteritis did not fit into this category in the UK during the fifty years between 1937 and 1986 (Galbraith *et al.*, 1987). Of the 34 outbreaks of waterborne disease recorded during that period, 30 were from private water supplies. The toll of disease involved over 1,900 cases. If the figures are looked at in isolation and the number of people who were made ill are compared with the total numbers drinking water from private and public water supplies, an interesting picture develops. The danger of contracting a disease by drinking water from private rather than public supplies can be calculated to lie in the area of twenty-two times more probable. If the figures from the final decade in the study alone are compared, the additional risk from private supplies goes up to over fifty times (Clapham, 1993). Private water supplies therefore appear to be intrinsically more dangerous than mains supplies. When this has been investigated more formally, results have been mixed. A study of 102 households (Meara, 1989) reported that there was no difference in self-reported illness between children who lived in rural areas on private water supplies and those on mains, although the rural children did take more medication.

It is important to remember that all supplies found to contain faecal indicator bacteria can theoretically contain pathogens, which may then lead to disease. Reports on the quality of small water supplies have found an 88 per cent failure for at least one parameter (Barraclough *et al.*, 1988); 84 per cent for coliforms alone (Humphrey and Cruickshank, 1985) and 89 per cent for coliforms and plate counts (Barrell, 1989). More recent studies by Kay (2003) and Lilly (2003) in Wales and Scotland have found similar failure rates.

If this is so, why were only thirteen outbreaks of waterborne disease from private water supplies recorded in those fifty years? It has been stated by Victoria *et al.* (1988) that although associations between poor sanitation and drinking-water quality and increased morbidity are generally assumed, in practice they are hard to demonstrate. It is well known that there is a general under-reporting of gastrointestinal illness. A study in the UK suggested that for every single case of infectious intestinal disease identified by the national surveillance system, another 1.4 were identified by laboratories and 6.2 faecal specimens were sent to laboratories for examination because disease was suspected. In addition, for this single identified case, a further 23 cases visited their doctors and another 136 cases occurred in the community, but did not visit their doctor or report it (Handysides, 1999).

This under-reporting is often thought to be even worse for illness associated with the consumption of water from small supplies. Private supplies are located in small rural communities, very often consisting of only one or two properties. Where outbreaks of disease occur, they will therefore normally be confined to a small number of people and except in particularly severe cases, it is unlikely that those affected will consult a doctor. In the event of a doctor being consulted, it is debatable that the cause will be identified, the appropriate authorities notified and the incident fully investigated. On a mains supply there are more people at risk and these numbers will cause reporting statistics to be skewed. A study in 1996 (by Dougan), found that people on private supplies consulted their General Practitioner less often than those on mains supplies, even if the amount of gastrointestinal illness was similar. Jones and Roworth (1996) also confirmed that the true incidence of disease from small water systems would not usually become apparent when they investigated a public water outbreak in a Scottish rural village. They noted that the large number of people with gastrointestinal illness (711 out of a population of 765) might have been expected to alert local doctors to the fact that there was a problem, unfortunately it did not. Only a few went to their General Practitioner and they were spread between three separate practices. If 93 per cent of a village becomes ill and it is not obvious, it must be even more unlikely that a few sick people on a private water supply would come to the attention of the medical authorities.

The second reason for under-reporting is associated with the nature of most waterborne infections. It is likely that a doctor confronted with a serious case of typhoid will notify the authorities, who will then investigate it. It is unlikely that milder outbreaks of gastroenteritis of unknown causative agent will be investigated so thoroughly. As the background levels of the major nineteenth-century diseases decline, the chances of their contaminating water supplies will diminish (Morgan, 1990). Modern outbreaks will be much more likely to be of unknown, viral or parasitic origin. Individual outbreaks of waterborne disease from private water supplies will not therefore normally attract much attention. The major cause of gastrointestinal illness amongst children under five, for example, is cryptosporidiosis (Casemore, 1991). This disease is particularly likely to be found in private spring supplies because of the levels of faecal contamination, but there has not been a large body of work investigating this relationship.

Epidemiological studies in Canada have also shown that there is a non-trivial endemic (i.e. habitually present in a certain locality or population) level of unreported disease due to the consumption of ordinary tap water (Payment *et al.*, 1991). The low level of severity of disease, however, results in few demands on the medical profession and thus does not often figure in disease statistics. Payment's work involved studying the gastrointestinal effects of drinking mains water and showed that levels of illness from water are much higher than is generally appreciated. He concluded that 35 per cent of the gastrointestinal illness found in his study was water related and, despite being caused by water treated to modern standards, were preventable. It can only be assumed that this endemic level of disease must be even higher amongst people on private supplies, where treatment is often missing or defective.

The smallest number of people on a private water supply where an outbreak was notified to the Communicable Disease Surveillance Centre (CDSC) for the half-century between 1937 and 1986 was 6, the mean was 146 (Galbraith *et al.*, 1987). There are very few private water supplies with a population that even approaches this higher number. Does this mean that the smaller supplies are safer? Patently not – in all studies for pathogenic contamination of water supplies, the smaller the supply the more likely it is to be contaminated. In fact, the majority of private supplies can never cause an outbreak. Single property supplies serve 45,000 people in England and Wales (Burfitt, 1997). However, no matter how bad the water or the total number of victims in the property, a disease associated with only one house can never, by its strict medical definition, be classed as an outbreak.

Microbiological considerations

Drinking-water containing pathogenic microorganisms does not strike down everybody equally. After a private water supply becomes contaminated, perhaps only one member of a family will be affected. Whether a person actually becomes ill is dependent on many factors. Different organisms require vastly different numbers to cause disease. With *Cryptosporidium* for example, although the actual minimum number of oocysts necessary to produce an infection is unknown, it is thought to be less than ten and may be only one. A *Yersinia* infection on the other hand requires up to 10^9 organisms to cause disease.

The general health of a person is also a factor, not only as to whether they become ill but also how ill they will become. The reaction of individuals will vary, even though they may all consume the same number of pathogenic organisms. Nutrition and fitness will affect the way a body's immune system is able to fight the disease. This will not always guarantee success of course, as even super-fit sports people can become seriously ill, but it is a factor to bear in mind when interpreting disease statistics. Some people consider that the reason for the increase in life expectancy throughout the eighteenth and nineteenth centuries was largely because of improved nutrition. The population kept getting the same diseases, but because of a better diet they (particularly the children) could withstand these disease attacks and eventually recovered, rather than dying. In developing countries, where child mortality remains appallingly high, thousands of children under the age of five die every day from drinking contaminated water. But they also die because they are probably ill from something else (such as intestinal worms), have limited access to medical facilities and are malnourished (a polite word for starving). When you talk about diarrhoea to people in the UK, it is usually subject to a certain amount of merriment. It is not such a funny subject in large parts of Africa, Asia or South America, particularly if your brother or sister has just died from it.

Whether a disease will make someone ill also depends on the immunity they have to it. Immunity can be acquired because a person has previously been ill from the disease and naturally produces antibodies to deal with subsequent re-infection or they have been immunized against it. Immunity takes two main forms; one is a long-term immunity as with yellow fever, where an inoculation provides protection for over ten years. The other is a short-term immunity that has to be maintained by the body being regularly challenged by the organism, this happens with malaria and possibly *Cryptosporidium*.

Small water systems that are used at caravan sites, camp sites and holiday cottages will have a transient population who may not have immunity to any of the contaminants that could be in the water. When they become ill they will probably have moved to another area of the country or returned home. As they contemplate the possible cause of their illness, there will be a range of potential candidates and they may not associate it with the water they drank on holiday. This is because in most developed countries people are used to public water supplies that are of consistently good quality and, as they will probably not been informed otherwise, they will have assumed that was the case with the water at the holiday accommodation. It is only when large or serious outbreaks occur that private supplies are suspected and there have been several recent UK outbreaks of *E. coli* O157 associated with campsites in rural areas served by unchlorinated private supplies.

Young children are most likely to be affected by pathogens in drinking-water, because the immune system takes time to develop. Babies who are breast-fed will acquire some immunity from their mother's milk, but will temporarily become more vulnerable when weaned. Cryptosporidiosis for example, is a disease that affects young children rather than babies. Adults usually have a fairly robust immunity to many common diseases but as people move into old age, the immune system starts to fail and they may begin to suffer from diseases that have not caused them any problems for decades. Some experts feel this decline has been overstated and that many people's immune systems stay in good condition well into old age. However, many diseases will kill old people where they do not have such a serious effect on younger people.

An important consideration nowadays is the increasing number of immuno-compromised people in the community as they are at risk from small water supplies for two reasons. The first is that a number of naturally occurring bacteria, referred to as heterotrophic bacteria, are not normally pathogenic in a healthy population (although there are some exceptions – see later sub-section on 'Bacteria') but they are much more likely to cause problems in the immuno-compromised. Water is not tested for these naturally occurring organisms, other than as part of a total (or heterotrophic) plate count. These bacteria, such as *Aeromonas* or *Pseudomonas* will grow in water and can multiply to give large number of individuals if the water is left standing for any length of time. For instance, bottled water that has stood on a shelf in a warm corner of a shop may have substantial amounts in it by the time it is sold. In one study of bottled still water, out of the seventeen bottles tested, seven had plate counts greater than a thousand and three had more than ten thousand. This is a lot higher than would be expected from tests at bottling plants and compares to an average of less than ten in tap water from UK public supplies (Fewtrell *et al.*, 1997). This is not a problem in a normal population, but bottled still water had to be banned from one ward in a hospital in the North of England. Visitors were bringing it in as presents, which led to problems for the patients because of their reduced immunity following major surgery. The numbers of these bacteria are rarely counted in private water supplies but, because of the levels of contamination and the general lack of treatment, they will be much higher than in public supplies.

The second problem for immuno-compromised patients is that an illness that is normally mild or self-limiting can become life-threatening. Cryptosporidiosis is the prime example. Normally it is a self-limiting, albeit distinctly unpleasant disease but in HIV/AIDS patients it was found to be a significant cause of mortality. Nowadays, because of the medical regimes for HIV/AIDS patients, particularly in the main cities, cryptosporidiosis is not as much a problem as it is for other immuno-compromised people, such as those suffering from leukaemia or recovering from major surgery (Farthing, 1999).

Once the many mechanisms involved are understood, several peculiarities of small water systems become clear. People may not get ill even though they have drunk contaminated water because they are adult or fit or only drink beverages using boiled water or have built up an immunity to the particular diseases that are being excreted by the animals on the land around their water source. I was once in discussion with the user of a remarkably contaminated water supply. The water was so bad that if boiled, it gave off a smell reminiscent of an outstandingly pungent farmyard. I was trying to persuade him to go on to the public water supply. As part of this discussion, I suggested that the water must have made him ill at sometime during the past. He replied that he had never, ever been ill. I countered that everyone has been ill at some time in their life but he retorted that all he had ever suffered from was a broken arm and that even I could not blame it on the water supply! I felt that with no evidence to the contrary, I must concede to this elegant line of argument. He and his family may well have built up immunity to all the pathogens they were regularly drinking in their water. I felt it pointless to point out that if the cows that were polluting his water supply picked up an exciting new disease, he may well be laid low for weeks. And, as he apparently felt he had won the argument and was quite smug about it, I felt it would serve him right.

Reports in the scientific press of waterborne outbreaks of disease linked to private water supplies are often associated with temporary populations. Is this because the water they are drinking is worse than that consumed by the static rural population or is that people who drink water and become ill while travelling are more likely to complain? As well as the reduced immunity levels, it is probable that it is easier to investigate an outbreak where the only common factor amongst the people affected is the water supply. This situation is similar to the popular view that people who attend wedding receptions are much more likely to get food poisoning than those who do not. Whilst some wedding caterers do leave a lot to be desired hygiene-wise, the real reason is that the outbreaks are easy to investigate and to draw conclusions over. The guests telephone each other to discuss the reception and the wedding presents. One of them also mentions numerous visits to the bathroom. The other person then relays their tale of frequent nocturnal toilet calls. Two and two are put together, the health officials are called in and all they have to do is find the menu for the one meal, ask everyone what they ate and conclude for example, that all the people who got sick consumed the chicken a la surprise. Compare this to a normal situation where illness has been caused by a small water system. The ill people will not be speaking to each other on a daily basis. The illness may not get reported for some time, if at all. When the water is tested, it will not show any sign of the pathogen; the slug of contaminated water will have left the system some time before. To add to the confusion, a number of people who drank the water will not become ill and several who did not drink it may (via food prepared with the water or when they brushed their teeth in it).

It is interesting to compare the suspected reasons for under-reporting disease occurrence in small water systems with the often over-reporting of figures in public water supply outbreaks. Outbreak numbers are normally assessed by a telephone survey of a proportion of the people in the affected area, where they are asked if they have had diarrhoea in the recent past (usually a month). This figure is then multiplied to give the total number of cases amongst the whole population in the affected area. However, what the people who are being telephoned seem to hear is not 'have you had diarrhoea in the last month' but 'have you ever had diarrhoea', or 'have you been ill with anything in the last three months' or 'you may be in for the chance of some compensation if you answer "yes" to this question'. What the investigators are looking for are people with a clinical problem or a debilitating disease that affects their day-to-day life, but the answer may still be 'yes', even if someone just goes to the toilet a bit more regularly or their faeces are not quite as solid as normal. A telephone survey of diarrhoea incidence, on its own, will be unable to differentiate between the normal level of the disease and the extra cases caused by the incident. A study in the Northwest of England (West, 1999) found that when a case control study was employed to assess the accuracy of this telephone method, the results were surprising. The control area, where there was no *Cryptosporidium* in the water, actually had a greater incidence of diarrhoea than the supply zone with the contaminated water. The study also found that when researchers asked their questions from a medical viewpoint there were fewer incidences reported than when the questioner suggested that financial compensation might be an issue – now there's a revelation. This one study will not provide definitive figures to cover all previous outbreaks, but it appears that some investigators may have over-estimated disease levels by a factor of ten. The famous outbreak of cryptosporidiosis in Milwaukee, Wisconsin is normally thought to have affected 403,000 people was based on a total of 613 telephone calls (Corso *et al.*, 2003). The Northwest researchers now consider that the true incidence was probably nearer 42,000 (Hunter, 2001).

There is a particularly interesting argument applied by some freethinkers to small water systems. It goes like this – a polluted private water supply is drunk by generations of a family and after getting a few childhood diseases, they build up a robust immunity to the water supply. An official from the local authority then comes along, tests the water and insists that some form of water treatment be installed. Despite their better judgement, the family spends its hard-won money on a treatment system. All goes well for a little while – the treatment system works and they gradually lose their immunity. Suddenly the treatment system fails and the whole family goes down with some dreadful disease. The conclusion to this story is that if they had been left alone, they would not have become ill. It is interesting that no one suggests the idea of introducing small but regular amounts of pathogenic material into public water supplies to build up a robust immunity amongst the community as a whole. To acquire immunity you have to have been ill in the first place and 'a few childhood diseases' are easy to dismiss (but perhaps not if you are the child). Neither is it suggested that the treatment system this mythical family installed should have been accompanied by effective source protection to keep the pathogens out, nor that the treatment system should have been properly maintained and fitted with a device that ensured it failed to safety. A point that should also be stressed is that this family, with their cast-iron stomachs, will not be the only people who drink the water. During Christmas or the summer holidays, loving grandchildren may visit. They will drink the water and having no immunity to it, may be subjected to an unpleasant bout of diarrhoea, at the least. The water will not be blamed because 'everyone has drunk the water'.

Microorganisms in water can either be viruses, bacteria, fungi, protozoa or parasites. Some have no effect on health whereas others will make people ill. The effects can range from a mild stomach upset to death. Pathogenic organisms use a variety of mechanisms to cause illness. Viruses invade and cause the death of the body's cells. Bacteria and protozoa produce toxins, releasing them either as part of their metabolism or as a result of dying and decaying. They can also initiate allergic reactions in the immune system that are so strong they can cause damage to the body's own cells. Parasites, particularly helminths (intestinal worms) will cause injury by their sheer physical presence. This can include malnutrition, caused by tapeworms consuming the person's food, or trauma as the parasites migrate through the body or lodge in an organ such as the liver, spleen, heart or lungs.

Many parasites spend part of their life cycle in a human host. They will not benefit by the host dying, so after millions of years of evolution the successful ones are unlikely to cause fatal illness. That is unless they are not in their natural host or another illness is present. Parasites can however, cause significant morbidity, particularly in children. Interestingly it is reported that tapeworm eggs were swallowed in Victorian times as part of a weight loss programme. Dieters swallowed the eggs, became infected by the worm (up to 7 m in length) and ate as much food as they wanted to. The worm took its share and when the person reached their desired weight, they took a little pill, the worm was killed and subsequently excreted. Simple and efficient – it's a wonder it is not being advertised nowadays by a complimentary medicine company.

Diseases associated with water have usefully been classified according to how the water helps cause the disease. Bradley developed this classification as discussed in White *et al.* (1972). The categories are given different names from time to time, but the basic system remains the same. The first category is 'waterborne'. Waterborne diseases are those in which the bacteria, viruses, etc. are carried in the water and the person ingests the organism by drinking water or eating food that has been prepared with water. Cholera, typhoid, cryptosporidiosis and rotaviruses are all waterborne diseases. There are many more. 'Water-washed' or 'water hygiene' diseases are those caused by poor hygiene practices and are usually a problem where there is insufficient water available to wash often enough or properly. The water that is used can be perfectly safe (although it very often isn't) but where it is scarce, washing has to take second place to drinking. These diseases will be most common in desert regions and amongst the very poor. Examples of water-washed diseases are typhus, scabies and ear or eye infections. Very often, sores or cuts, which would be washed clean in developed countries and thus heal naturally, can become severely infected. 'Water-related' or 'water habitat vector' diseases are those where, although the water does not cause the illness directly, it plays a vital part in the life cycle of the infectious agent, normally the host of a parasitic organism. Malaria is an example; the female mosquito lays her eggs in water, where the larval stage develops. The fourth and final classification is 'water touched' or 'water contact' diseases. They do not normally cause illness when ingested but are transmitted through skin contact. A serious water contact disease is schistosomiasis, otherwise known as bilharzia. The parasite spends part of its lifecycle inside an aquatic snail, but enters through the skin when people are bathing in open water. This classification also includes conditions such as skin disease caused by *Pseudomonas aeruginosa*.

Viruses

Viruses are the smallest microorganisms found in water and are little more than self-replicating parcels of RNA or DNA. The viruses that are of interest in water were identified by WHO (1993a). These are Norwalk virus, the imaginatively named small round virus or small round-structured virus, Norwalk-like virus (all these are now called Norovirus), Adenovirus, Enterovirus, Rotavirus, Hepatitis A virus and finally, the intricately named, enterically transmitted non-A, non-B hepatitis virus, otherwise known as hepatitis E. They are also nearly all classed by WHO as being of high health significance, with a low infective dose. Viruses are subject to occasional re-naming by scientists. Despite appearances, this is not done to confuse or annoy students who have only just learnt the old name. Re-naming will happen as more is found out about an organism or group of organisms.

Bacteria

Bacteria are a major cause of illness associated with drinking-water. Diseases such as cholera, bacillary dysentery and typhoid are very well-known examples. It is highly unlikely that these particular diseases will be found in private water systems in Europe and the US as they have largely been eradicated due to water and sanitation improvements since the nineteenth century. Because of these sanitary improvements, even the people who may be ill with them will not be excreting the organisms around springs or boreholes. Nowadays these ailments are usually only found in people who have recently travelled abroad. However, it is not impossible to imagine a scenario where typhoid, for example, could enter a small water source from tourists on a camping tour of the UK. Private water supplies are more likely to be contaminated with zoonotic bacteria (i.e. those that can be transferred between animals and man). Examples include *Campylobacter* and pathogenic *E. coli*, although many of these bacteria are more usually associated with food poisoning outbreaks. Nevertheless, it should be remembered that in many developing countries, raw water is not safe to drink, whether from a well or the public supply. A lot of people therefore only drink water that is bottled or privately treated (of course, the very poor in these areas do not have this choice). However, some untreated water will be consumed via washed salads or as an ingredient in inadequately cooked food. Throughout the world, up to 60 per cent of the illness caused by contaminated water supplies are thought to be consumed in food (Alegre, 1999).

Opportunistic or adventitious bacteria will cause illness in some circumstances but are usually harmless. *Pseudomonas aeruginosa* is an example; it occasionally causes skin or ear problems but with immuno-compromised people it will cause serious problems. Cyanobacteria, otherwise known as blue green algae, are another potential problem. Some can produce toxic substances, which if consumed, can cause very serious illness, even death. Fortunately, they normally only affect animals, such as cattle and dogs, drinking from lakes.

Protozoa

Protozoa are small unicellular creatures such as amoebae or *Cryptosporidium*. Very few protozoa are pathogenic although some can cause significant illness. As part of their lifecycle, some species form egg-like cysts, often called oocysts, and these can be soft-walled or hard-walled. Some species such as *Giardia* can produce both types. The hard-walled cysts enable the organism to survive in the environment for some time.

Helminths

Helminths are parasitic worms, usually with a complicated lifecycle involving two host species. Sometimes the worms can grow to several metres within the body, others remain small, such as the liver fluke, but can still cause a great deal of damage. Helminths can be ingested by a variety of methods, including being drunk in water or eaten in meat with the parasite or its cysts lodged in the tissue. They can also pass through the skin when the person is swimming or standing in water. Humans can be the primary host or a secondary host. The primary host is the one in which the sexual stage of the organism develops. A secondary host is one where the organism only resides within it during an intermediate stage. Most helminths are host specific but sometimes non-human species are ingested and the worm causes tissue damage as it moves through the body.

Indicator organisms

There are fifty-six potentially pathogenic organisms that can be found in drinking-water. It is impossible for a laboratory to look for every one, every time a sample of water is submitted. So, what is to be done? Do you look for the most dangerous or the most common? Neither – what is needed is an organism that will be representative of all of them, that is an indicator of potential problems. Unfortunately, the ideal organism for this task does not exist and the ones that are used are therefore subject to occasional criticism.

It is important to emphasize what we use indicator organisms for. The presence of faecal indicators in water points to the potential for pathogenic organisms to be present. The indicator may show that pathogens could be present somewhere in the supply or that they have not been removed by treatment. *E. coli* or faecal coliforms are not used to find out the number of *E. coli* bacteria in the water (although this has become slightly more of a problem with the growing importance of *E. coli* O157, causing people to identify the organism only in terms of its pathogenicity). Indicator organisms are there to tell us whether material from someone or something's gastrointestinal tract has entered the water because, if such an event has occurred once, unless something is done to prevent it, it can happen again.

Total coliforms, the other commonly used indicator organisms, indicate the efficiency of water treatment or of post-treatment contamination. Some people interpret their presence in water as showing the degree of general contamination, but they are particularly useful where good quality raw water undergoes treatment. If there are no faecal coliforms in the raw water and none in the finished product, you need a substitute to show that your treatment system is efficient; total coliforms are this surrogate. Total coliforms have been used for many years for this purpose and despite the occasional complaint, have served the water industry and consumers well. They are not particularly useful for untreated small water supplies however, where the water is neither regularly tested nor often of good quality.

What would the properties of the ideal indicator organism be? Because the vast majority of waterborne pathogenic organisms are spread via the faecal–oral route, indicator organisms must reflect this fact. Therefore the ideal indicator would have the following characteristics:

- Not be present if there were no possibility of pathogenic organisms being in the water. That is, if it cannot be isolated, the water is safe to drink.
- Survive in the environment longer than other pathogens and be less susceptible to disinfection. Thus any pathogens that were present in the water would have expired before the indicator.
- Survive for a reasonable length of time in the environment, so that remote pollution incidents will be detected.
- Be excreted by humans and all animals capable of excreting zoonotic pathogens.
- Not be found as naturally occurring organisms in the environment, thus if found, they have definitely come from a gastrointestinal tract.
- Be excreted in large numbers, so that even if contamination is slight and the dilution of the contaminant substantial, the organism can still be isolated in the laboratory.
- Be easy to grow and enumerate in the laboratory. This is so that highly specialized personnel are not required before the organism can be isolated. Enumeration should be simple and cheap so that the methods can be used all over the world.
- Be capable of being grown by a suitable methodology in the laboratory so that only that specific organism is isolated. Therefore, other organisms will not be suspected of being the indicator and falsely classify the water as being contaminated when it is not.
- Be safe to grow in the laboratory, so that not too many laboratory technicians become infected, therefore costing excessive amounts of money in training and job advertising costs.

There is also an argument that indicators should differentiate between contamination by human and animal excreta. This may be useful when trying to identify a source of contamination, perhaps following a disease outbreak or a previous sampling failure. Whether the contamination comes from a house toilet and sewage system or from a field of cows is important in order that the problem can be found and put right. Sorbitol-fermenting bifidobacteria have been identified as specifically indicating human faecal pollution. *Streptococcus bovis* is specific to animals in temperate climates and *Rhodococcus coprophilus* is specific to animals in tropical climates (Mara and Oragui, 1985). Specialized laboratories can also provide tracer organisms that will perform this task.

The difference from a disease point of view seems less important. A person with a private water supply that was contaminated with faecal coliforms once asked me whether the faecal material was from a human or an animal. The discussion centred on his belief that human excrement was intrinsically more dangerous. In fact, he said he didn't mind drinking 'the odd bit of dog s...t', but that if it was human, it was a different situation. This is not an argument that I find myself being particularly in sympathy with. I would suggest that it is not important which type of creature the faecal material comes from and that his preference for non-human excrement was aesthetic rather than scientific. It could in fact be argued that as humans are generally healthy, in most of the Western world at least, they will not excrete large amounts of pathogenic material close to private water supplies (Mara, 2003). Whereas animals grazing in a field may pick up many diseases and furthermore will be excreting them in much larger numbers. Thus animal excrement in drinking-water may be more of a risk than water containing the human sort.

What are the indicator organisms that we use and what are their characteristics? The primary one is *E. coli*, sometimes also known as faecal coliforms or thermotolerant coliforms. Scientific terms change over the years to reflect continuing accuracy or uncertainty and this test is one of them. Although the vast majority of faecal coliforms are in fact *E. coli*, a few of them are from other closely related genera but not the actual organism. Research has recently found non-faecal and non-*E. coli* coliforms in *E. coli* positive water samples. WHO has therefore proposed that the terms *E. coli* and faecal coliforms be replaced with 'thermotolerant coliforms'. To the water safety practitioner, these terms appear to be fairly interchangeable and they all indicate the potential presence of pathogens. The 'faecal' coliforms that are not strictly faecal in origin are found in organically rich industrial effluent and rotting plant material or soil. *E. coli* however, has been successfully used to identify faecal contamination for many years and has proved its worth in nearly all the 'ideal' categories' listed earlier. It is particularly easy and cheap to find and enumerate. It is also exclusively of gastrointestinal origin and is excreted in large numbers (10^9 per gram of faeces).

Criticism of *E. coli* as an indicator has centred on three problems. The first is that it does not survive in the environment as long as some of the pathogens – enteroviruses and protozoa can both be found in water where all the coliforms have died of old age or environmental stress. The second is that with the emergence of *E. coli* O157 and other toxin-producing members of the family, faecal coliforms are no longer without risk in the laboratory. The third problem is to do with the sensitivity of the tests used in the laboratory. There are three ways to test for coliforms. The first is the multiple-tube method; different dilutions of the water to be tested are incubated in tubes in a media where the bacteria metabolize and produce gas. The number of tubes that have gas in them give a 'most probable number' of coliforms in the water. This test is not used very often nowadays in laboratories testing private water supplies, although it is used every day at water treatment works (the standard for *E. coli* is zero so you only need to incubate one tube with 50 ml of water and if it has gas in it, the water has failed). One of its strengths however is that it can be carried out in the most basic of laboratories, so it is sometimes still used in developing countries. The multiple-tube method is also used as a back up; the two other tests are not as reliable with highly contaminated raw waters.

The second method, known as membrane filtration, is the standard European technique. The test samples are incubated at 37°C for total coliforms and 44°C for faecal coliforms. The difference in temperature utilizes the ability of faecal coliforms to survive at temperatures that would kill any non-faecal coliforms in the water. This is because they have recently been inside a mammal's warm gut and the non-faecal coliforms are accustomed to life at ambient temperatures. After incubation, the number of yellow colonies that have grown on the filter are counted and this gives a number for the bacteria in the water. Accuracy tends to drop with both very few and very numerous colonies. This stage of the process is known as the presumptive test. Although any colonies that grow are highly likely to be faecal coliforms, the laboratory cannot be certain until a series of additional tests to confirm the result have been carried out, at which point they are referred to as confirmed coliforms. To be absolutely sure that faecal coliforms are present can take over fifty hours from receipt of the samples to final confirmation. This is a long time to wait if drinking-water is contaminated.

A little-known problem with this method is in developing countries where confirmative testing may not occur. As discussed earlier, the initial testing includes incubation at a relatively high temperature so that the heat shock will kill bacteria used to European ambient temperatures. Unfortunately, in hot countries this incubation temperature may not be adequate to heat shock the non-faecal coliforms; Pinfold (1993), for instance, found that in Northeast Thailand, water jars that stored rainwater were reported as being contaminated with faecal material because the laboratories were finding yellow colonies on the growth medium. As the temperature of the water in the rain jars was generally 35°C or more he realized that the colonies were environmental coliforms and not faecal in origin after all. Previously to this work, the rainwater jars were being condemned as an unsafe method of water collection. Pinfold showed that the water in the jars was, in fact, of adequate quality, provided it was not contaminated post-collection.

The third technique, which is the standard one used in the US, is a simpler test known as the 'presence–absence' method. This indicates whether coliform bacteria are in the water and not the numbers per 100 ml. Presence–absence tests are proprietary systems and are therefore usually known by the trade name of the company that produces them. The most popular is the 'Colilert' system and laboratories usually refer to the test by this name. Membrane filtration was operated in the US until presence–absence was formally approved after several years of testing for accuracy by the Environmental Protection Agency.

The test period can be either eighteen or twenty-four hours. They are equally accurate and the laboratory decides which one to use based on when the sample is received. This is so that the results can be read during normal working hours. Any water sample containing coliforms turns yellow during incubation. Any samples that have turned yellow are then checked with an ultraviolet lamp in a dark room. If any fluoresce, faecal coliforms are present.

The literature comparing the two methods has indicated that presence–absence is the more sensitive (Gleeson and Gray, 1997) and is quicker. Where bulk analyses are being made, the cost of analysis is similar for both. Another advantage of the presence–absence test is that one examination provides results for both total and faecal coliforms. This is a major step forward as membrane filtration needs two separate samples to be taken and incubated at different temperatures.

Are there any advantages to the membrane filtration system? Is it better to know how many colonies there are in a water supply and thus the degree of contamination? In private supplies, it is a little naive to think that a large or small number of colonies will be truly representative of any particular water supply. The UK and US standards for total and faecal coliforms are the same, that is nil colony forming units (cfu) per 100 ml, so if any coliform bacteria are found, the sample has failed and the water is contaminated. Thus you only strictly need a test that says whether coliforms are present or not. It has been argued (Sartory, 2003) that a good laboratory can tell a lot from the nature of the colonies on a filter but this only holds good for a water supply that is examined on a regular basis. This will not of course, apply to the vast majority of small supplies.

For private water supplies therefore, the 'Colilert' system or one of the other proprietary brands appears to be the best method of identifying *E. coli*. As most public water samples prove negative, the time delay before getting a result may be a minor consideration, but as over 70 per cent of private water supplies are contaminated at some time during the year, the time difference before reporting may enable an intervention such as a boil notice to be applied more swiftly. A final consideration on this point is that presently, half of all negative private water supply samples would have been positive if the sample had been taken at another time, perhaps a more sensitive test would reduce this number of false negative results.

Because they have been so valuable over the years and WHO consider them to fulfil more of the categories for an ideal indicator than any others, thermotolerant coliforms will continue to be the primary indicator of choice for many years to come. *E. coli* in well water, for instance, has been found to be significantly associated with gastrointestinal illness as part of a one-year study of rural households in Canada (Raina *et al.*, 1999). It is unlikely that laboratories would want to abandon this well-tried organism, even though some criticisms have been levelled at it. It is possible however that Europe will follow the US lead, and start to use the presence–absence test. It will be difficult to abandon membrane filtration but I suspect that once it is done, everyone will wonder why it took so long to change over.

Because of the problems with faecal coliforms, particularly its longevity, other indicators should be used in conjunction with them to find evidence of pathogenic organisms that may otherwise be missed. *Enterococci* (previously known as faecal streptococci) are very useful for this purpose. Many studies, particularly those of Fewtrell and Kay (1996) and Clapham (1997), have found evidence of faecal contamination of small water systems using *Enterococci* where *E. coli* has not been isolated. *Enterococci* indicate contamination that occurred either at a greater distance from the source or further back in time than would be identified by coliforms. European and UK legislation has a standard for *Enterococci* (nil cfu per 100 ml) but the test is not routinely used for small supplies. It is however invaluable and should be an integral part of the basic suite of indicators in any monitoring system. During a study of private water supplies in the UK (Clapham, 1997), *Enterococci* were found to positively correlate with *Cryptosporidium* and *Giardia* oocysts, whereas faecal coliforms did not.

Another useful indicator organism for small water systems is *Clostridium*. *Clostridium perfringens* (formally *Clostridium welchii*) is a spore-forming organism that can survive for a long time in the environment; therefore it mimics some of the survival characteristics of the cyst-producing protozoa. In the study referred to previously, *Clostridium* was also found to positively correlate with *Cryptosporidium* and *Giardia*. *Clostridium* will be found in water from a further distance away in time or space than *Enterococci*. If found in water that has neither *Enterococci* nor faecal coliforms, *Clostridium* may indicate pollution from such a distance away that very few viable pathogens are still present in the water. Some people suggest that testing with other indicators should be carried out to confirm *Clostridium* results, rather than automatically applying a boil water notice or demanding that treatment is installed. Although confirmatory tests of a water supply are always useful, if a few *Clostridium* colonies are found then faecal contamination may be present, even if from some distance away. The legal limit for *Clostridium*, as for the other possible indicators of faecal contamination, is nil cfu per 100 ml.

To obtain a general indication of the quality of drinking-water, the total plate count assay is sometimes used. This is where a plated sample is incubated at a lower temperature than for coliforms; so that nearly all the bacteria in the water will form colonies. Environmental, but normally non-pathogenic organisms such as *Pseudomonas* or *Aeromonas* will be found by this method. It should only be employed to check individual supplies over a period of time because the test is used to identify significant changes in the number of bacteria present. A one-off test will not really tell you much – the water may contain any number of naturally occurring organisms without affecting health. The legislational standard for a plate count is therefore that the sample does not have a significantly larger number of colonies than measured in previous tests. If for instance, the count goes up by a factor of ten, there may have been a pollution event or a treatment failure. As most small water supplies are tested infrequently, a plate count is not a normal requirement of sampling. With some of the more regularly tested larger private supplies however, a plate count can be a useful addition to a sampler's armoury.

Outbreaks of disease associated with small water supplies

This section discusses a number of waterborne outbreaks associated with private water supplies in the UK. I have included them to show the nature and extent of the outbreaks that can occur from this source and to dispel the idea that private water supplies do not cause much disease. I would like to thank the authors of the original papers for their detailed reports and investigative work.

(1) Medieval Training Centre On 20 May 1993, a General Practitioner asked the Department of Public Health in Northumberland to investigate an outbreak of vomiting and diarrhoea associated with a private water supply. The supply served a medieval manor house used as a training centre. The supply served the manor, a cafeteria and a number of estate houses. The outbreak started on 16 May and caused 32 students, 6 tutors and 5 farm workers to become ill, 3 of whom were admitted to hospital. There were 120 people who lived or worked in the manor house, with a 32 per cent attack rate. The farm workers were permanently resident in other properties on the estate and out of a population of 80 there was only a 6 per cent attack rate. The supply had last been tested in February 1993 and found to be satisfactory. Faecal samples were taken from some patients and examined for *Salmonella*, *Shigella*, *Campylobacter*, *Cryptosporidium* and viruses. Some 11 of the 20 stool samples contained pathogens; 5 had *Campylobacter jejuni*, 4 *Cryptosporidium* and in 2 cases, both. Water samples were also taken and found to contain high counts of *E. coli*. When these results were received, a boil water notice was put on the supply and the owner de-sludged the tanks and disinfected the supply. Only one person became ill after the boil order was put on and that happened the day after, so they must have already been infected.

The water system was put in during the nineteenth century and came from several different springs. Pipework from the springs to a central large storage tank was mainly plastic. This had replaced some earlier Victorian cast-iron pipes. When the pipes were replaced, some of the springs were decommissioned. However, the original cast-iron pipes had been left in place and to seal it from the rest of the supply, a stopper was put in the collection chamber of one of the unused springs. Once the *E. coli* had been found, several other water samples were taken to try and locate the exact source of the problem. The sampling indicated it was the large storage tank or something further upstream. A search by environmental health officers found three dead lambs in the collection box from the spring that was no longer used. Unfortunately, the stopper that sealed it from the rest of the supply had perished and was allowing water through. As part of the investigation, another inlet to the storage tank was found but its source never located. To make matters worse, the field surrounding the storage tank and collection chambers had been sprayed with animal slurry five days before the outbreak. This was quickly followed by a period of unusually heavy rain. It is possible that the rain had washed the slurry into one of the chambers.

The treatment on site was with a 5 μm filter followed by ultraviolet disinfection. The ultraviolet light in the student block, where the highest incidence occurred, was coated with a brown deposit, probably iron and this would have seriously reduced its effectiveness. Manufacturers recommend replacement frequencies for their tubes but also say that they will be less effective if coated with deposit. The lamps were replaced at the recommended frequency, but the deposits were not attended to. In addition, one of the people who were ill drank water from a wash-hand basin that was supplied with untreated water.

In their comments, Duke *et al.* (1996) noted that they could not ascertain whether the dead lambs or the sprayed slurry caused the illness (the lambs were not tested for pathogens). They also said that the best way to prevent waterborne cryptosporidiosis is to avoid pollution of a water system with contaminated water or livestock waste. Why they should only refer to *Cryptosporidium*, I am not sure; this advice applies to all pathogenic organisms. The paper talked about the difficulty of locating and maintaining private water supplies and the investigation team pointed out that routine sampling, in line with legislative requirements, may not guarantee against 'episodic contamination'. It certainly did not in this case.

The outbreak exhibits several classic problems with small water supplies. There was the lack of a regular and accurate sanitary survey – three dead lambs could not have been that difficult to spot. Periodic monitoring was relied upon to spot problems, which it will not. There was a reliance on simple instructions, rather than taking an informed approach, for example when the lamps became coated with iron. This also points to a lack of maintenance, as does the need to have drinking-water tanks de-sludged. Inappropriate construction was also a factor – that is relying on a 'fail to danger' stopper and having a spring collection box that let several animals fall in and die. There is also evidence of inadequate protection against contamination, that is spraying the various storage chambers with animal slurry and lastly, the association with heavy rain. It was the classic accident waiting to happen. Interestingly, the legislative system in the US would probably have prevented this incident. The system would be classed as a community supply and thus require monthly testing, together with chlorination of the supply and daily testing for free chlorine residual. It would also probably have been classed as being under the direct influence of surface water, with extra requirements to ensure its safety. Finally, there would have been an annual inspection of the system by a certified engineer, who should have found the potential problems and had them attended to.

(2) *Research Laboratory, England* This incident occurred at a biological research institute in the South Cambridge area of southern England in 1995. The institute and the adjacent village had a mixture of both private and public water supplies. The private supply was sourced from two boreholes; one was in a boiler house, the other next to a car park. Each borehole was used for one week with one week off. Treatment consisted of water softening, chlorination, storage in two holding tanks and then pumping to header tanks for distribution. The institute and village also had a small private sewage treatment works. The institute's main sewer conveyed human, animal and laboratory waste to a sump and pumping station, 25 m from the car park borehole. One sewer passed within 1.5 m of the borehole. A new laboratory was being built on the site and heavy lorries had been driven on roads above the sewer between the borehole and the sewage pumping station.

In late May, the water from the private supply began to smell like the antiseptic TCP (Trichlorophenol). The borehole in use at the time was the one adjacent to the car park. There were no similar complaints about the mains water supply. The odour problem was quickly followed by diarrhoea and vomiting amongst people who used the private supply, both villagers and staff. Out of a population of 700 drinking the water, 58 (8 per cent) were ill. Boil water advice was issued on 23rd May and the car park borehole taken out of commission.

Environmental health officers were informed about the problem the day after and took samples of water from various points in the system. Twenty-four patients were visited and faecal samples were obtained. These were examined for a range of bacteria, viruses and protozoa associated with faecally contaminated food and water. Two people were found to have *Campylobacter* and two others *Giardia*. None of the other patients were found to have any pathogens in their stools. Samples of the water were analysed for *E. coli*, total coliforms and chemical contamination. The water was found to contain both coliforms and phenolic contamination. There were greater than 1,000 cfu per 100 ml for both *E. coli* and total coliforms. Chemical examination found xylenes, chlorobenzenes and hydrocarbons in the water, all below WHO Guidelines. However, 0.56 µg/l of total phenols were discovered, which is slightly above the UK legal limit and WHO Guideline Value of 0.50 µg/l. Consumption of phenols can cause vomiting and diarrhoea.

The owners of the supply quickly realized that heavy lorries must have damaged the sewer and contaminated the borehole. This theory was subsequently confirmed when excavations near the car park borehole showed that the area around the sewer was damp. It had not rained for some time and the adjacent areas were comparatively dry. It is possible that the sewer had been leaking into the borehole for a few days and the onset of illness merely coincided with a release of chemicals from the laboratory causing the smell problems. The environmental health officers served a notice requiring the institute to obtain its drinking-water from the public supply, to have tankers deliver safe water until the supply was reconnected and for a day nursery to be closed until the works were completed. Water for animal use remained on the private supply. Forty-one people became ill before the water was disconnected on 26 May; a further seventeen became ill up to the 13 June. It is probable, therefore, that pathogenic organisms were in the distribution system for some time.

The cases of disease were mild and there were no hospital admissions but there was the possibility of more serious problems amongst the young and old people on the supply. The outbreak control team felt that the siting of the sewer so close to the borehole was ill-advised and would not be acceptable for a public water supply. Furthermore, the paper (Reacher *et al.*, 1999) detailing the incident suggested that the legislation on private water supplies should be extended to require a system of accreditation based on a risk assessment by experienced water engineers. This is similar to the system of prior approval for small community systems in the US, where boreholes cannot be drilled within fifty feet (15.24 m) of any sewer. Because of the number of people supplied by the system, it would also be classed as a community system in the US. The authors of the paper considered that under the present UK legislative regime, owners of private supplies or environmental health officers had insufficient expertise to ensure adequate standards of engineering and operation of small water systems. This is an interesting observation as, apart from the dangerous proximity of the sewer, the supply was chlorinated and appeared to have been operated reasonably well; previous regular sampling had not identified any problems, but of course, that is no guarantee of safety. There may have been a problem with the chlorination device, or the massive loading after the sewer started leaking may have allowed some pathogens to survive. In the US, daily readings of free chlorine around the system would have identified this problem. There is also the possibility that the organism causing the problem was resistant to chlorination. This leads to the supposition that *Giardia* may have been the cause of the illnesses, but does not explain the presence of *Campylobacter*.

The people in charge of the supply appeared to take sensible decisions when the problems were brought to their attention. However, the call for a sanitary survey for small water systems is commendable and one that encourages a multiple-barrier approach to waterborne disease prevention. This incident reinforces the idea that spending time and money on preventing pollution entering a water supply by correct siting and adequate design and construction is much better than relying on a treatment system to remove any contamination once it occurs.

(3) Scottish campsite   Another outbreak of interest is discussed very briefly in the section on *E. coli* in Chapter 8, but a fuller discussion is useful. A small campsite on the West Coast of Scotland had an untreated, unprotected private spring-water supply that also supplied five local houses. In the summer of 1999, a sample of water was taken that contained 15 cfu of *E. coli* per 100 ml. This resulted in a 'boil water' notice being issued. A week later there were six cases of *E. coli* O157, all of whom were visitors to the campsite with limited exposure to the water. One child had only brushed its teeth in unboiled water, a 4-year old had had one glass of water and a 5-year old only drank the water over a two or three hour period. This indicates that the infectious dose for this pathogen is very low, particularly where children are concerned. None of the local residents become ill, even though they also drank the water and *E. coli* O157 was found in the water supply to one of the houses.

Sheep and deer grazed freely around the source of the supply and the probable cause of the outbreak was their contaminated droppings getting washed into the spring. Sheep and deer faeces collected from around the source contained *E. coli* O157 of the same phage type found in the water and the patients. The authors of the published report (License *et al.*, 2001) say that their findings raise questions over the acceptability of allowing the use of untreated private water supplies to temporary populations. In the US, this supply would be required by law to be chlorinated, as it would be classed as a transient non-community supply. As a minimum, the authors recommended that visitors to private water supplies be told about the nature of the water and the risks involved. Their report also brings into question the usefulness of present methods of ensuring boil water notices are sufficiently observed.

6 Waterborne diseases

Because small supplies are usually only tested for indicator organisms, it is rare that the authorities responsible for their safety know which specific pathogenic organisms may be present. It is important to be mindful of these organisms so their *modus operandi* can be studied and preventative measures put in place. The aim of Chapters 7–13 is to examine all the organisms in small water systems that are known to cause disease. Each chapter will discuss the organism, explain what it is, describe what its usual symptoms are and how it causes illness. A number of outbreaks associated with small water supplies will then be briefly discussed. These include both private supplies and smaller public supplies, where pertinent lessons can be learnt. This should give a reasonably thorough understanding of the ways that systems can fail and enable someone with responsibility for the control of small water systems to initiate preventative measures to stop similar outbreaks occurring.

The information will also provide a reasonable supply of useful facts when trying to encourage the improvement of water supplies through health education. Recent European legislation has allowed individual states to exempt any water supply serving less than fifty people (European Union, 1998). Where countries continue to require drinking-water to be of a particular standard for smaller supplies, they may consider formal action to be inappropriate and prefer to use persuasion to improve drinking-water safety. Presently, many authorities are loath to use strict enforcement for what is often seen as a personal preference to drink untreated water. Many people on private supplies are also proud of the fact that their water is untreated and free from chemicals. They consider this to be 'natural' and therefore the healthy option. Whilst in my opinion this argument leaves something to be desired – cholera and arsenic are both 'natural' – the belief is heartfelt and should not just be patronizingly dismissed. It is important to be able to work with people and persuade them to protect their supply from contamination. Enforcing authorities should be able to justify their requirements rather than merely relying on legislative powers to bring about improvements. Of course this will not always work, but formal enforcement should always be seen as a last rather than a first option.

This belief that formal enforcement is not always an appropriate approach is particularly true for the very small supply, particularly where the water is supplied to a single property. This type of supply (a 'Class F' supply in the UK) will normally only affect the household being supplied and there is a strong case against using legislation to force an improvement in water quality. If the householder is in full possession of all the facts then, the argument goes, they should be allowed to do what they like. A responsible authority must ensure that information for consumers is sufficient, accurate and specific to the type of supply. As I have discussed previously, however, the situation can be more complicated than just a matter of freedom from an interfering bureaucracy and personal choice. If the supply is well protected, the water is pathogen-free and chemically fairly unadulterated, with a responsible person maintaining it; there should be no problem. If the supply is one of the over 70 per cent in the UK that are contaminated with faecal material at sometime during the year, there may be a more serious problem. The chapters that follow should allow the reader to give detailed and specific examples of where illness has been caused by water from small supplies and why.

There are several water-related diseases that cause untold suffering in many of the hot countries of the world and millions of people are being killed or debilitated by them. They are not acquired from drinking-water but water is an integral part of the disease. Some of the parasites and vectors will breed in drinking-water sources such as open wells or surface supplies. To ensure that this book is comprehensive, I have included a brief explanation of these organisms, their life cycles and a description of the diseases they cause.

It should also be remembered when discussing contamination by specific pathogens that there is not a linear dose–response relationship between water consumption and pathogen intake. A recent study found that in a simulated model of a cryptosporidiosis outbreak, the majority of consumers never ingest a single oocyst. Of those that do, over half only ingested one a day. However, one in 10,000 consumers is exposed to very high densities of up to 150 oocysts (Gale, 2001). This is close to the number that would cause illness in half the people drinking the water (DuPont *et al.*, 1995).

In the writing of this chapter, I am particularly indebted to Professor Paul Hunter and his essential reference work, *Waterborne Disease: Epidemiology and Ecology* (Hunter, 1997), published by John Wiley and Sons. I have tried to keep the details of individual organisms specific to those encountered in small water systems and his book has been an invaluable help. Where briefly referred-to outbreaks are mentioned they are rarely referenced directly, as that would not be particularly helpful to the flow of the text. I would however like to earnestly apologize to anyone who feels that they should have been referenced but have not been. I would recommend that where any specific disease is encountered, Professor Hunter's book be referred to for a more complete picture.

7 Viral diseases

Waterborne viral diseases produce millions of cases of illness and are a major cause of gastroenteritis worldwide. Along with the protozoa they are much more likely to be the cause of recent outbreaks of waterborne disease in developed countries than bacteria. There are several classes for viral disease and they are each dealt with here.

Enteroviral diseases

Illnesses in this category are caused by enteroviruses within the picornavirus group (pico = very small). There are seventy-one different types and they were initially placed in four categories – Poliovirus, Coxsackie virus A, Coxsackie virus B and Echovirus. After the sixty-seventh type was discovered, virologists seem to have been tired of putting them in these categories and just called them 'enterovirus 68', 'enterovirus 69', etc. They are relatively easily cultured for viruses, are nearly always present in faeces and are often found in contaminated drinking-water. They are therefore sometimes used by virologists as indicators of faecal pollution.

Despite their ubiquity, they have not caused a great deal of documented waterborne illness and as they are always present in the gut, it is important to realize that they do not automatically cause disease. For those that do, illness follows ingestion of large numbers of the particular virus (normally in the region of 10^5 or 10^6). Different viruses cause different symptoms, the most notorious being the paralysis associated with polio. Surprisingly, only 0.01 per cent to 0.001 per cent of people who ingest poliovirus show any symptoms. For those that do acquire it, however, the illness can lead to a seriously debilitating condition. This can be fatal when the paralysis begins to affect respiratory functions. Other viruses in this group can cause serious problems such as meningitis, pneumonitis, pleurisy, hepatitis, conjunctivitis, encephalitis, croup and hand, foot and mouth disease. This last ailment is the one that is sometimes confused with the animal disease foot and mouth. However, there has never been a documented case of foot and mouth in a human.

Infection is acquired through the faecal–oral route, either from drinking-water, by person-to-person spread, in spray from coughing or via an intermediate object (also known as a fomite, which is defined as a substance capable of carrying infection, such as dust or hairs). Incubation periods and duration of the various infections depend on the disease. Hunter (1997) found no outbreaks associated with small water systems and only one public water supply outbreak. He considers that this is partly due to the general under-reporting of waterborne outbreaks of disease but also to the varied nature of the illness associated with this group of viruses.

Viral gastroenteritis

As they are thought to cause 10 million deaths per year worldwide and up to 5 billion cases of illness, diarrhoeal disease from this group of viruses is very important. They are considered to be the number one cause of acute gastroenteritis in nearly every country in the world, with increasing numbers of outbreaks in the temperate regions of the world. In the past, they have been difficult to identify in the laboratory and do not grow in culture cells, thus they are known as the 'fastidious' viruses. You might question however, how something that infects five billion people a year can be called 'fastidious'.

The types of virus associated with waterborne viral gastroenteritis are the Rotavirus (RNA, 72 nm), Adenovirus (DNA, 75 nm), Astrovirus (RNA, 28 nm), Coronavirus (RNA, >100 nm) and Norovirus (RNA, 30–35 nm). Norovirus was previously known as small round structured virus, Calicivirus, Norwalk virus or Norwalk-like virus and has recently been approved as the name for this group of viruses. I particularly like the name small round structured virus and occasionally picture the scientist who first discovered it. After the joy of discovery comes the question of naming. As the great brain whirls into action once again, the scientist thinks, 'well its definitely very small and it looks very round – I know, let's call it the small round structured virus'. Another interestingly named virus associated with illness is the sin nombre virus, this is merely the Spanish for 'no name virus'. Norovirus is considered to be the cause of 96 per cent of outbreaks of non-bacterial gastroenteritis in the United States (Boccia *et al.*, 2002).

Incubation time is very short, a few hours for Norovirus and about a day for Rotavirus. Onset is abrupt with vomiting, fever and diarrhoea. With Norovirus, patients are often unable to find a toilet quickly enough when the vomiting starts. The illness usually lasts a day or two, although diarrhoea in rotaviral infections can continue for up to five days. The illness stops by itself and medical help usually consists of the replacement of fluid lost because of the diarrhoea and vomiting. Spread of the virus is via person-to-person contact, in contaminated food or water and on projectile vomit. For some food-related Norovirus illnesses, a customer vomiting in the dining area has caused the outbreak, rather than bad practice in the kitchen. One study found an association between risk of contracting the disease and proximity to the ill person in the restaurant.

Incidents associated with small water systems are described here:

- In 1987, an explosive outbreak in an elementary school in Washington State was the first widely noted waterborne occurrence. Well water stored in a pressure tank was in a room flooded by sewage from an overflowing septic tank and it was thought that back-siphonage into a water overflow pipe caused the outbreak; 72 per cent of 490 teachers and pupils became ill.
- Lack of chlorination of well water at a rural summer camp in Pennsylvania in 1978.
- A poorly protected shallow well in Brazil in 1980.
- In Georgia, in 1982, a Norwalk virus outbreak was caused by a supply consisting of a well and three springs. One of the springs was inside a pigpen and surrounded by septic tanks. An unsurprised Hunter notes that a sample of the water in the spring revealed the presence of faecal coliforms. The outbreak was preceded by heavy rain.
- An ice-making plant in New England made ice containing Norwalk virus after heavy rain flooded the three wells and the septic tank at the plant in 1987.
- In Arizona in 1989, an outbreak occurred where treated sewage leaked through 60–80 m of fractured sandstone and limestone into water filled caverns supplying boreholes. This was considered to be a problem of a unique geological situation.
- Norwalk virus caused an outbreak at a caravan park in Australia in 1989. Untreated river water was kept in tanks that were contaminated by sewage. Nearly 27 per cent of the people on the site drank the river water although no one was supposed to, it was sign-posted as for washing purposes only.
- An outbreak reported in 1997 (Beller *et al.*, 1997) was caused by a well at a cafe used for tourists and travellers on bus journeys between Alaska and the Canadian Yukon. The café's septic tank contaminated the unchlorinated water supply.
- In March 1998, there was an outbreak of Norovirus in a small Finnish community. The cause was thought to have been inadequate chlorination (Kukkula *et al.*, 1999).
- A cross-connection between river water and the public supply was thought to have led to food poisoning for 130 people in South Wales and the South-west of England in 1999. It is thought that the contaminated mains led to problems in a bakery when the water was used to made custard for confectionery. The outbreak was caused by small round structured virus (Brugha *et al.*, 1999).
- A break in a long spur from a mains water pipe in a Southern Italian holiday resort was 'fixed' by connecting the resort's water tank to an unused irrigation system supply. The outbreak occurred in July 2000 and was the first reported outbreak of a Norwalk-like virus in Italy.

This is mainly due to the fact that there is no surveillance system for non-bacterial gastroenteritis in that country (Boccia *et al.*, 2002). Over 340 people became ill during the outbreak and it took the local health unit over two weeks to notify the central authorities about the outbreak. The measured attack rate was three times higher in staff than in visitors to the resort but this is likely to be because the staff were questioned about any illness they may have had, whereas many tourists with mild symptoms would not have bothered reporting their illness to their doctor. The outbreak had three peaks as new visitors each weekend became ill about two or three days after arrival. Although using water from the resort's storage tank was banned for drinking and in the kitchens, it was still used for showers and making ice. Techniques for disinfection of the water storage tanks were poor and may explain why the outbreak was so long-lived.

Viral hepatitis

Hepatitis means inflammation of the liver. There are several specific viruses known to cause hepatitis. They are classified alphabetically A to E. Hepatitis E is also sometimes confusingly known as non-A, non-B hepatitis virus. It is possible that another type – hepatitis F – will officially be added soon, but the final decision has not been made. The important hepatitis viruses in relation to water supplies are hepatitis A and hepatitis E. The A and E hepatitis viruses are also important because they are regularly the cause of food poisoning outbreaks. The food poisoning can be caused by contamination from someone with the disease preparing the food or through contaminated water being part of the preparation process. Infection is also acquired through the consumption of raw shellfish that have been grown in contaminated seawater. The viruses are regularly found in recreational waters that have sewage treatment outfalls or other sources of faecal contamination upstream.

Infection is through the faecal–oral route or person-to-person contact. Incubation times are slow: 10–50 days for hepatitis A and slightly longer for hepatitis E, at 21–60 days. Many people, particularly children, have subclinical infections and show no symptoms of the disease. Where the infection becomes apparent, symptoms start as a general fever and malaise, sometimes with nausea and stomach pains. A quarter of patients go on to develop jaundice due to the damage caused to the liver. The disease is usually self-limiting and most patients will recover with rest and painkillers. During the acute stages of both types of the disease there is no treatment available. Vaccination against hepatitis A is very effective and gives long-lasting protection, but there is no vaccine for hepatitis E.

Hepatitis A is a picornavirus so the virus particles are small (27–32 nm). It is excreted in large number and survives for a long time in water. It is highly infectious and common around the world. There are many outbreaks in temperate countries, but it is even more prevalent in hot countries. It is also one of the oldest diseases known to mankind (WHO, 2000b) and epidemics have been reported since the fifth century (Anon, 2002c). It is contracted at least a hundred times more often than cholera or typhoid and there are about 1.4 million cases per year. The only natural host of hepatitis A is man. Because it is so common it has had many names, including infectious hepatitis, epidemic jaundice, campaign jaundice and catarrhal jaundice.

Infection with hepatitis A usually occurs early in life and is associated with poverty and overcrowded living conditions. The disease is also associated with inadequate sanitation and contaminated drinking-water. In developing countries, most children will have had the disease before they are nine years old. As conditions improve, with better sanitation and drinking-water, age of onset becomes delayed. Then the disease changes from being endemic amongst young children (who go on to develop a long immunity), to being sporadically epidemic amongst adults. The disease can then spread quickly amongst an unprotected population from a single pollution event. The hepatitis A virus is rarely fatal (0.001 per cent of cases), but death is more likely the more older the patient is. Once jaundice has started in patients over fifty, it is very likely to be fatal.

Hepatitis E (27–34 nm) was only considered to be a separate disease in 1980 (WHO, 2001a). It was originally classified as a calicivirus, but according to WHO it is now unclassified. It is less infectious than hepatitis A and is only found in countries where faecal contamination of water supplies is common or amongst recently returned visitors. The disease it causes is officially known as enterically transmitted non-A, non-B hepatitis. The hepatitis E virus also has man as the natural host but there is a risk of zoonotic transmission as pigs, sheep, cows and rodents have recently been found to have the disease. Hepatitis E has four major strains, relating to specific geographical regions, these are China, the US, Mexico and Africa with parts of Asia. Hepatitis E is rare amongst children and usually attacks young adults from 15 to 40 years old. Death is more common with hepatitis E than hepatitis A, at 1–2 per cent of cases. Fatalities can also occur during pregnancy. There is no treatment for the acute stage of either of the diseases. Patients do not become symptomless carriers of hepatitis E.

Hepatitis A outbreaks of interest to small water supplies include the following:

- A Special School outbreak in 1969 in Australia where separate taps were provided for river water which was only supposed to be used for washing, and roof water which was for drinking. Children at the school probably drank the river water by mistake because they could not identify the correct supply due to their special needs. This appears to be a problem caused by insufficient time being spent on design considerations.
- Illness from a drinking-water supply to a college in the US, where the pipe ran alongside a sewer for some distance. The water pipe was

normally under pressure, but just prior to the outbreak in 1969, the supply was used for putting out a fire. This reduced the pressure in the pipe and probably allowed surface water and sewage in. Children with hepatitis A had recently played in this surface water. This is another example of design failure and further proof of the rule that 'if anything can go wrong – it will do so – at the worst possible time'.

- A spring water supply to a school that had general surface contamination and overflowing septic tank effluent running into it. This outbreak affected 49 children in Alabama in 1972.
- In Pennsylvania in 1980, 49 people were ill, probably due to a borehole in limestone being within 30 m of a septic tank. Limestone often has natural underground drainage caused by acidic rainwater eating into the rock (karstic) and this allows contamination to pass quickly though what sometimes appears to be solid rock.
- In Kentucky in 1982, tankers delivering untreated drinking-water from a set of springs situated in fractured limestone actually conveyed contaminated water. The limestone had sinkholes in it that collected rainwater. They also collected rubbish, faecal material and animal waste dumped by the local population. The water caused some illness in the community, but at the time the local doctors were 'too busy to report it', so the actual figures are unknown. What is known however is that the water contained 40,000 faecal coliforms per 100 ml and test dyes took two weeks to get from septic tanks into the springs (Bergeisen *et al.*, 1985).

Hepatitis E outbreaks occur in the Developing World and are frequently caused by sewage contaminated surface water (Kashmir – 1979) or from areas where leaky water pipes that are not under constant pressure, run alongside leaking sewage pipes (Madhya Pradesh, India – 1980 and 1989, Kolkata (Calcutta), India – 1980 and Sargodha, Pakistan – 1987). Public water supply related incidents are also associated with surface waters being given inadequate treatment, treatment failure or storage protection failure (Maharashtra, India – 1981, Medea, Algeria – 1980–81, Maun, Botswana – 1985, Delhi, India – 1986–87 and Kanpur, India – 1991). In this last outbreak 79,000 people were thought to have been made ill.

Outbreaks involving small water systems include ones caused by bad plumbing, for example, some toilets were connected to a drinking-water supply (Constantine, Algeria – 1983), shallow wells that were dug next to pit latrines in a Kolkata slum (1980) and unprotected wells in rural areas of Mexico (1986). In Somalia in 1988, 346 people died in an outbreak associated with drinking river water, pond water and well water. The death rate was highest in the well water drinkers at 8.6 per cent.

Insufficient resources are the main cause of this disease. Poor planning or maintenance can make many people ill or worse. However, this is often the result of reduced choices and no money, rather than a lack of knowledge or wilful mismanagement. Viral hepatitis can easily be controlled by simple measures such as chlorination of water supplies.

8 Disease-causing bacteria

The illnesses caused by bacteria in drinking-water stimulated the major improvements in water quality throughout the eighteenth and nineteenth centuries. Bacteria continue to cause serious illness in developing countries worldwide and despite nowadays being a minor problem in public water supplies in developed countries, they remain a serious one in private ones.

Campylobacter

Campylobacter is the most common cause of bacterial gastroenteritis in the UK and many other developed countries. There were over 56,000 reported cases in England and Wales in 2001. The disease usually produces mild, watery diarrhoea but can cause illness up to a severe inflammatory gastroenteritis. It has also been thought to be a factor of Guillain–Barre's syndrome, which is a paralytic condition like polio. The name for the organism comes from the Greek: campylo = curved and bacter = a rod. It is usually considered to be a food poisoning organism, particularly associated with poultry, but also travel to developing countries. The disease is zoonotic and people will become ill from contact with live animals and meat. People who work in situations where they are in contact with animals, either dead or alive, such as vets or butchers, usually become immune to the organism fairly rapidly.

It can be spread via drinking-water and several outbreaks have been associated with private water supplies or drinking from rivers and streams. A UK review found that 7 per cent of the patients studied drank from private water sources, although only 1 per cent of the general population has this type of supply (Frost *et al.*, 2000). The organism is frequently a cause of illness in developing countries where it is hyper-endemic, particularly in children. It is strongly associated with poor quality drinking-water and the absence of sanitation. A study in Bolivia, for example, showed that where stored drinking-water was chlorinated, all diarrhoeal diseases were reduced by 44 per cent and *Campylobacter* diarrhoea by over 80 per cent (WHO, 2000a). However, because it is not usually fatal in children over 6 months old, it is not often monitored in the poorer countries of the world.

WHO have calculated that the incidence amongst the under fives in developing countries is between 40 and 60 per cent. For comparison, in developed countries that figure is nearer three per thousand (0.3 per cent) and for the general population the worldwide figure is about nine per ten thousand (0.09 per cent) (Coker *et al.*, 2002). Under-reporting is obviously a factor for this normally mild disease and estimates vary from only 1 in 7 to 1 in 37 cases being officially reported. Because it is a food poisoning organism, it is a statutory requirement in the UK for doctors to report *Campylobacter* to the proper officer of the local authority under the Public Health (Infectious Diseases) Regulations 1988 (HMSO, 1988). Food poisoning is classed by the Department of Health as 'any disease of an infectious or toxic nature caused by or thought to be caused by the consumption of food or water'.

Campylobacter is normally found in wild fowl and domestic animals. A recent study found that up to 20 per cent of the outside of wrappings of chickens were contaminated with *Campylobacter* (Walbran, 2001). Another UK study found an average *Campylobacter* contamination level for poultry of 50 per cent (63 per cent in fresh chickens and 33 per cent in frozen ones) (Food Standards Agency, 2001). Although the organism is particularly associated with poultry eaten outside the home (Tauxe, 2000) raw milk is another vector. When Scotland introduced compulsory pasteurization of all milk in the 1980s there was a dramatic reduction in campylobacteriosis. *Campylobacter* will survive in water for four weeks at 4°C, but for shorter period at higher temperatures. It also lives longer in clean water than in rivers or sewage; this may be because of pressure from other organisms in highly contaminated water. Chlorination and ultraviolet disinfection will inactivate *Campylobacter*, and reverse osmosis will remove it.

The first well-documented case occurred in 1938 in Illinois and was a milk-related outbreak of diarrhoea. However, the microbiological pioneer Escherich isolated, what now appears to be, *Campylobacter* in 1886 and he thought that it was responsible at the time for several diarrhoeal-related deaths in children. *Campylobacter* was only recognized as a common zoonotic pathogen when Skirrow (1977) introduced a simple laboratory method for its identification. The first officially recognized waterborne outbreak occurred the year after, with 3,000 cases in Bennington, Canada (Vogt *et al.*, 1982). Prior to the introduction of the improved laboratory method, many food and faecal specimens taken as part of an outbreak investigation came back from the laboratory as negative – it is now thought that many of these probably contained *Campylobacter*.

Campylobacter has two main subgroups containing fifteen distinct species. One of the subgroup species is also called thermophilic (some like it hot). This is the group most likely to affect humans and contains *C. jejuni*, which is the most common species isolated in outbreaks. *C. jejuni* itself has two subspecies. The other thermophilic *Campylobacters* are *C. coli*, which is regularly found, *C. lardis* and *C. upsaliensis*. Other species can cause illness in man but the ones of public health significance are *C. jejuni* and *C. coli* with over 99 per cent of cases attributed to these two organisms. Subgroup two includes *C. fetus, C. hyointestinalis, C. concisus, C. mucsalis* and *C. sputorum.* The genera *Arcobacter* and *Helicobacter* are also considered to be part of the Family *Campylobacteriacea.*

The infective dose can be as low as 50 organisms but is usually of the order of 10,000. The organism normally colonizes the small intestine and is sensitive to stomach acid, so it prefers conditions when food is in the stomach, providing the buffering that helps it survive the low pH. Incubation is normally two to four days but can be from one to seven. Diarrhoea and stomach-ache are the most common symptoms and the abdominal pain can occasionally be so bad that it can be mistaken for appendicitis. Vomiting is not normally associated with this disease. Campylobacteriosis normally only lasts two to three days but can be longer. The disease is usually self-limiting and replacement of fluids is generally the only treatment necessary. It will however, respond to antibiotics.

Hunter (1997) has described the majority of outbreaks associated with drinking-water. Causes of contamination from small supplies have included the following:

- A chlorinated, unfiltered surface supply with large amounts of animal waste near where the water was extracted and heavy rain before the outbreak: 3,000 people were ill. This was the first recorded waterborne outbreak (Canada, June 1978).
- A supply using a temporary reservoir that was open to animals. The reservoir was small, stagnant and unchlorinated (Israel, July 1982).
- A private, unchlorinated, borehole supply feeding a UK boarding school, with an open storage tank in a roof space where roosting birds were probably defecating into it (UK, May and June 1983).
- An outbreak in Florida that caused an 879 per cent increase in the sale of anti-diarrhoeal medicine (a figure I find wonderfully accurate): 865 people were made ill, with one case of Guillain–Barre's syndrome. This was probably caused by broken pre- or post-treatment chlorinators failing to remove contamination from birds in open-topped reservoirs. The deep well community system had numerous deficiencies including an unlicensed operator as well as chlorination failure (Sacks *et al.,* 1986) (US, May 1983).
- An outbreak at a power station with 166 cases, where faulty plumbing caused contamination from a lake to enter the mains supply (Canada, March 1985).
- A surface water supply in a rural area that was only chlorinated when the river was high. Even this safety mechanism was delayed for a few hours after the river began to rise (New Zealand, March 1986).
- Seventy-five out of eighty-eight soldiers on manoeuvres became ill after drinking from a stream. The conscripts drank up to 4l of water before

dinner – it must have been very hot and beer supplies must have been very limited (Finland, July and August 1987).

- Chlorinated water from a private borehole where work had recently been carried out on the pipe work. The chlorination levels at the point where the work took place must have been unable to cope with the additional loading (Finland, November and December 1986).

- An untreated surface supply where the collection point was badly designed and insufficiently protected (Canada, June 1987).

- Water from four wells on a campsite that affected 44 staff and campers. The water was neither filtered nor chlorinated. The outbreak occurred a few days after heavy rain and there were farm animals in the surrounding fields (New Zealand, August 1989).

- An unchlorinated rural supply, where the outbreak was preceded by heavy spring run-off (Canada, Spring 1990).

- Untreated water from upland lakes where sheep had access and the spring melt could have washed their faeces into the water (Norway, Spring 1990).

- An outbreak caused by contamination from the faeces of wild animals, probably birds, bats or possums in rainwater storage tanks. This resulted in 23 members of staff at a holiday resort becoming ill. (Merritt *et al.*, 1999) (Queensland, Australia, June 1997).

Recent UK outbreaks from private water supplies include the following:

- A child's birthday party on a farm in 1993. The only people who became ill were children who had drunk from the farm's unchlorinated supply (Furtado *et al.*, 1998).

- A joint outbreak of campylobacteriosis and cryptosporidiosis occurred in an unchlorinated surface water supply at a training college in 1993 (Duke *et al.*, 1996). This is discussed more fully in the previous examples of outbreaks associated with small water supplies.

- Also in May that year (1993) an outbreak occurred at a boarding school in the North West of England. They had had an outbreak of cryptosporidiosis the year before. *Campylobacter coli* was found in a reservoir storing water from a borehole. Chlorine levels were low and faecal coliforms were found in the tap water (Furtado *et al.*, 1998).

- In April 1994, a holiday resort in East Anglia had an outbreak of campylobacteriosis. *Campylobacter jejuni* was found in the drinking-water. The private supply was an unchlorinated borehole with an inadequately maintained ultraviolet system for disinfection (Furtado *et al.*, 1998).

- Eight out of nine people living in flats in a large rural house became ill from *Campylobacter* in September 1994. They used an untreated surface water supply that was found to contain faecal coliforms (Furtado *et al.*, 1998).

- An outbreak in 1995 was caused by leakage from a sewer into a borehole supply (Furtado *et al.*, 1998). This is also discussed more fully in the section on examples of outbreaks associated with small water supplies.
- In June 2002, an outbreak occurred at a campsite in North Wales with an untreated private water supply (Anon, 2003).
- A joint outbreak with *E. coli* O157 happened on another campsite in North Wales in August 2002 (Anon, 2003).

Hunter comments that the majority of these outbreaks were rural and involved unchlorinated water supplies. He also points out the most of the outbreaks associated with this organism were caused by contamination from animal or bird faeces. Furtado *et al.* (1998) note that most of the outbreaks could probably have been avoided if appropriate treatment and monitoring procedures had been followed.

Cholera

Cholera is mainly spread via water and food contaminated with infected faecal matter, although soiled bedding and flies also can transfer the bacteria. The disease produces profuse watery diarrhoea (rice water stools) and can be fatal, particularly in children or the malnourished. Death from cholera can often be avoided if the patient drinks adequate amounts of clean water (with suitable amounts of salt and sugar added) provided they are comparatively healthy to start with. Its cause has been known since 1855 and the German microbiologist Robert Koch identified the organism, *Vibrio cholerae*, in 1884. There are two known serogroups of cholera – O1 and O139. It has spread throughout the world in a series of pandemics (worldwide epidemics) and we are still in the middle of the eighth one, which started in 1991. In fact, WHO considers that the pandemic is increasing rather than abating. The latest one is caused by the El Tor biotype of serotype O1. It has caused outbreaks in South America, Africa and Asia (Anon, 2002c). A recent finding has highlighted the reason why the once harmless cholera bacteria became pathogenic. A tiny thread-like virus known as CTX infects *Vibrio cholerae* and it inserts some of its own genetic material into the bacteria. This includes the gene that produces the toxin that paralyses the intestines of victims (Anon, 2000).

Because its eradication can be achieved by simple public health measures, the history of how a country combats cholera is usually considered to be the history of its public health movement. I have therefore included more detail about this aspect of cholera than for the other diseases examined here. *King Cholera: The Biography of a Disease* by Norman Longmate, published by Hamish Hamilton, has been an invaluable source of information for this section on the history of this terrible but fascinating disease.

It has been suggested that Asiatic cholera or Cholera Morbus as it was known, has been recognized in India since 400 BC, however the history of the modern disease began with the first pandemic in August 1817 (Longmate, 1966). It originated in India near Kolkata (Calcutta), probably following a religious festival coinciding with a newly virulent strain of the disease. Before that the disease would have been endemic in a milder form. It quickly spread throughout Asia, across to Europe and then America, killing millions as it passed. Cholera was a terrifying disease that killed more people in a single episode than any other disease since the Black Death in the thirteenth century. Because the cause was unknown but the approach of the disease was well publicized, people panicked in its wake. In many countries, people were shot if they tried to run away from infected areas because the officials were frightened that they would spread the disease. In Persia, soldiers shot into the air to try to kill the approaching disease. In Germany, they thought the disease was an excuse used by the British to explain the deaths of Indians they had killed. The German peasants thought their aristocracy had taken up this idea and were using it as an excuse to kill them, so they liberated the hospitals and took the patients away.

At the end of 1831, the disease arrived in the UK in the north-eastern port of Sunderland and within a few days many of its inhabitants were dead. Cholera then moved through the whole country leaving thousands dead. Many people claimed that the cause of the visitation was because the people had been sinful and it was a message from God. ('Its special prey is the drunkard, the fornicator, the filthy in body and the filthy in soul' – *The Times*, 1848). Others thought that the disease arose in a poisonous gas (miasma) that was produced by the piles of refuse or the filthy streams and ponds that were a feature of the towns of the period. The people who believed that the noxious vapours caused illness when they were breathed in were called 'miasmists'. Many famous early pioneers of public health, including Florence Nightingale and the sanitary reformer Edwin Chadwick, were miasmists. Although they were eventually proved wrong by the 'contagionists', their theories of disease prevention, which involved removing the accumulations and cleaning up the streams by providing sanitation, went a long way towards removing the causes of ill health.

The reason cholera took such a hold on Britain was mainly a by-product of the Industrial Revolution. Between 1800 and 1830, the population of many towns had grown two or three times. Cities that were villages in 1800, unchanged since the time of Elizabeth the First, were unrecognizable by 1832. Work in the cities increased as newly invented machines enabled huge leaps in commerce. Many poor people left the land where the living was meagre and tried to get work in the new towns. Although the population had grown, the infrastructure had not. The streets were unpaved and sufficient adequate housing was not available. Fresh clean water was virtually unknown, drinking-water was often obtained from rivers or wells and delivered in containers or piped to a standpipe without treatment. Washing facilities had to be shared, often by a whole courtyard. Sewers had only been built for the very rich, the flush toilet had not yet been invented and privy middens were used by up to one hundred people each. Streams and rivers were used for the disposal of human sewage and foul industrial effluent. In Bradford, for example, little boys used to set fire to the canal for amusement, the amount of methane given off from the rotting material was enough to support a flame. People were poor and undernourished and therefore resistance to disease was low.

The second and third pandemics came to Britain in 1848 and 1853. During these pandemics, not only was the cause of the disease unknown but cures were unknown as well. Experiments were conducted to find out why it was that some people were afflicted and others left untouched. A survey by a Doctor Roberts in Scotland at the time found that people with dark hair often had the disease, blond people less often and redheads were seldom stricken. It was not until later that someone pointed out that there were a lot more dark-haired people about. Doctors puzzled over the disease and because the patient went into a fever before they died, two possible causes of death were identified, either overheating or freezing to death (the fever being the body's response to the cold produced by the disease). Many cures were invented to counteract these causes, using heat, boiling water and steam or plunging patients into baths full of cold water or ice. Traditional remedies also had their adherents; these included the use of leaches and bleeding. Tobacco enemas were popular as were medicines containing poisons and common recreational drugs such as opium. Another option was a 'cholera belt' made of red flannel. This was tied firmly around the stomach to protect it from the disease.

After the cholera epidemic arrived in the UK, many people tried to discover what caused the disease. Several eminent men undertook tours of the country and produced reports on the sanitary conditions of the poor. Some wanted to help the suffering of their fellow man or were inspired by political ideas, others by a wish to reduce the burden of costs on the ratepayers. Several reports went into graphic detail about the terrible conditions in large towns. They often seem to vie with one another to find the worst place. When James Smith of Deanston wrote in 1837 that he thought Bradford was the filthiest town in the land, another, Angus Reach, wrote in 1842 that this was only because he had never been to Halifax. It was Edwin Chadwick however, who produced the most important report of the time. He was an unpopular man but very hard working and determined that public health must be improved. He produced a document for the Poor Law Commissioners called *The Report on the Sanitary Condition of the Labouring Population* (1842). One of the most damning statistics to come from Chadwick's report was the average life expectancy of a range of city dwellers. A businessman in Derby in 1842 could expect to live until he was 49 whereas a labourer in Leeds had a life expectancy of 19. It is important to realize however that these figures may be slightly misleading. The reason the figure is so low for the poorer classes was partially the high infant mortality rate. Because mothers were ill and undernourished, and conditions were unsanitary, their children did not often survive childhood, so these tragic deaths lowered average life expectancy. Chadwick's report proved to be a major contributor in drawing the attention of Parliament and the middle classes to the dreadful conditions of the poor of the large towns.

The first major breakthrough in the fight against cholera came in 1854 when Doctor John Snow had the handle of the Broad Street pump removed. As well as being a landmark in the history of cholera, it was the first example of the importance of epidemiology and its ability to identify the cause of a disease using scientific methods. The importance of water treatment was recognized by the middle of the century and many towns took over the role of provider of water. This ensured filtered water was eventually piped to the majority of the population and was arguably the major turning point in the fight against waterborne disease. One of the other important factors was the provision of sewers. This was accompanied by the introduction of the flush toilet that allowed water to transport human faecal matter rather than the night soil men having to remove it. This took the contamination away from people's houses but it increased the total amount of contamination entering rivers, as the sewers at the time merely emptied into watercourses. However, the building of 'sewage farms' soon followed. The people of Victorian Britain also gradually improved their diet, providing a healthier population. This was helped by the provision of school meals around the turn of the century. At the time of the Boer War (1902), many conscripts from amongst the working class were unfit for soldiering due to their having such a bad diet. School meals were instituted to counter this problem. A similar measure was the introduction of allotments in 1902.

I would not like it to be thought that this stream of improvements was brought about in a systematic way or with the nation's leaders acting in an altruistic, if paternalistic, manner throughout the second half of the eighteenth century and the first half of the nineteenth. Many improvements only came about after fierce campaigning and debate. The arguments over contagion versus miasma for instance, continued long after the cholera *Vibrio* was isolated. Improvements were often a matter of economics and not benevolence. For example, healthier employees worked harder than sickly ones. The invention of the clay drainage pipe, which replaced brick sewers, allowed much longer lengths to be laid for the same price. Politicians were also often elected on their promises of public health improvement. However, they were often more interested in the power that went with the position rather than the provision of benefits for their fellow men.

By the 1930s cholera was virtually eradicated in Great Britain. Most houses were provided with fresh treated water and flush toilets. Diet had been improved and medical services enhanced by the National Health Service's provision of free treatment after 1947. Inoculation to prevent disease had become commonplace and the only cases of cholera were from immigrants or travellers who had returned from overseas. This situation remains the case today but vigilance is always needed. Modern precautions in the UK involve compulsory testing for cholera, TB and other infectious diseases for all immigrants and people returning to the country after an absence greater than six months. Cases that occur in visitors and returning nationals must be notified to the authorities by law. If people are ill on board ship or aeroplanes, the transport companies should inform the appropriate authorities. British people travelling abroad are also encouraged to be immunized and are given information to reduce the risk of their contracting any waterborne diseases.

Outbreaks around the world occur in areas that have living conditions similar to nineteenth-century Britain. This includes poor quality water, insufficient excreta disposal, a population explosion, and intolerable destitution. In Peru for example, there was a large outbreak of cholera in 1991. This was thought to have been started by bilge water being pumped into the harbour at Callao from a ship newly arrived from the Bay of Bengal (the cholera bacterium can survive in seawater quite happily). The organism was then consumed in raw seafood taken from near the shore. It was further spread by water-tanker drivers who instead of paying for treated water from the state water company, obtained free water from contaminated rivers. They then delivered this water to the poverty stricken occupants of the 'pueblos jovenes'. These are the shanty towns that continue to spring up around Lima and (with other euphemistic names) all the major cities in the developing world. Permanent elimination of the disease will only be achieved through the provision of adequate amounts of treated drinking-water and suitable sewage disposal. The cost of such infrastructural measures is, of course, very high.

Escherichia coli

This organism is perhaps the most important waterborne bacterium after the cholera *Vibrio*. Not only is it used as the main indicator of pathogenic organisms in water but it is also increasingly becoming a pathogen in its own right. In environmental health circles, interest in *Escherichia coli* (invariably known as *E. coli*, mainly due to the difficulty of saying the word Escherichia) has been high for some time, particularly following a major food poisoning outbreak in Lanarkshire, Scotland in 1996, which caused twenty deaths. This outbreak was associated with the toxin-producing type of *E. coli* known as O157:H7, which has gone on to become infamous worldwide. However, there are many other strains that will produce toxins once ingested and WHO recently identified over one hundred zoonotic toxin-producing *E. coli* serotypes (WHO, 1998a). Although rapidly becoming recognized as major causes of serious illness, surveillance for non-O157 strains is almost non-existent. The various illnesses caused by the organism are watery diarrhoea, stomach cramps, haemorrhagic colitis, septicaemia and haemolytic uraemic syndrome (HUS). HUS is very serious and causes acute kidney failure. It is the condition that affected those who died in the

Lanarkshire outbreak. Once it begins there is no formal treatment for it other than a general care regime.

Dr Theodor Escherich discovered *E. coli* in Germany in 1885. *E. coli* is sub-divided into six categories based on its virulence. These are Enteropathogenic *E. coli* (EPEC) and Enterotoxigenic *E. coli* (ETEC) that cause traveller's diarrhoea and enteritis in infants. Diffusely adherent *E. coli* (DAEC) causes childhood diarrhoea and Enteroaggregative *E. coli* (EaggEC) causes persistent diarrhoea in children. Enteroinvasive *E. coli* (EIEC) causes Shigella-like dysentery, with blood in the stools, and Verocytotoxigenic *E. coli* (VTEC) also known as Shiga toxin-producing *E. coli* (STEC), causes Shigella-like dysentery and HUS. For the technically minded, a vero cell is a kidney cell of the African Green monkey, which is used routinely in tissue culture because of its similarities to the human cell. Thus vero-toxins are compounds that are toxic to the kidney cells of African green monkeys.

An *E. coli* strain may therefore be referred to as for example, an ETEC or an STEC and this will tell you how serious the potential illness will be. Various strains of *E. coli* are also referred to by specific antigens found on the surface of particular parts of the organism. The initials used are O (outer membrane), H (flagella – which is the tail-like appendage that allows the bacterium to move) and K (capsule i.e. the extracellular envelope). Thus *E. coli* O157:H7 refers to a specific antigen on the outer surface and another on the flagella. There may then be phage types to further identify the exact strain of the organism. Rather than being confusing, the system is designed to clarify the exact organism involved and allows for the strain found in a water supply to be linked with the one causing the illness.

The vast majority of *E. coli* are not pathogenic and it was exactly this quality that made the bacteria suitable for use as an indicator organism in the laboratory. Recently, however, scientists are becoming more aware of the pathogenicity of different strains with more of them being identified as laboratory techniques improve. The normal infective dose is thought to be between 10^5 and 10^{10} depending on type, however, it may be much lower for certain serotypes and down to less than 100 for some. Onset times range from eight hours to four days depending on type, as does the duration of illness, which varies from twenty-four hours to several weeks (Bell and Kyriakides 1998). Younger and older people appear to be much more likely to develop HUS if infected, but the reasons for this are not yet fully appreciated.

The main reservoir for *E. coli* O157 is the intestines of healthy cattle. The first time this was recognized was 1982, although the organism may have been present earlier and was only recognized at that time due to improved detection, surveillance and reporting. *E. coli* is mainly a foodborne organism normally associated with meat or meat products and in the US it is known as the 'burger disease'. The investigation and response to a recent serious outbreak in Canada was hampered because the medical authorities initially assumed it to be foodborne rather than waterborne (O'Connor, 2002) Outbreaks have also been associated with apple juice (known as cider in the US), soft cheeses and sprouting vegetables. Raw milk from cows and goats has caused outbreaks and the general recommendation of the UK Task Force on *E. coli* O157 is that all milk should be pasteurized for safety (Reily, 2001). Visits, particularly of children, to open farms or petting zoos have also led to people becoming seriously ill. The UK Health and Safety Executive has produced guidance for such farms to reduce the risks of infection (HSE, 1998). In Sweden, children under five years of age are advised against visiting farms and this measure has reduced the incidence there (Reily, 2001). Contamination therefore comes from faecal material and raw meat and it should generally be assumed that all cattle will excrete the organism at some time during their life. The recent study reported by Reily (2001) reported that 10 per cent of housed cattle and 5 per cent of cattle at grass were infected at any one time with *E. coli* O157 and an average 23 per cent of all herds had at least one animal shedding the organism. Of course, rates will vary between different herds but it is the deposition of infective faecal material near the source of a private water supply that is a worry to those responsible for their safety. Fortunately, O157 and the other STECs are no more resistant to chlorination than any other *E. coli* and typical free residual chlorine levels of at least 0.5 mg/l following standard water treatment will inactivate them as will correctly fitted and maintained ultra-violet irradiation lamps. Nevertheless, It is a worrisome organism because many private water supplies have no disinfection and the potential effects of infection are so serious. The UK Task Force on *E. coli* was concerned that single property private water supplies in the UK are not required to be tested for coliforms. In their report they referred to a study of private supplies to single domestic properties in Aberdeenshire in Scotland, which found that over the course of a year, 89 per cent failed to comply with microbiological drinking-water standards (Reily, 2001).

Those responsible for small supplies have to be careful that the spreading of slurry does not contaminate the water and that farm animals are kept away from the source. The code of practice covering pollution prevention on farms recommends that sewage sludge and other organic wastes should not be spread within 10 m of a watercourse or 50 m from springs, wells and boreholes (SOA, 1997). The UK Task Force also considered that fencing around a source to prevent farm animals getting too close should be a legal requirement for all private water supplies. Their recommendation for transient private supplies, such as campsites, is that visitors should be advised to boil all water unless it is known to be of good quality. The Task Force emphasized that where water is sampled infrequently, the testing will not reflect the possible fluctuation in the quality of the supply (Reily, 2001). They were also keen that boil water notices were employed whenever high levels of coliforms or *E. coli* were found in single properties, campsites and holiday homes. Their final comments commended the risk assessment approach as a significant way forward in improving the integrity of water from private water supplies and they proposed that it should become a regulatory requirement.

Hunter (1997) lists several outbreaks of *E. coli* O157 associated with drinking-water exposure:

- The first outbreak definitely identified as being transmitted by water affected 243 people in Missouri in 1989 and 1990. A public water leak occurred and overflowing sewage may have entered the pipe and contaminated unchlorinated water after it had been filtered. Thirty-two people were hospitalized, four died and two developed HUS. Shortly before the peak of the outbreak, 45 water meters were replaced and 2 mains water pipes ruptured during the operation (Swerdlow *et al.*, 1992).

- A small outbreak occurred in 1990 in Grampian, Scotland where the supply was fed by a different reservoir to the one normally used. This was due to excessively dry conditions preventing the normal one being available. The new reservoir was fed by a land-drain system that may have been contaminated with cattle slurry.

- In Fife in Scotland, there was a joint outbreak with *Campylobacter* in March 1995: 6 people had *E. coli* O157 (two developed HUS) and another 627 had campylobacteriosis. An illegal cross-connection with a sewage-polluted stream at a vegetable processor contaminated the distribution system. Because the food processor's own borehole pump had stopped working, they pumped water from a stream and reconnected to an old mains connection. The pressure of the pump taking water from the stream was greater than that in the mains and the stream water entered the village drinking-water supply (Jones and Roworth, 1996).

- A report in 1998 detailed the case of a seriously ill young child (16 months old) living on a farm in Ontario. The probable cause was a contaminated shallow well containing the organism of the same phage type and toxin as the one isolated from the child. It was found that 63 per cent of the cattle on the farm were excreting *E. coli* O157. The possibility that the disease was transferred by direct contact with the animals on the farm was discounted because the child had had no contact with them and did not drink unpasteurized milk or eat ground beef. The wellhead was defective below ground level, allowing surface water, contaminated by cattle waste, to flow into it (Jackson *et al.*, 1998).

- Another outbreak where a transient population was more likely to be ill than residents, occurred in Alpine, Wyoming, US in the summer of 1998: 157 people became ill and 3 developed HUS needing hospitalization. Alpine is a tourist town of about 500 people. Visitors to the town had a 50 per cent chance of becoming ill if they drank the water, whereas for residents it was 23 per cent. The supply was an unchlorinated spring where the water was collected in perforated pipes buried

2–3 m below ground. The water was then stored in an underground storage tank and piped directly to the consumers. Risks to the supply had been identified by sanitary surveys in 1992 and 1997. The surveys had shown that contamination was possible from surface water and animal droppings (mainly deer and elk). Sampling had also shown that coliforms were regularly present in the water. During the investigation, *E. coli* O157 was not found either in animal droppings or the water tank, but it must be remembered that this organism has a low infectious dose and it is not likely that the bacteria would automatically be found in follow-up sampling. The investigators noted that in eighteen outbreaks of waterborne *E. coli* O157 in the US between 1982 and 1998, unchlorinated supplies caused five. They also stated that because of under-reporting and under-diagnosis, reported outbreaks probably represent a small fraction of the true numbers of *E. coli* O157 outbreaks associated with drinking-water (Olsen *et al.*, 2002).

- On the West Coast of Scotland in the summer of 1999, an outbreak was traced to an untreated, unprotected private water supply at a rural campsite. The campers were not aware that the supply was private and therefore did not boil the water or arrange for alternative supplies such as bottled water (License *et al.*, 2001). See also the section on outbreaks in private water supplies, where this outbreak is discussed in more detail.
- A large outbreak at a state fair in New York in 1999 occurred where the unchlorinated shallow well supply was contaminated by cow manure after a large rainstorm.
- The most famous recent case of waterborne *E. coli* O157 poisoning was from a public supply in the town of Walkerton in Canada in 2000, in which 7 people died, including a 2-year-old child, and 43 people had HUS. The infection was caused by slurry spreading close to a shallow borehole subject to 'the direct influence of surface water'. The supply should have been chlorinated but due to a variety of circumstances it was not. Because of its importance and the lessons that can be drawn from it for private water supplies, the report should be sought out and studied by everyone interested in public health and water. It is available on the Internet at the web site of the Ontario, Canada's Attorney General's Office (Attorney General, 2003). The full report is by O'Connor (2002).
- At Inverkip, Scotland, another campsite outbreak of *E. coli* O157 occurred in June 2001. This time it was a Girl Guide camp and 15 out of 24 guides became ill; 14 were subsequently found to be excreting the organism. The water supply to the site and the nearby cottage was from a private supply, and both, the water supply and faecal samples from the local cattle were found to contain *E. coli* O157.
- In July and August 2002, again in Scotland, a contaminated treated private water supply caused a small outbreak to a caravan park and campsite (Anon, 2003).

- A joint outbreak of *E. coli* O157 and *Campylobacter* from a small campsite in North Wales occurred in August 2002. There were 16 confirmed cases from an untreated private supply after unusually heavy rain had fallen the week before (Anon, 2003).

Salmonella typhi

Typhoid is another of the world's major waterborne diarrhoeal diseases. There are over 12 million cases of typhoid and paratyphoid ('para' means resembling) per year, causing approximately 600,000 deaths annually. It is particularly prevalent on the Indian subcontinent but people living in, and travellers to South America, West Africa and South Asia can also contract it. Typhoid got its name because it produces symptoms similar to typhus, although lice rather than water spread that disease. In 1837, William Gerhard of Philadelphia showed that typhoid was different to typhus and in 1880, Carl Eberth and Edwin Klebs finally identified the causative organism, *Salmonella typhi* (Anon, 2002c). Typhoid is therefore a salmonella infection and there are many other types of salmonella that will cause human disease. Previously it was considered that salmonella needed a huge dose to cause illness – something in the order of 10^9 organisms. However, for certain serotypes the infective dose is now considered to be less than one thousand (e.g. typhoid) and for others possibly less than ten. Incubation is 12–48 hours for most salmonellae but longer for typhoid at 10–14 days. Salmonella infections are usually self-limiting with natural recovery after a couple of days. Typhoid however, can be a serious condition with severe illness, infection of the lymph nodes, bloodstream and gall bladder and occasionally perforation of the intestine. Untreated cases will have a 15 per cent fatality rate.

In developed countries, typhoid is mainly associated with food poisoning. In the UK, there have been several famous outbreaks associated with food, particularly one in Aberdeen in 1964. This was caused by the contamination of a tin of corned beef imported from South America. About 500 people were ill. However, only 50 people became ill from the corned beef, the others ate cooked meat that had been cross-contaminated by meat-slicing machines. The disease affected a disproportionate number of young women, as a popular diet of the day recommended a lot of cold, sliced meat. Food poisoning is, of course, a guaranteed way to lose weight, so the pundits may have been correct in their claims. Another interesting story associated with this outbreak (which may be true or may be popular myth), was that a UK Government Ministry tried to sell all the other tins of corned beef seized at the time to Pakistan. The Pakistanis however smelled a rat and refused to buy it. This outbreak, although classed as food poisoning, did have a water connection. The tins were cooled in contaminated river water after processing and, as the meat cooled and shrank, it produced a slight vacuum that sucked infected water in through pinprick holes in the seals caused by manufacturing defects. The previous major outbreak was in Croydon, England in 1937. A typhoid carrier was mending raw water inlet pipes leading to a treatment works. He defecated whilst repairing the pipes and contaminated the water. At the same time there was a shut down of the chlorination plant at the works.

Typhoid has been associated with spread by water since the third quarter of the nineteenth century, and it was during an outbreak of this disease that the world's first chlorination plant was installed at Maidstone in Kent, in 1897. Because filtration and chlorination are successful against waterborne typhoid, it is now virtually unknown in developed countries. It much prefers to infect humans rather than animals, and small water systems are unlikely to become contaminated in the Western World. It is highly likely that other salmonellae will be present however. *Salmonella* serotypes infect the gut of many animals and birds and are fairly ubiquitous in surface waters. A particular problem is salmonella in poultry.

The causative agent of typhoid is normally thought of as *Salmonella typhi*, however the complete picture is a bit more complicated. *Salmonella* is one species technically known as *Salmonella cholerae-suis*, this is sub-divided into seven subspecies (including one confusingly called *cholerae-suis*) with further subdivisions based on O and H antigens, as *E. coli* has. So, for example, what is generally known as *Salmonella enteritidis* is in fact, *Salmonella cholerae-suis* subspecies *cholerae-suis* serogroup 9 serotype *enteritidis* – so now you know.

There are four recent outbreaks of waterborne typhoid of interest to people with responsibilities for small water systems. These are as follows:

- In 1973, a labour camp in Florida used wells that were susceptible to surface water contamination. There were leaks in the sewerage system and the water tanks were found to contain faecal material. Their locks were broken, indicating deliberate contamination. The camp's chlorination system was inadequate and coliforms had previously been found in the water.
- An estimated 9,000 people were affected by perhaps the largest ever outbreak in 1975/76 in India. Although this was a public supply, fed by two wells, only one had a chlorine gas cylinder providing disinfection. When that ran out, it had to be taken off-site to be refilled and the plant operated with bleaching powder. During this procedure sewage overflowed over the well area.
- In 1988 in Kashmir, an outbreak occurred at a village where the treatment plant had broken down. The supply was provided intermittently and this produced a small vacuum in the pipes when the supply was turned off. This sucked in contaminated groundwater through cracks in the pipe and infected 230 people.

- A private reverse-osmosis desalination operator sold water by the gallon in Tabuk in Saudi Arabia. In 1992, the filter had not been changed for two years. The water came from an aquifer that was underneath an area where the city's sewage collected. This was another accident waiting to happen and when it did 81 people became ill.

Shigella

Dysentery is a widely used general term that basically means serious, bloody diarrhoea (or bloody serious diarrhoea). Hippocrates first used the term and at that time it meant diarrhoea with mucus and blood in loose stools. There are two types of dysentery; amoebic – which is caused by *Entamoeba histolytica* and is discussed in the section on protozoa; and bacterial – which is caused by various forms of *Shigella*. In fact several organisms cause bacillary dysentery but *Shigella* is by far the most important. There are several groups and sub-types of *Shigella* but *Shigella flexneri, Shigella sonnei* and *Shigella dysenteriae* are the most well known. Bacterial dysentery is therefore also known as Shigellosis. Shigellosis is a major worldwide disease that affects 2 million or more people a year of which about 650,000 will die because of it. An epidemic in Central America that began in 1968 caused 20,000 deaths from over 500,000 cases in a four-year period (WHO, 1996b). Although in the UK cases *of Shigella sonnei* infection have fallen to a 53-year low, it is now showing signs of antibiotic drug resistance, which is a worry for the future.

Bacterial dysentery has a normal incubation period of 2 or 3 days, although it can take up to 7 days. Man is the only host, although the organism will live for some time in fresh water. The infective dose is less than 200 organisms in healthy adults and patients can excrete 10^9 per gram of faeces, so it can be very infectious. It is spread by the faecal–oral route and is associated with poor hygiene practices leading to person to person spread, food poisoning and water where there are inadequate water treatment and sanitary facilities.

The disease can be a very unpleasant gastrointestinal disease with stools that eventually become mostly mucous and blood, as the surface of the intestine is attacked. Pain, fever, headache, tenemus and unsurprisingly, malaise are symptoms. Tenemus is the word that describes a painful straining whilst trying to defecate. This can lead to haemorrhoids and rectal prolapse. *Shigella sonnei* infections usually cause a milder self-limiting version of the disease. In fact, most infections from this type of *Shigella* have no symptoms but if diarrhoea does occur, vomiting and potentially, dehydration usually accompany it. This is the most common type found in industrialized countries and is nearly always a childhood disease. Infection reduces in the over-fives as toilet behaviour and hand-washing improves. *Shigella dysenteriae* on the other hand, often causes a much more serious disease with a lot of very watery stools. It can also cause severe pain and gangrene of the bowel wall, which can be fatal. Other symptoms include arthritis, lesions of the mouth and genitals and severe kidney damage. *Shigella dysenteriae* and *Shigella flexneri* are more common in hot countries where water and foodborne spread is common. Flies that have been in contact with faeces (due to inadequate sanitary arrangement) can also be vectors of *Shigella*.

Hunter details several outbreaks associated with water. The ones associated with small water systems are as follows:

- A well continuously contaminated by a leaking septic tank which made 1,200 people ill in a town of 6,500 in Florida in 1974. The people became ill after a chlorination failure lasting 48 hours.
- A shallow well supplying a private school in Argentina caused illness in about 70 per cent of its pupils. The supply was hand-chlorinated, which is not a particularly reliable method. Rotavirus and *Shigella sonnei* were isolated from the water.
- Faecal pollution of a well in southern India caused six deaths and 546 cases of dysentery in 1972. One interesting aspect of this case is that the poorer people in the village did not get the disease as they were not connected to the water pipes fed by the well.
- A private campsite with a cross-connection between the sewage system and the drinking-water system in Arizona caused illness in 1,850 people in 1979. The pipes to the water and sewerage systems were identical and the water system was poorly designed.
- In Israel in 1985, 8,000 cases of gastrointestinal illness were probably caused by an interconnected well system without routine chlorination. One well in particular had been dug in porous soil and was sited 7 m from a large mains sewer. The population served was over 100,000 and of those who were ill, 1 boy died and 90 went to hospital. The sewer leaked in several places and was also thought to be the cause of a subsequent typhoid outbreak.
- The final outbreak concerns illness caused by *Shigella flexneri*. This was in Northern Thailand where unboiled river water was drunk. There was no chlorine in the water pipes and faecal coliforms were isolated from the water. The local villagers admitted to dumping faeces into the river and the system was not disinfected because the village had run out of chlorine. A total of 242 people were ill, 65 per cent of cases within a four-day period in August 1991.

As Hunter points out, shigellosis is caused by water with human faeces in it, which is not properly chlorinated. Where small water supplies are adequately protected from contamination by human sewage and they are adequately chlorinated or otherwise disinfected, no cases of bacillary dysentery will occur.

Yersinia

The *Yersinia* family contains many different species but only *Yersinia pseudotuberculosis* and *Yersinia enterocolitica* have been linked to water-borne outbreaks of disease. *Yersinia enterocolitica* causes gastroenteritis, mainly in children under five years old. Diarrhoea and stomach pains characterize the disease. In adults the organism may also cause arthritis and occasionally septicaemia (blood-poisoning) in the old. *Yersinia pseudotuberculosis* causes a condition known as mesenteric adenitis, which has symptoms similar to appendicitis. *Yersinia* species are found all over the world and have been isolated in well water, water treatment plants, rivers and lakes. The most famous member of the *Yersinia* family is *Yersinia pestis* (which used to be called *Pasteurella pestis*), which is the causative organism for Plague otherwise known as the Black Death. This deadly disease, spread by rat fleas, decimated the world's population in the Middle Ages. However, as the Black Death is not spread by water and has never been found in a private water supply, I shall discuss it no more.

There have been several reports of waterborne outbreaks associated with *Yersinia* but only one was from a public supply. The cause of that outbreak, reported in 1993, was contamination of uncovered water storage tanks in the roof space of a hospital. A lot of debris was found at the bottom of the tanks and *Yersinia enterocolitica* was isolated from it. However, the four people who were ill were only found to have *Yersinia frederiksenii* in their stools.

- The first noted waterborne outbreak of *Yersinia* was in 1974 in the US. The source was an untreated well water supply to a ski lodge in Montana. *Yersinia enterocolitica* and coliforms were recovered from the water. Thirty people out of a population of fifty became ill. It has been noted (Hunter, 1997) that this outbreak occurred before several commonly found pathogens, notably Norovirus and *Cryptosporidium*, were suspected of causing waterborne illness. It is possible therefore that one of those organisms, rather than *Yersinia*, may have been the cause.
- *Yersinia enterocolitica* was also found in a smaller outbreak in Canada in 1986 where a family drank well water. It was found in both the water and stool samples from patients.
- Two outbreaks occurred in Japan in the 1980s and the causative agent in both cases was thought to be *Yersinia pseudotuberculosis*. The first outbreak involved water from a mountain stream used for drinking, where 260 people were ill out of a village of 1,400. The other involved a well where 11 out of 17 children who drank the water were affected. In both incidents, children were much more likely to become ill than the adults. Again, it is possible that another un-recognized organism caused this illness.

Other bacteria of potential concern

The majority of these organisms have occasionally been associated with drinking-water but have not been found to be of particular importance in small supplies. They are included here for completeness.

Helicobacter

Helicobacter pylori is associated with the incidence of stomach ulcers, gastric cancer and heart disease. It is a recently recognized organism and was only found to be pathogenic in 1980. There are no recorded water-borne outbreaks, but it appears that infection may be associated with both drinking water and recreational contact, particularly in developing countries and especially in children. A recent study of surface waters and shallow private water supplies in the US (Hegarty *et al.*, 1999) found *Helicobacter* in 60 per cent of surface waters and 65 per cent of shallow groundwater supplies. Unfortunately the researchers found no statistically significant correlation between *Helicobacter pylori* and *E. coli*. Coliforms were not found in four of the wells containing *Helicobacter pylori*. The report considered it possible that incidence was proportional to the distance between the water supply and septic tanks. One well where the organism was isolated was sited within 5 m of a septic tank, even though local regulations required a minimum distance of approximately 30 m. The organism has never been found in public water supplies and is unlikely to survive conventional treatment. Levels of chlorine that will kill coliforms will destroy *Helicobacter pylori*. It is possible however that it may be found in small water systems where treatment, particularly disinfection, is absent.

Leptospira

Leptospirosis is a zoonotic disease (also known as Weil's Disease and before that Black Jaundice) caused by contact with the urine of infected rodents, cattle or dogs. The organism gets into the body through damaged skin or mucous membranes. It is traditionally a work-related disease afflicting people who may come into contact with rats, for example farmers, riverbank workers and sewer workers. Occasionally the disease is acquired through recreational activities such as canoeing or rowing. Golfers have also become infected when retrieving their balls from water hazards (ones that were obviously more hazardous than they thought). Leptospirosis starts with influenza-like symptoms and is usually a mild disease. More serious damage may be caused to the liver and patients can become jaundiced (particularly, you would expect, against rodents). Leptospirosis has a 10–20 per cent fatality rate amongst patients who develop the second (post-fever) stage of the disease. People normally recover naturally and treatment by antibiotics is often unnecessary. A total of 67 cases occurred in the UK between 1981 and 1992. People who have regular contact with surface water or may have work-related contact with rats are advised to inform their doctor if they get 'flu-like' symptoms. *Leptospira icterohaemorrhagiae* is the most commonly found species in the UK and is the one found in the rat population.

There has only been one known drinking-water outbreak, and that was in Italy in 1984. The water most likely to have caused the disease was from a village drinking fountain. On closer examination of the system, a 'new' storage reservoir was discovered; when it was opened it was found to contain a dead hedgehog.

Mycobacterium

This bacterial family contains many pathogenic species, including the ones that cause tuberculosis and leprosy, but there has been no evidence for those two diseases being spread through small drinking-water systems. *Mycobacterium avium* subspecies *paratuberculosis*, sometimes known as 'MAP', is naturally found in the environment. Another term for this organism is *Mycobacterium avium* complex (MAC) and it has recently been given a high priority for research by the American Water Works Association as an important emerging pathogen. This is because of its wide distribution, ability to cause serious disease, its resistance to disinfection and its ability to re-grow within biofilms (AWWA, 1999). MAP has been known to cause a gut infection in ruminant animals called Johne's disease. Recently there have been questions asked about the possibility of this organism causing an inflammatory bowel condition in humans. It has been suggested that treated drinking-water could contain MAP and thus be responsible for some cases of the disease. The UK Drinking Water Inspectorate think this is highly unlikely (DWI, 2001a). They suggest that because contamination would be via faecal material from animals, contaminated sources would tend not be used for public drinking-water supplies. Standard water treatment works would also remove the organism and finally, it is improbable that water would be responsible for recent increases in bowel disease when mains water quality is improving year on year. This was confirmed by research by the Public Health Laboratory Service (now part of the Health Protection Agency), when they failed to find MAP in any of the supplies they tested. Of course private water supplies are regularly contaminated by faecal material from cattle and sheep and often do not have any form of treatment. There will have been little or no research on the incidence of MAP in private water supplies.

Other types of mycobacterium have occasionally been associated with illness from water supplies. *Mycobacterium xenopi*, which causes chest disease, has been isolated from tap water during the course of several public water outbreaks. Two public water supplies, one to a hospital and the other feeding a block of flats, were found to contain this organism when they were sampled. *Mycobacterium fortuitum* was suspected of causing problems in two hospital ice-making machines, although the organism was also found elsewhere in the water systems. *Mycobacterium marinum* has been associated with contact with water in aquaria or swimming pools. *Mycobacterium ulcerans* causes infective ulcers, called Buruli ulcer disease, but is restricted to developing countries, where most of the people who are affected live near surface waters.

Tularaemia

Tularaemia is caused by *Francisella tularensis*, a virulent organism found throughout most of the Northern hemisphere. The infective dose is up to 10^8 organisms if taken orally, but much less (ten to fifty) if infection is acquired through the skin. It has been associated with waterborne outbreaks in Italy, Turkey and possibly Kosovo. It is a zoonotic disease spread by a wide range of rodents, lagomorphs (rabbits and hares) and birds; outbreaks have been associated with both hares and voles in Bosnia. It is also known as 'deer fly fever' and 'water-rat trappers disease' (Hunter, 1997). I am somewhat suspicious about these names as I have never seen a flying deer (no matter how fevered) and consider the profession of water-rat catching to be of such limited financial remuneration, that I doubt it exists. The disease however, is an unpleasant one with diarrhoea, sore joints, vomiting, ulceration and fever. Onset is sudden; the incubation period is normally three to five days. If untreated by antibiotics, it is fatal in 15 per cent of cases. The breakdown of public health measures during wartime seems to cause increased incidence of the disease. The largest outbreak, for example, was on the Eastern Front during the Second World War.

After the incidence of anthrax in America, which followed the dreadful terrorist attacks on New York on 11 September 2001, the US Government identified $1.5 billion for basic research into 'select agents' that could be used by bioterrorists. This was an increase of five times over the previous amount. As well as anthrax, smallpox, botulism and plague, tularaemia was considered as one of these Class A agents (MacKenzie, 2002).

- The first known waterborne outbreak was in 1982 in a small medieval town in Italy. The town had two separate water supplies and the one causing the illness was very basic in construction. It was unchlorinated and in a poor state of repair. Ironically, the people drinking the water apparently preferred its 'natural spring water' qualities to the other supply. The source was open to the atmosphere with signs that animals were present in the area. A sick hare was found near the source, infected with the organism.
- An outbreak in postwar Kosovo in 1999 occurred in a rural area and was preceded by a large increase in the numbers of rats and mice seen by the villagers. This in turn was followed by an increase in the numbers of weasels. This explosion of the rodent population and thus their predators was caused by the breakdown of public health and sanitary

measures. The sanitary facilities were latrines, and water was obtained from badly damaged and unprotected wells. Rats were found living in the wellhead stonework and several rodent carcasses were recovered from the water at the bottom of the wells. There were more than 900 suspected cases of the disease. Researchers were unsure whether the outbreak was foodborne, caused by contamination of food by infected rodents or waterborne, due to the contamination of the wells. Because the disease has such severe symptoms, those studying the outbreak had to consider whether the organism was deliberately introduced into the water or food supply as an act of biological warfare. The investigators eventually discounted this theory and concluded that it was a naturally occurring event in unnatural circumstances (Reintjes *et al.*, 2002).

9 Environmental or adventitious bacteria

There are several species of bacteria that are widely found in the aquatic environment but do not normally cause illness in the immuno-competent. They are not therefore particularly associated with health problems from drinking-water. It is important to be aware of them nevertheless, as they have occasionally been associated with disease where people may already be ill with other conditions or their immune system is reduced and unable to cope (Dufour, 1990). They are usually known as environmental bacteria, but I have also come across the terms adventitious or heterotrophic in this context (although heterotrophic strictly means they get their source of energy and cellular carbon from the oxidation of organic material, that is, by feeding on plants or animals – rather than photosynthesis). Where laboratories carry out plate counts, it is often these bacteria that are cultured. There will be many different types of environmental bacteria but the important ones for drinking-water safety are listed here.

Aeromonas

Aeromonas are commonly found in both fresh and salt waters. There are several species, each one favouring a particular environmental niche. *Aeromonas hydrophila* is found mainly in clean river water, *Aeromonas sobria* in stagnant water and *Aeromonas caviae* in marine water. They are so common that people have tried to use them in rivers as indicators of pollution. They are known to cause diarrhoea and infection in soft tissue where damaged skin comes into contact with contaminated river or lake water.

Aeromonas caviae is the one most commonly associated with diarrhoea. Diarrhoeal infection is usually mild, although more severe symptoms have occasionally been known, including bloody diarrhoea and chronic colitis (inflammation of the colon).

Aeromonas have been found in treated chlorinated water and sometimes there is re-growth in the distribution pipes. Chlorine only appears to have a temporary effect on them and this may mean that it stops them from reproducing but does not kill them. If left (presumably so they can get their breath back and have a bit of a rest after the chlorine attack) they can continue as normal. Hunter (1997) was unable to find any confirmed waterborne outbreaks of disease directly associated with *Aeromonas*, either from drinking-water or via recreational use. There have however been individual cases where water, either via immersion, contact or consumption of untreated water, has been strongly associated. It is possible that the lack of proof of causation of a waterborne outbreak is a result of absence of evidence rather than evidence of absence.

One interesting fact about *Aeromonas* is that *Aeromonas hydrophila* has a symbiotic relation with leeches and lives in their gut, where it helps digest its food (blood) by producing enzymes. Apparently up to 20 per cent of those people who have been treated with leeches develop an infection. One would have thought that being treated with leaches is bad enough without going on to develop further complications from *Aeromonas*.

Burkholderia

Burkholderia pseudomallei causes the disease known as melioidosis. It is included here for completeness and has been associated with very few outbreaks of disease from drinking-water. Until recently it was considered to be a *Pseudomonas* and is commonly found in tropical soils and surface waters, particularly rice paddies. It is a particular problem in Thailand, but is also endemic in Northern Australia, where the disease peaks in the rainy season. Melioidosis is acquired via contact with infected water or soil through damaged skin. Its main symptoms are purulent (pus-filled) abscesses. They are found in different parts of the body and can become quite large.

One outbreak, associated with drinking-water, occurred in 1997 in a rural community in Australia. There were five cases and three people died as a result of getting the disease. The supply to the community was a series of boreholes where the water was treated and chlorinated. The cause of the outbreak is thought to have been an aerator used in the treatment process to raise the pH of the water. It probably became contaminated after soil entered the aerator or by infiltration from a leaky or damaged water pipe. The aerator merely provided somewhere for the bacteria to grow. When it was taken out of service for repair, the sediment and biofilm growths were probably disturbed thus releasing the organism into the water supply. The investigation at the time of the outbreak also thought that chlorination failure might have contributed to the problem (Inglis *et al.*, 2000).

Plesiomonas

These cause gastroenteritis, which ranges from mild symptoms to bloody diarrhoea. They have also been known to be associated with non-gastrointestinal illnesses, such as septic arthritis and meningitis. They are common in fresh waters and have been found in both surface and well water. Laboratory techniques for identifying them are not as well developed as those for other organisms and they are not routinely looked for. Only two waterborne outbreaks have been reported, in 1973 and 1974, both from Osaka in Japan. The first outbreak affected 978 people who had stayed in a youth centre. *Plesiomonas shigelloides* was found in one tap sample and in the stools of twenty-one patients. There is some question as to whether this was, in fact, acquired from the water supply. There is even more doubt as to whether the second outbreak was associated with water. The organism was not found in any water samples taken during that outbreak.

Pseudomonas

Pseudomonas is frequently found in the environment and has been isolated from surface water, seawater, soil and both tap and bottled water. Apparently, they are also extremely fond of spa-pools, where they are often found, particularly where disinfection practices are not up to scratch. There are over one hundred species of *Pseudomonas* and several are pathogenic to humans, the most important being *Pseudomonas aeruginosa*, which unfortunately is also resistant to many antibiotics. Although in healthy individuals it is usually harmless, when it comes into contact with a site of injury or an existing illness it can cause infection, for example inflammation of the eye or inner ear. Other diseases associated with *Pseudomonas aeruginosa* are meningitis, brain abscesses and joint, skin, urinary tract and gastrointestinal infections. Outbreaks of illness associated with *Pseudomonas aeruginosa* are not therefore associated with drinking-water but are usually skin or ear infections from contact with contaminated water in swimming pools, spa-pools or whirlpool baths.

10 Protozoa

Protozoa are unicellular organisms that cause waterborne illness wherever poor quality water and inadequate sanitation have been found. Recently, they have become to be seen as more of a problem in developed countries as laboratory techniques have improved, making them easier to isolate. They are therefore being found to be the causative organism in outbreaks that would previously have had unknown aetiology (origin or cause). Several species are also very difficult to treat using normal chlorination methods as they have a very robust outer cell wall.

Amoebic dysentery

There are two types of dysentery, one is bacterial and caused by various species of *Shigella* and has already been discussed. The other is amoebic and is caused by the protozoa *Entamoeba histolytica*. This disease is found throughout the world and the number of people affected is huge. It has been claimed that annually, 40 million people develop intestinal disease or liver abscesses from this organism, with 40,000 of them dying from amoebiasis. Some claim it is the third leading cause of morbidity and mortality due to parasitic disease in humans (after malaria and schistosomiasis). Others have estimated that it is responsible for between 50,000 and 100,000 deaths every year. In many publications, *Entamoeba histolytica* is cited as either infecting one-tenth of the world's population or 500 million people. Examination of 50,000 faecal samples in Bombay by Amin *et al.* (1979) showed that 16 per cent contained *Entamoeba histolytica*. Hiagins *et al.* (1984) examined 1,387 stools from Flores, Sumatra and Java and *Entamoeba histolytica* affected about 50 per cent of the population, although it is probable that only 1 in 10 of those infected would have been ill during a single year. The numbers vary because of the different ways disease statistics are collected and so will never be completely accurate but it is clear that amoebic dysentery is one of the world's major diseases and that it causes untold misery.

The protozoa are transmitted in faecally contaminated food or water via cysts that are about 12 μm in diameter. Amoebic cysts can easily survive for

30–90 days in moist conditions and mild temperatures. They will live for several weeks in the external environment because of the protection from their tough outer walls. Within the cyst are small daughter organisms called trophozoites. They can also be passed in stools but because they do not have the protection of the outer wall, they are rapidly destroyed once excreted. If ingested they will not survive exposure to stomach acid. If the outer walls of excreted cysts have been stressed, they are also destroyed by stomach acid. If the cysts survive these natural defences, their walls dissolve in the small intestines and the trophozoites are released. The trophozoites multiply by binary fission and produce further cysts and the life cycle continues.

Any untreated water that comes from surface supplies may be contaminated. In developing countries, anyone who drinks untreated or unboiled water or eats food made with untreated water is at risk. In industrialized countries, problems are associated with poor hygiene practice. Risk groups include travellers, recent immigrants and institutionalized populations. Hands can also become contaminated with the parasites when a person changes the nappy of an infected infant. Amoebiasis can also be transmitted by sexual contact but this is a minor route of infection. Infected individuals are the only reservoir for the parasite. Humans are the only important host although dogs, cats, pigs and rats have occasionally been found with the cysts.

Symptoms vary according to the patient. Factors that affect this include the physical and emotional condition of the host when the immune system is challenged. Infants are especially vulnerable to these pathogens because of their underdeveloped immune systems. The symptoms usually appear 2–4 weeks after infection, but the onset may range from a few days to a few months. In most people, the symptoms of amoebiasis are intermittent and mild (various gastrointestinal upsets, including colitis and diarrhoea). Sometimes there are no symptoms at all. Nevertheless, treatment is essential; 40 per cent of all untreated cases eventually have further problems such as amoebic hepatitis or hepatic abscesses. In severe cases, dehydration and anaemia result from the loss of fluids and blood. Once ingested, the amoeba secretes enzymes that can destroy the tissues that line the large intestine. This leads to a perforated colon and peritonitis (an inflammation of the lining of the abdominal wall). These complications can be fatal. The liver and lungs may also become sites of infection. Black, solid, bloody, mucous covered, fatty stools characterize acute amoebiasis. Many chronic cases are asymptomatic and there have been instances where infection has lasted for up to forty years.

Humans can be host to at least six species of *Entamoeba*, in addition to several amoebae belonging to other genera. In recent years, it has been recognized that there are in fact two species of what has previously only been known as *Entamoeba histolytica*. It has long been thought that many of the people apparently infected with *Entamoeba histolytica* never develop symptoms. This was thought to mean that the parasite had very variable virulence.

However, in 1925, Emile Brumpt suggested an alternative explanation, which was that there were two species; one capable of causing disease and one that did not. He called this other one *Entamoeba dispar*. Although the theory was initially dismissed, in the 1970s, data started to appear that supported it. In 1993, a formal re-description of *Entamoeba histolytica* was agreed, separating it from *Entamoeba dispar*. *Entamoeba histolytica* is the species that can cause amoebiasis and dysentery; *Entamoeba dispar* is not pathogenic. *Entamoeba dispar* is the most commonly encountered species, although the two are very similar and differentiating between them relies on complicated laboratory analysis.

Cryptosporidium

Cryptosporidium is thought to be annually responsible for up to 500 million infections worldwide. The organism is a small (4–6 μm), protozoan parasite of the coccidia genus. *Cryptosporidium* has a worldwide distribution and is found in a vast number of hosts including birds, fish and mammals. There are several species of the organism but only one initially appeared to infect man, this was *Cryptosporidium parvum*. Recent research has led to the discovery of two genotypes of this species. Genotype one is exclusive to humans, genotype two is found in both livestock and humans (Peng *et al.*, 1997). Previously they were known simply as type one and type two, but recently they have been given separate names, although this still has to be made official. Type one is known as *Cryptosporidium humanis* to reflect its purely human infectivity and type two retains the name *Cryptosporidium parvum* (Smith, 2003). The relative incidence of the two types has been found to be roughly one-third *humanis* and two-thirds *parvum* (McLauchlin *et al.*, 1999). Other forms of *Cryptosporidium* have been found to infect HIV-positive people including *Cryptosporidium felis*, which has infected a man in Italy and *Cryptosporidium muris*, a man in Kenya.

There is, as yet no specific treatment for cryptosporidiosis. In healthy patients, the organism causes an unpleasant, self-limiting, gastrointestinal illness that may last up to 28 days, but is usually of 7–14 days duration. Amongst the immuno-compromised however, it can be life-threatening. Patients will get seven or eight separate symptoms on average, the most common being diarrhoea (93 per cent) and loss of appetite (84 per cent) (Casemore, 1991a). Two-thirds of the people in the UK contracting the illness (cryptosporidiosis) are children. The infective dose is unknown, but is considered to be below ten and some experts think it may be as low as one. Healthy volunteers in the US were found to have a median infective dose of 132 oocysts, although one participant became ill after receiving 30 oocysts (DuPont *et al.*, 1995). To confuse the situation slightly, it has been found during tests in the UK that densities of 2.86 oocysts per litre of drinking-water have failed to cause any detectable illness in the population (Craun *et al.*, 1998). Many water treatment facilities in the UK continuously monitor for *Cryptosporidium* and the legal maximum is one oocyst per 10 l. There has been no evidence that illness has been caused by water containing these levels.

Cryptosporidium is popularly, but wrongly, thought to be an exclusively waterborne agent, associated with mains water from surface supplies. In fact cryptosporidiosis is commonly associated with zoonotic spread, mainly from calves or lambs. This may be by direct contact with the animals or their faeces, for example when children go on educational visits to farms or indirectly via raw milk, which may be contaminated with faecal matter or raw water from private supplies. Human to human spread is now considered to be the most common cause of cryptosporidiosis either through the classic faecal–oral route or via fomites (Casemore, 1991b). *Cryptosporidium* is excreted from hosts in cyst form, known as an oocyst. Infected young animals can excrete up to 10^{10} oocysts a day and a human can excrete 1 billion oocysts in 24 hours. The oocyst is very hardy and can live in the environment for many months. Once a suitable host ingests the oocyst, it releases four sporozoites into the small intestine. They then begin the asexual stage of the life cycle and eventually initiate the symptoms of the disease.

Oocysts have often been found in drinking-water supplies and 28 per cent of treated-water supplies in the US have been found to contain them (Rose, 1990), although at quite low levels (0.002–0.009 per litre). This does not necessarily mean that populations are at risk; in fact it may show that these levels are not likely to cause an outbreak due to their contributing to a low endemic level of disease or to an increased level of immunity in the community. Oocysts have been found in all types of water sources, including protected boreholes that had never shown any previous contamination (Morgan *et al.*, 1995). Because the wall surrounding it is very tough, *Cryptosporidium* can withstand massive doses of chlorine (20,000 mg free residual chlorine per litre), doses that would easily decimate other organisms. Because of the large numbers excreted, the relatively small number needed to initiate the illness and the failure of chlorination as a control method, *Cryptosporidium* has become a problem for people responsible for water safety across the globe.

Cryptosporidium was first discovered in 1907 by a Doctor Tyzzer in the stomach lining of a mouse. It was considered at that time to be of minor interest to human health and consequently very little work was done on the organism. It was only found to be infective to humans in 1971. The first official human case of cryptosporidiosis was in 1976 and afflicted a 3-year-old girl in Tennessee. The first recorded outbreak was in 1983. The first waterborne outbreak in the UK was in 1987 in Ayrshire, where 27 people were made ill (Maddison, 2000). In England and Wales, the annual number of laboratory results for cryptosporidiosis now ranges from 3,700 to 5,700. In England and Wales, cryptosporidiosis is the fourth most commonly identified cause of gastrointestinal disease following *Campylobacter, Salmonella* and *Rotavirus* (Anon, 2002b). The highest incidence is in toddlers and young adults and is twice as common as *Salmonella* in the 1-to-5 age group. The infection now appears to have become endemic.

There is normally a slightly higher incidence of the disease in rural areas. This may be because of the greater contact with farm animals but it could also be because that is where the private water supplies are. Temporal and seasonal distributions of cryptosporidiosis vary from country to country. In the UK, there has been found to be an increase of cases in the early spring (mainly *C. parvum*) with a second lesser peak in October and November (where there is higher incidence of *C. humanis*) (Chalmers, 2003). These peaks coincide with times of high rainfall and the spring peak is also the time of farming events such as slurry and muck spreading, lambing and calving (Grimason *et al.*, 1990). Interestingly, the autumn increase in *C. humanis* cases may be somewhat swimming pool related as UK residents return from abroad infected with the organism and continue swimming as they did on holiday.

There appears to be a corresponding association with general contamination of private water supplies during the spring and autumn. Because of the poor protection of many small supplies, during times of heavy rain, particularly those following dry periods, faecal matter containing *Cryptosporidium* oocysts will be washed into the water system. One study in the United States showed that 87 per cent of raw waters were infected with *Cryptosporidium* (Rose *et al.*, 1991).

A problem with *Cryptosporidium* has been the difficulty of finding oocysts in the laboratory and determining whether they are viable or not. There are two techniques for identifying the parasite. The first essentially involves filtering large quantities of water through what is either a tightly compressed sponge or a tightly wrapped tube of special tape. The filter is then removed and any trapped particles are carefully freed from the filter. The second technique uses a laser to remove all suitably sized particles from a stream of water. This is followed by staining the particles with a selective medium so that any oocysts can be seen under a powerful fluorescing microscope. This is highly technical work and requires specially trained technicians; particularly those working at the microscope because they need to differentiate oocysts from other similarly sized and shaped particles. Isolating bacteria is much simpler and consequently much cheaper. The problem of identification of the organism in water is, unsurprisingly, also a problem when enumerating them in stool samples. In the past, for example, air bubbles have been mistakenly diagnosed as *Cryptosporidium* oocysts (Maddison, 2000). Looking at an oocyst through a microscope does not always tell you whether it is viable (capable of causing the disease) or not. Sometimes viability will be fairly obvious, such as when the oocyst is empty or the wall has split to allow the sporozoites out. If, however, the oocyst is intact, you cannot be sure about its viability without using animal studies (which should be avoided where possible) or other complicated scientific processes.

Even when an oocyst has been identified and looks viable (and may indeed be viable), it could be a species of *Cryptosporidium* that is not infectious to man. A new species that appears only to be infectious to sheep has recently been identified (Hunter, 2002). If this is the case, water that is found to contain oocysts may not make anyone ill. Any costs associated with reacting to the discovery of these oocysts would therefore be wasted.

Viability is an important consideration when assessing the success of any treatment that claims to inactivate the oocyst, but not physically remove it from the water. Ultraviolet disinfection is becoming a popular method for *Cryptosporidium* deactivation; it can be successful in both medium and low-pressure units. However, because ultraviolet treatment damages the DNA of microorganisms rather than physically removing them, any oocysts found in water would make it appear to be unwholesome, even if it was perfectly safe. Because of these difficulties and the cost of isolation and identification, routine monitoring for oocysts in private water supplies is not carried out. Research has been undertaken on *Cryptosporidium* in small supplies, but only rarely. Routine monitoring for oocysts is restricted at present to mains water treatment plants where there is a reasonable risk of the organism entering the plant in the raw water.

Several highly publicized waterborne outbreaks have occurred in the UK since 1976. The rate is fairly low, but because they are mains supplies, a lot of people can become ill. These incidents have had a variety of causes, including breakdown in plant or poor treatment processes, artificially high levels of oocysts in the feeder supplies or post-treatment faecal contamination. This contamination has also had a number of causes, including heavy rain washing sheep faeces into pristine surface waters that only have disinfection as a treatment system and leakage of cattle manure into inadequately protected aqueducts. The most famous *Cryptosporidium* incident was the Milwaukee outbreak, which was reported to have caused 403,000 cases and several deaths. Despite these numbers probably being an overestimation (see earlier section on microbiological considerations), it was a huge outbreak attracting much publicity. The cost has recently been calculated as US$96.2 million. Even when the over-estimation is accounted for, it cost about US$57.2 million (Corso *et al.*, 2003).

Infections arising from contamination of private supplies are on a much smaller scale and would not normally attract much media attention. It is likely however that the organism may contaminate many such small supplies. This is due to a variety of reasons including the proximity of young farm animals to many supplies in remote rural areas, their inadequate protection, the lack of treatment and disinfection and the inability of most small-scale treatment systems to remove the organism. It has been argued that people who regularly drink water from a small private supply will become immune to *Cryptosporidium*. However, even if this is the case (which I am not convinced about), people visiting farmhouses, campsites or other isolated dwellings using small water supplies will still be at risk, even if other organisms, such as faecal coliforms, have been removed through disinfection.

There is another reason why there will generally be a lower risk to people drinking from public water supplies than those using individual small supplies, particularly surface water or those under the influence of surface water. This is because non-pristine public water sources are usually treated using a multiple-barrier approach. No conventional water treatment can guarantee the complete removal of oocysts, although there may be a sufficient reduction in numbers to reduce the health hazard to almost zero. In public water treatment plants, flocculation using iron or aluminium salts, followed by carefully run filtration processes (particularly meticulous control of backwashing operations and ensuring correct depths of filter beds) will effectively remove *Cryptosporidium* by up to 99.99 per cent (this is a four log reduction – see the section on disinfection in Chapter 18 for an explanation of 'log removal'). Because they are so small, oocysts will not be completely removed by settlement. It may take eight weeks for a single oocyst to settle to the bottom of a 5 m deep reservoir (Ives, 1991).

Because local authorities do not routinely sample for *Cryptosporidium*, research has been carried out to establish that it can be present in private water supplies. There have been three studies recently that have looked into this problem. Because each study gives a detailed insight into *Cryptosporidium* in small water supplies, it is useful to discuss their findings in some detail. The first was in the Yorkshire area of the UK and 15 separate sites were sampled 10 times each (Clapham, 1997). The aim was to show that contamination of private water supplies by *Cryptosporidium* was a reality; this had never been proved before in the UK. Supplies were therefore selected that had particularly poor source protection and animals in the surrounding area. The water was also tested when the supplies were most likely to be contaminated, that is following rain during the autumn period: 9 of the 15 supplies were found to contain *Cryptosporidium* (60 per cent). In addition, 8 of the supplies also contained another protozoan parasite, *Giardia lamblia* (53.3 per cent); 98.6 per cent of the samples showed some form of faecal contamination and the levels of faecal indicators were almost uniformly high with a faecal coliform geometric mean of 224.5 per 100 ml and a maximum of 6,000 per 100 ml. The study proved that private water supplies without reasonable source protection could be considered to be at risk when animals are present.

Other parameters of water quality were looked at during the study to try and establish a link between them and *Cryptosporidium*. It would be useful when monitoring private water supplies to have an indicator that correlated with oocyst contamination but was cheaper and easier to isolate in the laboratory. The study found that there was a statistically significant correlation to *Cryptosporidium* for *Enterococci* ($r = 0.6783$; $p < 0.001$) and *Clostridium perfringens* ($r = 0.6268$; $p < 0.001$). Unfortunately, there was no significant correlation between *Cryptosporidium* and coliforms (either total or faecal), turbidity or conductivity. Sometimes turbidity is thought of as an indicator of a recent pollution incident and therefore *Cryptosporidium*. Unfortunately, this study showed that it is not yet possible to rely on turbidity as a substitute for *Cryptosporidium* testing. Even more unfortunately, there was no correlation between *Cryptosporidium* and the scoring of a sanitary survey. One explanation for this could be that there was a defect in the weighting of the sanitary survey scoring and shows that further work needs to be done on ensuring that sanitary surveys reflect the actual incidence of contamination, particularly for *Cryptosporidium*.

The results of the study were also interesting when looked at from a monthly point of view. During the November sampling, 23 per cent of samples were positive for *Cryptosporidium* compared with 15 per cent in December and 11 per cent for January and early February. When November was considered separately, the first half of the month gave 24 per cent positive samples compared with 12 per cent for the second half. This probably shows that the first half of November is the downward slope of the autumn peak and that the winter month results are the background level. There could of course be other reasons for this difference, although factors, such as time of day and rainfall events, were largely negated by judicious control of the sampling process.

The second study (Maddison, 2000) was perhaps the first to use geographical information systems to link incidence of cryptosporidiosis with private water supplies. In the area she studied (part of the West Midlands), Maddison found that over a four-year period the number of cases of cryptosporidiosis within 100 m of a private water supply were higher than would be expected in the general population. She found 29 cases but would have expected to find 17. In her thesis she discussed various issues surrounding the practicalities of her study and most of them tend to make the results under-estimate the problem. These included the following:

- Under-reporting of illness by people on private supplies.
- The source of some supplies being further away than 100 m from the properties they serve.
- Some people being made ill by supplies that serve transient populations. Thus they would have gone home by the time the illness developed and reported them to their home authority rather than the one being studied.
- People on private water supplies may have a higher level of immunity than the general population.
- Because of the low numbers of people who would be affected by an outbreak from a private supply, it is less likely that an investigation would be carried out.

Although she was grateful for the help she received from local authorities, she did not get as much assistance as she should have done for such an important study. She pointed out that for some authorities this, 'may be linked to poor organization of private water monitoring'.

The third major study on *Cryptosporidium* in private water supplies in the UK was carried out in 2000 (Watkins *et al.*, 2001). The investigators looked at seven medium-sized supplies across the UK. They sampled each site daily during two periods – one in the spring and one in the autumn. Samples were taken for *Cryptosporidium* and *Giardia* as well as a range of indicator organisms and *Campylobacter*. The sites of the private water supplies ranged from a holiday camp to a hotel. The amount of *Cryptosporidium* found varied from near absence at one site to being present in 75 per cent of samples in another (50 per cent also contained *Giardia*). All seven sites however had positive results for *Cryptosporidium*. The number of indicator organisms also varied between not being found at one site to their presence identified in 98 per cent of samples at another. Six out of the seven sites were positive for indicator organisms at some time. Once again, this study showed that *Cryptosporidium* is to be expected in all but the very best private water supplies.

Another result of this survey reinforced the random nature of normal sampling regimes and the importance of rainfall. At one site, during sampling in the spring period, only one sample was positive for protozoa but in the autumn period a substantial 56 per cent of the samples were positive. At another site, only one sample was positive for *E. coli* in the spring but in the autumn, 93 per cent of daily samples were positive. There was occasional heavy rainfall during the sampling period and in four of the sites the quality of the water was seriously affected. There was a substantial rise in the counts of both the indicators and the pathogens being studied. One site also had *E. coli* O157:H7 in the water during the spring sampling period. This site had chlorination and filtration installed before the autumn sampling period. Even with the treatment in place, during severe weather, the system failed and pathogens were found in the water. The report's authors stated that low frequency sampling programmes would not detect such dramatic changes in water quality unless sampling is targeted at heavy rainfall events. They concluded that sites that are well-protected following sanitary surveys and have suitable treatment will have no microbiological problems, but high microbiological contamination will periodically occur at sites in rural areas where there is no control over source water quality, where water treatment is absent and there is low-frequency monitoring.

Outbreaks of cryptosporidiosis associated with small water systems include the following:

- An outbreak in a UK boarding school in April 1992. A reservoir supplying the school was thought to have been contaminated by slurry being spread nearby. The supply was chlorinated, but several pupils became ill. The last case coincided with the removal of the supply and the use of emergency tankers (Furtado *et al.*, 1998).

- Two deep unchlorinated wells caused an outbreak in 1994 in Washington State, US. Water from a treated sewage irrigation system was found dripping along one of the well's outer casing. Eighty-six people were ill.

An immuno-compromised child died of cryptosporidiosis in the Yorkshire area a few years ago; this was not a pleasant death, with the child suffering about five years of continuous, serious diarrhoea, day and night, before finally succumbing to it. It is therefore a false (and insulting) proposition to say that it is not important for small water supplies to be kept free of *Cryptosporidium*.

Cyclospora

Cyclospora cayetanensis is a protozoan parasite that is only known to cause disease in man. It is a 'new' organism and was first reported to cause illness in 1977 in New Guinea. It was not investigated again until 1985 when Charles Sterling and his team from the University of Arizona carried out further work in Peru. *Cyclospora* looks like *Cryptosporidium* but the oocysts are slightly larger at 8–10 μm in diameter. It has been suggested that they also look like certain cyanobacteria. The organism causes diarrhoea; other symptoms can include nausea, anorexia, abdominal cramps, fatigue and weight loss. It has no treatment. *Cyclospora* infections are fairly ubiquitous and have been found in the Americas, the Caribbean, Africa, India, South-east Asia, Australia and England (Stirling and Ortega, 1999). It is possible that *Cyclospora* is endemic in many developing countries and when people routinely start to consider *Cyclospora* as a potential source of waterborne illness, it will probably start to be found more regularly.

Because of difficulties with studying the organism, its life cycle is not fully understood. However, it is known that to complete its life cycle, the oocysts must spend some days or weeks in the environment, thus preventing any direct person-to-person spread. Methods used to identify the organism are similar to those used for *Cryptosporidium*, but these are not ideal for *Cyclospora*. Organisms resembling *Cyclospora* have been found in primates, dogs, ducks and chickens but as yet man is the only confirmed host. The infective dose is not known, but as with other protozoan parasites, it is thought to be low.

Infections have shown temporal fluctuations with disease coinciding with warm weather and, as with most waterborne organisms, rainfall. The oocysts survive well in water and prefer moist conditions. Waterborne spread is therefore possible in small water systems if a person with the disease excretes near an unprotected water source or uses a sewer that leaks close to a source or its distribution system. Once infected it is believed that a certain amount of immunity can be built up. As with other protozoa, *Cyclospora* may have some resistance to chlorine, but because outbreaks are usually only associated with developing countries, it is strongly suspected that suitable water treatment and normal disinfection processes will be sufficient to remove it (Stirling and Ortega, 1999).

Cyclospora outbreaks have mainly been associated with food poisoning, particularly from raspberries and basil originating in developing countries. Several well-publicized outbreaks in the United States have implicated raspberries from Guatemala. In 1996, for example there were 1,465 cases. It is thought that the oocysts may be spread by contamination with polluted water via irrigation or in water mixed with pesticide used to spray the berries. In Lima, Peru, the organism was isolated in water in sewage lagoons, which is often used to irrigate crops.

Most reported waterborne cases have been in developing countries, or in travellers returning from them:

- An outbreak at a Chicago hospital in 1990 involved at least 11 cases and was thought to be from a stagnant rooftop drinking-water storage reservoir.
- In Nepal in 1992, there was an outbreak associated with either drinking untreated water or milk reconstituted with it.
- An incident, reported in 1993, was associated with swimming in a lake near Chicago.
- A case reported in 1993 was thought to be caused by 'oral siphoning', to empty a saltwater fish tank.
- Another outbreak in Nepal, this time in 1994, was amongst British soldiers who drank chlorinated water, which was a mixture of public supply and river water. The infections ceased when the water was boiled.
- In Massachusetts, borehole water was implicated in an incident where three people became ill.

Giardia

The second most important waterborne protozoan parasite after *Cryptosporidium* is *Giardia lamblia*, it is also known as *Giardia duodenalis* or *Giardia intestinalis*. This organism causes a severe diarrhoeal food poisoning-like disease called giardiasis. The first waterborne case of giardiasis appears to have been reported in 1974 and concerned a woman who drank water from a rainwater collection system on a farm (Brady and Wolfe, 1974). Cysts of *Giardia* are passed out in faeces and swallowed via water or food in the classic faecal–oral route. Once in the body, the cysts pass to the duodenum where two trophozoites are released. These are 12–18 μm long, with two flagella at one end and a concave disc at the other. The disc attaches to the walls of the small intestine. As they pass through the intestines they produce more cysts, ready to be excreted.

Symptoms have been described as acute onset of diarrhoea – which is often explosive, abdominal cramps, bloating and flatulence (Hunter, 1997).

Malaise is often common and sulphuric belching is quite characteristic. This therefore is an organism to keep away from, and if you get it, people will be advised to keep away from you. Hunter also describes an interesting method of diagnosis of the disease. As well as microscopic examination of faeces, patients can swallow a small weight with a fine line attached to it – the other end is taped to the patient's cheek. After a few hours the weight is recovered and trophozoites can be seen attached to the thread. Unlike cryptosporidiosis however, giardiasis can be treated.

Many of the properties and problems of *Giardia* are similar to those of *Cryptosporidium*. It is zoonotic and found in many animal species. It is more resistant to treatment by chlorine than coliforms and is capable of survival in the environment for long periods of time. Incidence of giardiasis has often been associated with drinking from surface waters, but it is also likely to be found in poor quality small water supplies. *Giardia* is considered to be especially common in the US, but as shown in the study of *Cryptosporidium* in private water supplies discussed earlier (Clapham, 1997), it can also be present in a number of private supplies in the UK. This difference may be that the organism is merely looked for more often in the US. The study by Watkins *et al.* (2001) also found that the presence of *Giardia* was much more prevalent in the autumn than the spring. Unfortunately they were not able to identify a reason for this.

Interestingly, several outbreaks of giardiasis in the rural hinterlands of the US have been associated with beavers. The disease is sometimes known as 'Beaver Fever' because of this connection. *Giardia* is a common cause of traveller's diarrhoea due to backpackers drinking unboiled water from lakes and streams away from civilization – presumably not thinking that wildlife has probably defecated into it. Hunter (1997) has comprehensively described the reported outbreaks associated with *Giardia*. Of the 19 outbreaks he details, 14 are either associated with people drinking from small water systems or surface water, or ones that have problems similar to those of small systems.

Some of the reasons for these outbreaks make interesting reading:

- An outbreak on a US farm involved water supplied from a brick cistern that collected rainwater (reported in 1974).
- There was an outbreak associated with a camping trip in Utah in 1974 where the only risk factor for developing the disease was not drinking untreated water before the trip, that is the only people who did not get ill were those who had previously become immune.
- Also in 1974, a small public system in New York State had an outbreak where human waste in the watershed could have contaminated water that was chlorinated but unfiltered. Interestingly in this outbreak, people in the same town who drank from private water supplies were not affected.
- Inadequate flocculation, coagulation, sedimentation and the deterioration of water treatment filters caused a problem in a rural town in the US in

1976. This was in addition to finding three beavers below the water intake, with *Giardia* in their stools.

- In 1980, an unfiltered, marginally chlorinated water supply in Montana, caused illness in nearly 800 people after heavy rain had made the water very turbid.

- An autumn outbreak in Aspen, Colorado was reported in 1984, where there was inadequate chlorination, lack of coagulation and damaged filters. Only new residents in the area (less than two years) were affected, presumably because this had happened before and existing residents had become immune.

- Another autumn outbreak occurred in Nevada in 1983 when a beaver built its lodge in a pipe leading to a reservoir. The outbreak stopped when chlorination was stepped up and the beaver removed.

- Chlorinated but unfiltered water was suspected in a trailer park in Vermont in 1986, where the chlorination was thought to be inadequate during peak times. The outbreak followed the destruction of a beaver dam upstream that led to a large release of water. Perhaps the beavers decided to get their own back.

- In Massachusetts in 1985/86, an outbreak was caused where water from a reservoir recently brought back into service was chlorinated but not filtered. The report on the incidence said that the reservoir had evidence of beaver activity (yes – them again) and graffiti. This is the first report I have read where graffiti appears to have been suspected of being a cause of disease. The graffiti presumably indicated human activity and the *Giardia* problems might have been more associated with toilet habits than vandalism.

- A creek in British Colombia in 1986 had high turbidity following spring run-off of snowmelt. Four beavers and two muskrats excreting *Giardia* were found in the creek. The creek was closed and the incidence of disease started to decline. The town's officials however decided not to issue a boil-water notice because it might have affected their tourist trade. The creek was soon used again, but with an inlet further upstream. When the original inlet was re-opened sometime later, another giardiasis outbreak occurred. This appears to be a little like the 'Jaws' scenario – perhaps the film about it will be called 'Buck Teeth', to reflect the beaver connection.

- Two outbreaks, five years apart in Creston, British Colombia were reported in 1993 and 1994. Beavers upstream from the water intake were found to be excreting large numbers of cysts.

- There was deliberate contamination with human faeces of a water tank in a block of flats in Scotland in 1990. Four people were affected.

- In the UK in November 1991, a West Midlands rural community using a private supply taken from a groundwater source became ill with *Giardia*. The storage reservoir was thought to have been contaminated with animal faeces from livestock in the area. The supply was not regularly chlorinated (Furtado *et al.*, 1998).

- A large outbreak occurred at a hotel in Greece in May and June 1997. This was a mixed outbreak with 107 probable cases of infectious diarrhoea of which 58 were confirmed *Giardia lamblia*. Although the actual cause of the outbreak was not found, guest said that the water in their rooms was discoloured and smelt of sewage. The water was examined and various pathogens were found. These included rotavirus and small round-structured virus (Norovirus), but *Giardia* was not one of the pathogens tested for. The water supply was chlorinated but not filtered (Hardie *et al.*, 1999).

As well as drinking-water causing outbreaks of giardiasis, recreational use, where small amounts of water are swallowed, has been known to cause the disease. Like recreational cryptosporidiosis, incidents are mainly associated with faecal accidents in swimming pools.

Naegleria

Naegleria is a type of amoeba that feeds on bacteria. They live in warm freshwater and so the organism is common in surface waters and soil in warm countries. Humans are only thought to be affected by the species *Naegleria fowleri*. Infection is rare and normally only acquired by contact with surface waters when swimming. There have only been 144 recorded cases worldwide. The organism enters through the nose and moves to the brain or spinal column. It can cause primary amoebic meningoencephalitis. This is a terrible disease that starts with a headache and fever followed by vomiting, a worse fever and a coma. The disease is often fatal and death occurs between seven and ten days after infection. Most cases have been in the US (63) followed by Australia and Czechoslovakia. One case in the UK followed contact in a spa pool. Of course, the disease may be a lot more common in hot countries in the developing world but lack of suitable laboratory facilities and knowledge of the organism may have prevented it being identified.

11 Helminth infestation

There is only one waterborne helminth infestation discussed here. It is the nematode (roundworm or threadworm) infection known to cause Guinea Worm disease or dracunculiasis.

Dracunculiasis

Dracunculiasis is endemic in Africa. The disease also used to be common in India and the Middle East, but is now confined to rural areas in 13 countries in Africa, between the Sahara and the equator (Ruiz-Tiben *et al.*, 1995). Worldwide numbers of those infected fell from over 3 million in 1986 to 220,000 in 1994. This disabling condition is caused by the roundworm parasite *Dracunculus medinensis*. Its intermediate host is the copecod water flea. Copecods are very small and can barely be seen in a glass of water if is held up to the light. People become infected when they drink stagnant, dirty water containing the water fleas. When they enter the stomach, gastric juices kill the fleas but release their larvae. The larvae burrow through the stomach wall and migrate around the body. The worms, which are about the thickness of a piece of spaghetti, grow to nearly a metre in length. They mate, the males die and the pregnant females migrate to the skin. People do not know they are infected until the worm actually reaches the surface of the skin, which can take up to a year. Here they form a blister that eventually ruptures. At this time there are other symptoms such as nausea, diarrhoea, vomiting and dizziness. Because the blister feels hot (another name for the creature is 'fiery serpent' because of this hot blister), people bathe in cold water to cool down. This causes the female to release millions of immature worms in a milky white fluid. The water fleas eat them, which starts the process off again. The worms emerge mainly through the legs and feet but in 10 per cent of cases they may come out through the arms, genitalia, trunk or buttocks.

When the worm first appears, people should begin to draw it out slowly and carefully, to prevent pulling its head off. If the head is removed by mistake, a painful swelling reaction occurs all along the length of the remaining piece of worm. Traditionally, the affected person winds the worm around a stick, a few centimetres per day, until it is completely removed. This can take several weeks. During this process, because of the lesions and damaged skin around the site, other infections can occur. Some people think that the medical sign, still used by some vets and doctors, with a picture of a snake coiled around a stick or sword, is in fact an ancient representation of this process of encouraging the guinea worm to leave the body.

There is no cure for dracunculiasis and people do not develop immunity to this disease. Eradication programmes involve simple measures such as education about the lifecycle of the worms, keeping infected people away from water and using cloths to filter the water fleas.

12 Water-related diseases

The definition of a water-related disease is one where water is the habitat for the vector of the disease during part of its life cycle. This chapter is included for completeness as many open wells and watercourses in developing countries will provide that habitat. Even in developed countries with sophisticated surveillance techniques, public health measures may not be able to cope with rapid spread caused by modern travel. West Nile virus is a perfect example; a couple of cases brought into the US from another country have spread like wildfire.

The vectors that carry these diseases are usually insects, particularly mosquitoes. The illnesses spread by mosquitoes are thought to be the biggest causes of death from any of the infectious diseases. The female mosquito feeds on blood to ensure her eggs develop. When she injects her mouthparts into the bloodstream of an infected individual, she can also ingest malaria, dengue fever, etc. The infection is passed on when she feeds again.

The vectors do not, of course, restrict their activity to private water supplies. Streams, rivers, lakes, sewage works, ponds, puddles and even small areas of collected rainfall in trees or discarded tyres can provide a home. Different insect vectors require different types of water to breed in – some prefer still waters, others only use streams or rivers. The mosquitoes that spread malaria and dengue fever usually prefer clean water but the mosquito that spreads filariasis breeds in highly polluted water and overflowing pit latrines (WHO, 1992).

Interventions to reduce these diseases should take a multiple-barrier approach. If one part of the fight fails, the others may still be successful. As well as treating individual patients and generally reducing vector numbers, if you can break part of the life cycle of the insect or parasite it will not reach full maturity and the disease cannot be passed on. Sometimes removing or altering the nature of the habitat water can achieve this. Open wells should therefore be kept covered or otherwise protected to prevent the vector gaining access. Other methods include trying to remove standing water to reduce places where the mosquitoes can breed or well-organized pesticide spraying programmes of the water or the insides of houses where the adults frequent. The female mosquito lays her eggs in water and when they hatch, the larvae live in the water until they pupate. The larvae have breathing tubes connected to the surface of the water and another method of eradication is to pour light oil over the top of the water – this blocks the tubes and kills the larvae. Personal protection, such as using mosquito nets (particularly ones soaked in insecticide) and insect-repellents, can also be useful. Large areas of the Camargue in the South of France have been rendered malaria-free by intensive and intelligent use of all these techniques. Fortunately for the French and the visitors to the area, it was a sound financial idea to carry out and maintain these expensive vector control programmes. Unfortunately for many people in the developing world, the economics do not add up for them.

I will now briefly describe the major water-related diseases.

Dengue fever

Dengue fever is a tropical disease, mainly concentrated in Asia and the Americas, but it is increasingly a problem in Africa. It is one of the most common and widespread insect-borne infections with about 50 million people afflicted annually and 24,000 deaths (WHO, 2002a). Symptoms vary from fever to fatal haemorrhaging (serious bleeding from blood vessels). About 5 per cent of the people who get the disease die from it – mainly children. The mosquito is the vector for dengue fever and the main species that spreads it is *Aedes aegypti. Aedes* are the type of mosquito that breed in rain-filled waste cans, tyres, etc. This species also transmits yellow fever.

The disease is caused by four serotypes of the *Flavivirus* arbovirus and each strain causes different symptoms. Once attacked by one, the patient (if they survive) will get life-long immunity to it, unfortunately this does not prevent them getting attacked by another. The disease is also known as 'breakbone fever' because of the pain associated with it. It is of course important that people with the disease are kept away from mosquitoes to prevent the disease spreading further.

Some experts quietly admire the deadly *Aedes* mosquito and consider it well adapted to its lifestyle. It is fast and has a relatively quiet hum, so as not to wake you up or attract attention before it bites. The mosquito attacks from behind or below, mainly on the ankles or feet. It lives close to humans (normally within a hundred metres) and feeds on man in preference to any other animal. Some people even think that when pottery was invented thousands of years ago, the *Aedes* adapted to develop a strain that could live in water in man-made containers all year round, rather than just in puddles and holes in trees during the rainy season. With modern forms of transport, dengue is spread more quickly, particularly by people who have been bitten but have not started to show signs of the disease. Modern control methods rely on reducing the places where the mosquito can breed (Gluber and Clark, 1995). As the mosquitoes can breed in very little water, the effect of increased production of such things as non-returnable glass bottles provides many more potential habitats.

Lymphatic filariasis

This disease is spread by the *Culex pipiens* mosquito. Other names for the infection are Bancroftian filariasis and in severe cases, elephantiasis. It affects over 120 million people in 73 countries of the tropics and sub-tropics. When the mosquito feeds, it ingests microscopic thread-like round-worms (200–300 μm) from the blood of an infected human host. They are passed on to subsequent victims when the mosquito feeds again. The worms then travel from the skin to the lymph nodes where they grow up to half a metre in length. They can live for up to seven years and after mating, will release millions more microscopic worms into the blood stream. It is thought that you need repeated bites over many years to get filariasis. Short-term visitors to endemic areas therefore have little chance of getting this disease.

Breakdown in public health measures and insufficient mosquito control are the cause of the growth in the incidence of this disease. Control measures include dosing whole communities to kill the microscopic worms. This does not kill all the adult worms but reduces them to acceptable levels and has to be repeated annually. Because it is not completely effective, infected people remain a risk if mosquitoes bite them. Changes to surface water systems can also increase the number and size of breeding places for the mosquito.

Malaria

In 1898, it was discovered that malaria is accidentally injected into the blood stream when certain types of the *Anopheles* mosquito impale their victims to obtain blood. The organism that causes malaria is called *Plasmodia*, which is a single-celled protozoan parasite. The symptoms of malaria are severe fever, headache and nausea. The parasite invades the red blood cells causing damage to organs and anaemia. It can also cause kidney failure and obstruct the circulation by damaging blood cells and turning the blood to what is sometimes described as 'sludge'. Another name for malaria is black water fever. There are now 300–500 million cases of malaria per year, with 1.5–2.7 million deaths in 91 countries around the world (WHO, 1996d). About 75 per cent of these deaths are in children under 5 years old. It appears that in sub-Saharan Africa the infection rate is already so high that it does not appear possible for the situation to get worse.

There are several reasons for the increased incidence of the disease, including a reduction in the amounts spent on eradication, the effects of civil unrest and increased urban migration. Many of the prophylactic (protective) drugs are becoming less useful as resistance to them spreads and others are found to have severe side effects. Probably the main cause, however, was ending the use of cheap, effective, persistent insecticides because of environmental worries. There were many successful eradication campaigns in the 1950s and 1960s, mainly using DDT. Whilst countries that have abandoned spraying have had large increases in the disease, Ecuador, which has increased DDT spraying, has had a 61 per cent reduction since 1993. Robert *et al.* (1997) suggest that pressure from active environmentalists in rich countries to have this chemical banned will, for a variety of reasons, result in millions of new cases of malaria in the world's poor countries. They suggest that perhaps it should not be banned until the known detrimental effects outweigh the illness it prevents. Dr Robert and his co-authors stated, 'we are now facing the unprecedented event of eliminating, without meaningful debate, the most cost-effective chemical for the prevention of malaria. The health of hundreds of millions of persons in malaria-endemic countries should be given greater consideration before proceeding further'. The Commission for Environmental Cooperation, a North American organization, is now trying to get DDT banned in Mexico before 2006. It seems that the imperialist approach of the seventeenth and eighteenth centuries to people in hot countries – forced religious conversion and the wearing of clothes – has been replaced by the need to take on white middle-class attitudes to environmental issues. DDT, by the way, is the abbreviation for dichlorodiphenyltrichloroethane (a mixture of the isomers 1-(2-chlorophenyl)-1-(4-chlorophenyl)-2,2,2-trichloroethane and 1,1-bis(4-chlorophenyl)-2,2,2-trichloroethane).

One preventative for malaria is quinine, obtained from the bark of several types of South American tree. It is considered by some experts to be the drug that has benefited more people than any other in fighting infectious diseases (Reiter, 2000). The trees, originally found in the Andes, produce a powder called cinchona. Introduced into the Old World in 1640, it was widely used in mainland Europe but, because it came from the Catholic areas of Peru and Bolivia and had not been mentioned in ancient Greek texts, Protestant England took some persuading that it was a useful medicine. Quinine is a constituent of the tonic used in gin and tonic. Many people do not realize that when they drink gin and tonic it is really a medicine to keep malaria at bay.

There is some talk that due to global warming, malaria may return to Europe. Although an attractive idea if you want to get a job as a doom-monger, it is not really a possibility. Malaria was active in Britain and a lot of the rest of Europe when the temperature was even colder than it is at present. There have also been malaria epidemics as far north as the Arctic Circle. During the Little Ice Age, about 1550 to 1750, at its coldest the Arctic pack ice extended so far south that Eskimos landed their kayaks in Scotland. Nevertheless, during this time of severe cold, malaria was common in England and caused many deaths around the marshlands where the mosquitoes lived. The disease started to decline in the nineteenth century, but not because Europe was getting colder, because it wasn't – it was getting warmer. As well as the success of quinine, the reason for the decline has been attributed to a variety of reasons. These include drainage and land changes removing mosquito habitats, additional growing of turnips and mangel-wurzels that increased the numbers of animals on the land thus reducing the chances of man being bitten by a feeding female mosquito (Reiter, 2000). The only real risk of getting the disease is to travel to those areas where it is endemic. It is a great pity that it cannot be eradicated there as well.

Yellow fever

The yellow fever organism belongs to the group of viruses known as *Flavivirus*. The disease is spread by the *Aedes aegypti* mosquito and is a serious endemic problem in Africa and the tropical Americas. As well as the mosquito passing the virus by piercing the skin to feed, it can infect its offspring via its eggs, which will transmit the virus to man when they become adult. The disease has been known for over 400 years, but it was not until 1900 that is was found to be transmitted by mosquitoes. At present there are thought to be 200,000 cases per year and 30,000 deaths. Very effective vaccination has been available for sixty years and can give protective cover for up to ten years.

After being bitten by the mosquito, the incubation period is three to six days. Some people do not show any symptoms, but for those that do, symptoms include high temperature, fever, pain, diarrhoea and jaundice (thus the name yellow fever). This condition normally lasts for about four days and the majority of people then get well. However, for about 15 per cent of cases, a second attack of fever quickly follows. This time symptoms also include vomiting, bloody stools, haemorrhaging, particularly around the mouth and nose, encephalitis and renal problems. Death may follow.

There are several other insect-borne water-related diseases which mainly have no cure or vaccine and include Eastern and Western Equine Encephalitis, Japanese Encephalitis, Onchocerciasis (also known as River Blindness), Rift Valley Fever, St Louis Encephalitis and West Nile Virus.

13 Water-contact diseases and cyanobacteria

This chapter discusses cyanobacteria, which can cause a variety of problems and only one water-contact disease: schistosomiasis. As well as being a significant cause of illness worldwide, it has been associated with problems with drinking-water from surface supplies such as lakes and streams when people enter the water to fill their water containers.

Schistosomiasis

Schistosomiasis is also known as bilharzia. This is because a German called Theodor Bilharz first identified the parasites in 1851 in Cairo. This disease has been known to cause illness in humans since the days of the ancient Egyptians. It has also been known in Japan for a long time. Very old Japanese folk songs from the Yamanashi area refer to the effects of the disease in the snappily titled, *'I do not want to marry a man in Magoshima; It's a hardship to drink the polluted water of Nozouike pond'* and *'If you want to marry into Nakanowari village, I shall present you a funeral dress and a tomb'* (Toyoda, 2002).

After malaria, schistosomiasis is the second most important disease in terms of human beings infected and the suffering it causes. It affects over 200 million people in 74 countries in the developing world and 20 million suffer severe consequences from the disease (WHO, 1996c). Although it is not as fatal as many other tropical diseases, about 20,000 people a year will die from it. Again, this is a disease that affects a great proportion of the world's children and about 80 per cent of those affected live in Africa. Having schistosomiasis can have a chronic, severely debilitating effect on the body. Problems include anaemia and impaired development, both mental and physical. Cases are normally identified by blood in the urine or faeces. As with many diseases of hot countries, migration caused by poverty or war brings people with the disease to new areas. This has happened recently in Somalia and Djibouti. Changes to rivers caused by damming also spreads the disease to new districts. Around the man-made Lake Volta in Ghana, 90 per cent of the children are now affected.

The disease is caused by fluke-like creatures called schistosomes. They are transmitted through fresh water and canals, lakes, rivers, streams or springs can all be infested. Where the disease may be present, it is recommended that raw water is always boiled. Infected people can pass the eggs of the parasite in their faeces and urine. If the eggs hatch, they produce small miracidia that swim and penetrate the water snails that act as the intermediate host. The life cycle of the parasite within the snail is complicated but eventually thousands of tiny thin fish-shaped organisms called cercariae are produced. They then leave the snail and swim to the human host. They penetrate the skin in seconds (and lose their forked tail) and in doing so become schistosomes. Schistosomes are also known as digenetic blood trematodes, waterborne flatworms or blood flukes. Once they have left the snail the cercariae can survive for about two days. Therefore, where this disease is endemic, water that is used for washing should be stored for over two days before use. Transmission can only take place if people use open fresh water as a toilet and other people enter the water to work (such as rice farming or fishing), bathe or gather water for drinking.

Once in the human host, the schistosomes grow up to 20 mm long within 30–45 days, after which the females begin to produce eggs. Although early symptoms are often absent, itchy skin caused by the irritation of the cercariae entering the body is common. Fever, chills and a cough can occur up to two months after infection. Symptoms of schistosomiasis are caused by the body's reaction to the eggs, not by the worms themselves. The females lay from 200 to 2,000 eggs per day for up to five years. The eggs are spiny and they tear and scar tissues. Where someone has been repeatedly infected over many years, there can be damage to the liver, intestines, lungs, female reproductive organs and bladder. Death can be caused by complications of bladder cancer, renal failure, liver fibrosis or hypertension.

There are suitable medicines available that will quickly treat the disease and reverse the conditions it causes, if treated early enough. Repeated infections will cause irreversible damage. The WHO strategy for dealing with this disease is through giving medicines to infected people, discouraging defecation in open waters and the provision of sanitation and safe drinking-water (WHO, 1993b). The provision of a proper water supply will reduce infection levels as it stops people going into surface waters to fill their water jars (Hunter, 2002). It is also possible to alter the water snail's habitat or use molluscicides to remove them, but this may upset the food chain balance in areas where fish are an important part of the diet. It is also said that hard rubbing with a towel of any areas of skin that have been in contact with suspected surface waters will prevent infection. This is not to be relied on.

Cyanobacteria

Cyanobacteria can cause problems for environmental health practitioners around the world, yet these dangerous but remarkable organisms (vitally important to life on this planet) are often little understood. They not only have two contradictory names, but there is some confusion as to whether they are plants or animals. For the purposes of clarity, I will refer to them throughout as cyanobacteria, but they are often better known as blue-green algae. Although they are normally associated with recreational water problems and have not particularly been associated with small water supplies, this is possible and that is why they are included here. For this section, I found the book entitled *Toxic Cyanobacteria in Water: A Guide to Their Public Health Consequences, Monitoring and Management*, edited by Ingrid Chorus and Jamie Bartram, published by E & FN Spon, to be an excellent and very detailed account of their pathogenicity and aetiology. Again, where briefly referred-to outbreaks are mentioned they are rarely referenced directly and I would apologize to anyone who feels that they should have been referenced but have not been. Chorus and Bartram's book should always be referred to where further information is needed.

So, what are these fascinating creatures? Cyanobacteria make up the largest group of simple non-nuclear single-celled microorganisms. But, are they animals or plants? There was much discussion on this subject before it was determined that, despite their botanical appearance, and the fact that they contain chlorophyll, they are bacteria. However, the boundaries are not that clear and some confusion remains, so don't be too surprised if this definition changes. They photosynthesize, are filamentous (which basically means hairy) and colony forming. They need only water, CO_2, inorganic substances and light to live; even so, they also survive for a long time in the dark. The amount of light that they need is so low that they can often out-compete other organisms in highly turbid waters.

The reason why cyanobacteria are of public health interest is that some produce a variety of poisonous organic chemicals, known as cyanotoxins. The toxins can enter drinking-water systems if a lake or river is used as a raw water supply. There is normally only a risk when the cyanobacteria multiply to form large mats (blooms) or scum on the surface of the water. The toxins are normally kept within the cell walls, so they only become a problem when the cells are ingested or if the cells die and start to disintegrate in the water. All cases of gastrointestinal disease following consumption of cyanobacteria have been associated with decomposing cells. It is thought that the amount of free toxin in the water is normally 10–20 per cent of the total amount of toxin present. This figure may be greater in ageing populations as more of the cyanobacteria begin to die-off. If there is an outbreak of acute gastrointestinal illness, it is best to consider viruses, bacteria or protozoan parasites as the possible cause, however if an obvious culprit cannot be found, the drinking-water comes from a surface source and it is late summer, then cyanobacteria should be considered.

The first properly recorded case of acute cyanotoxin poisoning was in Australia in 1878. Numerous animals that drank the water from Lake Alexandrina died very quickly afterwards. However, over a thousand years ago, a Chinese general reported loosing troops when they crossed a river that looked much greener than normal. As they negotiated their way across, the soldiers drank some of the water, subsequently fell ill and died. People in many cultures have had to learn that when they take water from lakes, they must dig holes close to the bank and wait for them to fill with water. Cyanobacteria are largely filtered out by this basic bankside filtration.

Cyanobacteria are found virtually everywhere. On land they have been discovered in soil, on desert sand, in volcanic ash and on rocks. They are a normal constituent of freshwater but thrive in salty water where they form a large percentage of marine plankton. Cyanobacteria cope with extreme environmental conditions and live in polar regions and hot springs, even in the thermal vents from undersea volcanoes. They can form symbiotic relationships with animals, plants and fungi and have been around a long time. Some fossilized cyanobacteria have been uncovered that are 3,450 million years old.

It is believed that cyanobacteria were instrumental in the most important change to life on earth – from unicellular to multicellular organisms. Originally, life on earth consisted of simple, single cells with no nucleus (prokaryotics). The theory is that after millions of years, one cyanobacterium engulfed another one nearby and somehow they became a single, complex microorganism with a nucleus (eukaryotic). This eventually led to an explosion of growth, including plants, animals and fungi. Eventually this huge family included human beings – but just because they are possibly your cousins, does not mean you can relax when they are around.

Bill Bryson, the gifted writer and linguist waxed lyrical about cyanobacteria in his book *Down Under* (2000). Cyanobacteria form coral-like constructions called stromatolites, which have been unchanged for 3.5 billion years (more than three-quarters of the way back to when life first emerged on the planet) but are nowadays only found in three places on earth. He noted, when he found one, that occasionally you see tiny bubbles of oxygen rising from the large cowpat-shaped blobs. He went on to say that, 'this is the stromatolite's only trick and it isn't much, but it is what made life on this planet what it is today'. They each take only CO_2 and light energy to form 'the faintest puff' of oxygen. They were the only form of life found on the planet for two billion years, but because they were found in such large numbers, they produced enough oxygen over the millennia, to raise the level in the atmosphere from virtually nothing to 20 per cent. This was 'enough to allow the development of other, more complex life forms' – Mr Bryson included.

There are thousands of species of cyanobacteria, some produce toxins and some do not. Even individual species may contain toxic and non-toxic strains. Unfortunately, it is not possible from microscopic examination to tell whether a particular strain is toxic. Some cyanobacteria produce blooms and some do not. Some are blue-green and some are not (some are red – they produce the dangerous 'red tides' in tropical seawaters). Some live in cloudy water associated with excessive pollution, others like clear water. Some species form mats on the surface of underwater rocks, others prefer to live close to the surface of the water. Others contain gas-filled vesicles that allow them to move up and down in the water, so they can find the most advantageous ecological niche at any one time.

As they are so ubiquitous, should they constantly worry us? The answer is no (with qualifications). I could only find one outbreak where acute lethal poisoning had been associated with drinking-water thought to contain cyanobacteria. But there have been many cases of serious illness associated with recreational water use. The only other recorded deaths were during an incident in Caruaru, northeast Brazil, when water that contained cyano-toxins was used in kidney dialysis machines. Sixty patients died when inad-equately treated water from a local reservoir with excessive algal blooms was used during a serious drought in 1996 (Pouria *et al.*, 1998). There have been numerous animal deaths recorded, but they are much more likely than humans to drink from the sides of lakes or rivers containing blooms or scum. In fact, in most animal poisoning cases there has been an obvious scum at the drinking point. The number of cyanobacteria in water that humans would find acceptable to drink would not generally contain enough cyanotoxins to cause illness. Some cyanobacteria will also cause water to taste musty or earthy and this has sometimes alerted authorities to a prob-lem. However, it is important to stress that many cyanobacteria do not cause any taste or odour problems and just because these problems are absent, it does not mean the water is safe. Boiling the water will kill the cyanobacteria but not remove the cyanotoxins.

Cyanobacterial blooms are mainly found in lakes and slow moving bodies of water. This is because in a swift-flowing river, the physical movement of the water will tend to prevent them forming large mats. To form blooms, cyanobacteria also need waters that are high in nutrients, so they can grow and multiply. Waters that are high in nutrients are called eutrophic. The process of becoming eutrophic is called, unsurprisingly, eutrophication and was first recognized in 1969. It is often defined as – the enhancement of nat-ural processes of biological production caused by increases in the levels of nutrient, usually phosphates and nitrates. A body of water with normal amounts of nutrients is called oligotrophic. If it is unusually low in nutrients it is called mesotrophic. Nitrates are a normal constituent of sewage and are found in treatment works effluent and in fertilizer run-off. Nitrates can also come from industrial production or natural sources such as cows. Some cyanobacteria can even fix nitrogen directly from the air. Phosphates are nat-urally excreted (two to four grams per person per day), applied as fertilizer and found in washing powder and detergents. Because of the nitrogen-fixing ability of cyanobacteria it is normal to try to reduce the phosphorus content of the water to limit their growth. Reducing man-made sources of phospho-rus can be successful, but even totally removing phosphorus from detergents will only decrease the amount in sewage effluent by about half. This is because the other half is being excreted. Reservoir storage will naturally remove 50–65 per cent of the phosphorus in water.

In temperate parts of the world the blooms are seasonal, occurring at the end of summer and in the autumn, as the water temperature builds up. This warmth encourages the cyanobacteria to multiply. In eutrophic conditions, temperatures normally need to be above 25°C. In Mediterranean and tropical regions, blooms can occur year round and will last longer than the two to four months normally found in temperate climates.

When blooms form, they contain millions of the organisms, the wind can then blow them to the edge of the water, where they accumulate and form scums. A bloom will contain several species of cyanobacteria and different ones will dominate as the bloom ages. This will not be apparent on a daily basis, but they will change from week to week. Blooms are a nuisance for water management, as well as being potentially dangerous. The blooms will also become unpleasant when the cyanobacteria die and start to rot. Mats that grow on the bottom, either in the mud or on rocks, can also be a problem. This is because when they photosynthesize, they produce oxygen; this gas lodges in the algal mats and causes parts of it to tear off, rise to the surface and die.

Die-off within cyanobacterial blooms can be natural or caused by man when pesticides (normally copper sulphate) are sprayed on the water to remove them. If the correct application of copper sulphate is made the death of all cyanobacteria will occur within three days. Other algaecides that have been used to control blooms have been; copper citrate, potassium permanganate, chlorine and copper triethanolamine. Alternatives to copper sulphate are being sought because it is poisonous to other life forms. In some cases algaecide usage itself has caused a problem. Where blooms have been sprayed, the cyanobacteria die and start to decompose, thus releasing their toxins. When considering using an algaecide where water is abstracted for drinking purposes, it is important to use it as a preventative measure rather than a cure, that is to stop the build-up of a bloom when numbers (and toxins) are low, rather than when they have become a major problem.

Biological control has been introduced as an alternative to chemical dosing. Fishing or adding predators can remove plankton-eating fish and this allows the cyanobacteria's natural competitors an extra edge. Aquatic plants can be introduced that take up phosphorus as they grow. As with all biological control this process cannot be an exact science and there are many examples where species have been introduced to solve one problem and ended up causing an even bigger one.

Another 'natural' control that has been suggested is the application of decomposing barley straw. The theory is that fungi on the straw are antibiotic, possibly due to hydrogen peroxide production when it rots. Some studies have found a decrease in cyanobacteria; others have not. The studies that have been done were not without their criticisms. The use of barley straw is cheap and highly visible, so it can look as if the authorities are not only doing something about the problem but are also using an environmentally sound option. However some experts (Chorus and Bartram, 1999) consider the practice is dubious, particularly as it may increase the problem by introducing more organic, oxygen-consuming material into the water.

The toxins produced by cyanobacteria can be separated into groups by the way they act on the human body. The most common are the hepatotoxins. These cause severe damage to the liver within twenty-four hours of being ingested. The second group is the neurotoxic alkaloids that affect the central nervous system. A third group causes skin irritation and are called the dermatotoxic alkaloids. The final category is the irritant toxins. They cause problems with internal linings of the body, such as in the nose or throat.

The first documented cases of human gastrointestinal illness associated with cyanobacteria were along the Ohio River in 1931. The course of the illness seemed to follow a toxic bloom as it flowed downstream. Another outbreak was in 1966 in Harare which had an annual problem caused by water from a particular dam. In other parts of the town, which used water from another source, there was no illness. The worst outbreak was in Bahia in Brazil: 2000 people were ill and 88, mainly children, died after drinking water from the newly flooded Haparica Dam in 1988. The water was later found to contain *Anabaena* and *Microcystis* cyanobacteria. As well as gastrointestinal problems, there have been recorded outbreaks where liver damage has been caused. In 1979, 150 people (of which 140 were children) were hospitalized after drinking water from a reservoir on Palm Island, off the coast of Queensland. The reservoir had recently had an algal bloom problem and been treated with copper sulphate at 1 μg/l. The people affected had various symptoms of liver damage. In New South Wales in 1981, there were more people with evidence of severe liver damage after drinking water from another reservoir that had also had a *Microcystis* bloom treated with copper sulphate. During the dialysis incident in Caruaru, 117 patients developed cholestatic liver disease and 47 of them died. The livers of the patients and the filters on the dialysis machine were found to contain microcystin toxin. Chronic effects from cyanotoxins have mainly been recorded in China, where they cause liver damage. Until recently a lot of the population drank water from ponds and ditches, which have, unsurprisingly, been found to contain cyanobacteria. The Chinese government has been carrying out campaigns to reduce the problem and have replaced traditional surface water drinking sources with wells and boreholes.

For drinking-water safety, the best way to prevent problems is to stop the cyanobacteria getting into the water supply in the first place. Providing suitably fine microstrainers and baffles can do this or ensuring the supply's intake is properly designed and situated. If filtering is used to physically remove the cyanobacteria, the organisms should always be carefully removed, as killing them or rupturing their cell walls will release the toxins and it is much easier to remove the cyanobacteria than their toxins. Cyanotoxins are extremely stable and most treatment systems use activated carbon filters that work by adsorption. Wood-based powdered activated carbon (PAC) is effective at removing some, but not all. Granulated activated carbon (GAC) is more effective, but needs a contact time of fifteen minutes or more. Regular replacement of the filter medium is necessary and if the medium is used up, the filter will fail to danger. Ozonation is effective and uses powerful oxidation to break down the chemicals, but it works better in treated water than raw. Potassium permanganate, another powerful oxidizing agent, is also effective for treated water. Chlorination can remove some cyanotoxins but not all. Ultraviolet irradiation at doses two orders of magnitude higher than normal has also been reported as being capable of removing some cyanotoxins (Carlile, 1994). All these methods are complicated to operate however and should not be relied on without a suitable back-up system.

There are guideline values for cyanobacteria in both recreational water and water used as a source of drinking-water. Guidelines in drinking-water take two forms. One is the actual amount of toxin in the finished water. There has only been sufficient research carried out on one toxin, microcystin, for WHO to give a provisional guideline value. This is $1.0\,\mu g/l$ for the total amount of toxin both in the water and still within the cell body. The other guideline is the amount of cyanobacteria in the intake water. At 200 cells/ml vigilance is needed. At 2,000 cells/ml the people responsible for the water supply should consult with the health authority, inform hospitals about potential dialysis problems and take samples to identify the cyanobacteria and which toxins, if any, are present in the water. The next level is 100,000 cells/ml – when this is reached alternative sources of water should be provided. The water can still be used for washing, laundry and flushing the toilet, so cutting off the supply is not advised unless absolutely necessary.

I would not want you to run away with the idea that cyanotoxins are all bad. If you can forget for a moment that these toxins can do a very good job of spoiling your day at the bathing hole, there are several substances produced by cyanobacteria that have been found to be anti-tumour, antiviral, antibiotic and antifungal. Research laboratories are trying to find ways to use these substances for medical purposes. Cyanobacteria are also sold as dietary supplements, hopefully under strict control. I believe this idea started in California.

14 Chemical contamination

Microbiological contamination and waterborne disease are not the only problems associated with private water supplies. There are a number of possible chemical contaminants and people who are responsible for drinking-water safety must be aware of them and what risks they may present. Some chemical contaminants, such as lead and arsenic, are acknowledged as causing damage to health and this danger has been known about for some time. They should be removed from water or reduced to an absolutely safe level. There are other, less common or less well-known chemicals, where questions have been asked about their safety, but absolute proof is missing of their hazardousness at the levels normally found in drinking-water. There is a lot of research presently taking place to find out what problems they may cause and what a safe level for them would be. It is often prudent to adopt a precautionary approach to these chemicals until more information is available, but sometimes the worries prove unfounded after a great deal of money has been spent on their eradication.

Other chemicals are a problem because of their aesthetic or nuisance causing properties. Hardness is an example as it damages plumbing fittings. Others will produce taste and odour problems or cause staining to clothes or sanitary ware. Manganese, iron and aluminium are examples of aesthetic parameters regularly found in small water supplies. Even though aesthetic parameters may not cause ill effects, their presence should be reduced to a minimum, as any contamination of water is unwelcome.

The methods used by the UN and WHO to obtain safety levels for chemicals in drinking-water are based upon an Acceptable Daily Intake (ADI). The ADI is defined as 'an estimate of a daily exposure to the human population (including sensitive subgroups) that is likely to be without an appreciable risk of adverse affects' (USEPA, 1989). ADIs are derived from long-term feeding studies on laboratory animals with assessments being made on the carcinogenic, mutagenic (producing mutation of genetic material), teratogenic (substances causing deviations from normal growth and development of the foetus), neurotoxic and reproductive effects. The dose level causing no toxicologically significant effects on animals is referred to as the 'no observable adverse effect level' (NOAEL). The NOAEL is usually divided by a safety factor of 100 to become the ADI. This is to take account of a tenfold variation in susceptibility in animals and a further tenfold variation in humans (Hurst *et al.,* 1991). ADIs are expressed in terms of milligrams ingested per kilogram of bodyweight. Individual countries then use this to calculate maximum residue levels (MRLs) in food and water based on the amounts that are consumed in that individual country, either for the general population or for specific subgroups such as infants, ethnic groups or vegetarians. In hotter countries, water is consumed in greater amounts and therefore for the same ADI the MRL of any chemical in the water should be less. WHO has produced a comprehensive set of Guideline Values for all the important contaminants of water (WHO, 1993a). The organization stresses however that these are only guidelines and each individual country should adapt them to take account of its climate, population and their particular dietary habits. The European Union has taken these guidelines and incorporated the vast majority of them into the standards laid down in the Drinking-water Directive (European Union, 1998).

The general processes for the contamination of water have already been described and a variety of mechanisms, both natural and anthropogenic (manmade) can add chemicals to drinking-water. Some may even be introduced as part of a treatment process designed to improve the water quality. Chapter 15 and 16 discuss the potential contaminants of particular interest in small or private water supplies and their effect or otherwise on health.

15 Inorganic contaminants

Aluminium

UK maximum concentration allowed: 200 µg/l.
WHO Guideline Value: None set – complaints will be expected at 200 µg/l.

Aluminium (Al) is a metal and is the third most common element in the earth's crust. Its chemistry is complex and the element can occur in many soluble forms known as *species*. Aluminium is commonly found combined with silicon as aluminosilicate minerals in clays, rocks and soils. It is a general rule that the lower the pH of the water, the greater will be the amount of metal that can be dissolved from the soil and geological deposits. Thus, where rainwater is acidic, the water should be expected to contain some metal contamination. In private water supplies, because of the ubiquitous nature of aluminium, it should be encountered on a fairly regular basis.

In the past, there has been controversy over aluminium's connection to the biological damage associated with acid rain. There have also been questions over its safety in drinking-water because of a possible relationship with Alzheimer's disease. In some areas of the world, where there are naturally high levels of aluminium in the water, there are raised levels of Alzheimer's disease or other conditions with similar symptoms. Until recently however, the element was considered totally harmless, with no evidence that it could cause disease (Packham, 1990). WHO still classes it as an aesthetic rather than health-related contaminant (WHO, 1993a). They state there is little evidence that orally ingested aluminium is acutely toxic to humans, despite the widespread occurrence of the element in foods, drinking-water and many antacid preparations. In most of the USA, aluminium is not even included in the list of chemicals that are looked for when testing water from small supplies. They know it is there, because it is such a common element, but they do not consider it to be a health problem. Neither is it an aesthetic problem, as it is never found in high enough concentrations to cause taste problems. So, the argument goes, why bother looking for it?

Aluminium can form strong complexes with organic and inorganic substances. At normal pH, as an aluminate, it is highly insoluble and generally not available for participation in chemical or biological reactions. At lower pH values, however it becomes increasingly soluble (Gjessing *et al.*, 1989). Simply put, if the pH of water is raised the dissolved aluminium will precipitate out of it; if the pH is lowered it will dissolve into it. Of particular interest to questions about acid rain is the fact that aluminium sometimes binds to naturally occurring humic and fulvic acids in soils and peat. Although there may be high levels of aluminium in waters from moorland areas containing these substances, it is not easily released from this binding effect and consequently is not available to become toxic to aquatic life (Gjessing *et al.*, 1989).

The highest concentrations of aluminium in surface waters occur at high flows, when pH is at its lowest. At these peak flows, surface waters contain a lot of aluminium-bearing soil (aluminium is found in the soil as part of the weathering process). Normal flows are more likely to be composed of groundwater from lower soil zones, where the acid is partially neutralized by inorganic weathering and the aluminium is thus precipitated out. This increased acidity during high flows can also result in enhanced levels of aluminium from recently precipitated aluminium hydroxide or other sediments from the bottom of the stream, dissolving once again in the water (DoE, 1988). Aluminium has now been shown to be quite toxic once it is actually released in the aquatic environment. It was the acid in acid rain that was originally thought to be the cause of damage to flora and fauna, but it is now believed to be the aluminium released by the lowered pH that causes the problems (DoE, 1988).

Aluminium has no known health benefits (Gjessing *et al.*, 1989) but is not normally present in man because the gastrointestinal tract is a proficient barrier against it. If it does enter the body, it is efficiently excreted via the kidneys (Costa, 1991). However, once it has managed to pass into the blood stream in sufficient amounts, due to a medical problem or because of excessive quantities being present in the water, aluminium has been shown to have some toxic effects (Wisniewski, 1991). Some time ago, there were several studies linking Alzheimer's disease with aluminium in water supplies (Flaten, 1989; Martyn *et al.*, 1989; Michel *et al.*, 1991). However, at least one of the authors (Martyn) has stated that none of the studies were straightforward to interpret and the results should be construed cautiously. Other scientists have criticized the data as being inadequate (Walton, 1991). Alzheimer's disease is characterized by two growths within the brain; these are called senile plaques and neurofibrillary tangles. Elevated levels of aluminium, as alumino-silicates, have been found, both in the plaques (Candy *et al.*, 1986; Edwardson, 1991) and the tangles (Perl and Good, 1991). Research however, has indicated that when studying the plaques, the aluminium may have been introduced during the staining process used to identify them (Landsberg and Watt, 1992). This research however did not study aluminium deposition in neurofibrillary tangles. The worries about aluminium are that it either causes the plaques and tangles or it is responsible for mental degeneration following its deposition in them. A counter argument is that, just as the build-up of calcium in certain degenerative abscesses is a consequence, not a cause of the problem, so aluminium in the brain is a consequence of Alzheimer's disease (Wisniewski, 1991). This hypothesis is corroborated by the deposition of aluminium in the brains of people with Down's Syndrome and those affected by *dementia pugilistica* (punch-drunk syndrome) (Hardy, 1991).

Birchall (1991) on the other hand looked at the relationship between aluminium take-up and the presence of silicon in the diet. He stated that a major role for silicon is to actually reduce the bioavailability of aluminium. With a silicon-deficient diet, animals show a pathological response due, not to the lack of silicon, but the increased amount of aluminium in the body. Birchall argues that the correlation between aluminium in drinking-water and Alzheimer's disease is less significant than an inverse relationship of the disease to silicon in water. Just as soft acid waters contain high levels of aluminium, they will also contain significantly reduced levels of silicon. This research shows the problems associated with scientific studies equally as clearly as it has the protective effect of silicon. Another important study was by Crapper McLachan (1991). He looked at the toxicological data on aluminium concentration in the brain and its dementing effects. After examining the published epidemiological studies, he concludes that there is no evidence for or against aluminium *causing* Alzheimer's disease. He states however, that there is considerable evidence supporting the idea that aluminium plays an important role in the cascade of events, which eventually lead to neuronal malfunction and neurone death.

With all the interest, how do you decide whether aluminium in private water supplies is safe? A major survey by Wilcock (1991) concluded as follows:

- There is no proof that aluminium causes Alzheimer's disease.
- There is some evidence to suggest that aluminium contributes to cognition and excess levels lead to mental impairment.
- Some epidemiological evidence shows an increased level of Alzheimer's disease can be found in some areas of the world where there are higher concentrations of aluminium in drinking-water.
- It should not be forgotten that human beings ingest aluminium by means other than drinking-water. Aluminium pans for example, will leach aluminium into the food during cooking.

Another important study (Walton, 1991) determined the following:

- Aluminium is found in many forms and some of these speciations are more bio-available than others.
- There is some evidence to suggest that Alzheimer's is a series of separate, but linked diseases, which would explain some of the incongruous results of some scientific studies.

- Any evidence linking aluminium to the aetiology of Alzheimer's disease is wholly circumstantial. There is no convincing evidence however to suggest that the European Union and WHO limit of 200 µg/l should be reduced.

A further study in Australia tested healthy volunteers with drinking-water containing the commonly used water treatment chemical, aluminium sulphate. It showed that the aluminium was not a risk factor for Alzheimer's disease, reassuring the world's water treatment engineers that they could continue using this effective chemical (Anon, 1999). A few years before, due to pressure from environmental groups worried about the unproven link with Alzheimer's disease, several water treatment plants in the UK replaced aluminium with iron as a coagulation agent (flocculant). In the early days of this changeover, plant workers reported several problems with its use. These included difficulties with control of the floc blanket and inaccuracies with dosing levels. Because of the continuing use of iron as a flocculating agent and improvements in its manufacture and use, these problems have largely disappeared.

The EU originally set maximum allowable concentrations for aluminium based on WHO Guidelines issued in 1970. At that time 'orally ingested aluminium compounds did not appear to have deleterious effects on normal individuals'. The Guidelines nevertheless reported that the presence of aluminium compounds in water used in kidney dialysis had been associated with neurological disorders (WHO, 1970). Despite these findings and because aluminium sulphate is an extremely effective treatment option for water, WHO suggested a Guideline Value of 200 µg/l. WHO has more recently reviewed the evidence for and against aluminium and concluded there is insufficient evidence for reducing levels although they state that a good treatment plant should be able to achieve levels of 100 µg/l. They kept the guideline level at 200 µg/l because a significant reduction would rule out the use of aluminium sulphate as a flocculating agent, particularly in small water treatment plants (WHO, 1993a). This would lead to an increased worldwide risk from diseases that are at present kept under control by using it (Patel, 1992). The UK government has occasionally allowed much higher levels in water. One report (DoE, 1991) stated that the government considered there was no public health hazard associated with aluminium up to levels of 3,300 µg/l. Higher amounts can be reduced to acceptable levels with treatment systems that raise pH and cause the metal to precipitate out. This can then be removed by physical filtration or settlement. Similar methods are recommended for use in iron and manganese removal.

Perhaps, more important than any long-term potential for chronic illness from aluminium are the problems associated with acute accidental poisoning. This was most famously highlighted by the incident at Camelford Water Treatment Works, in Cornwall, UK on 6 July 1988. A relief driver poured 20 tonnes of aluminium sulphate into the contact chlorine reservoir by mistake (Rowland *et al.*, 1990). There were many reported ill effects following the incident, but several epidemiological problems have made it difficult to decide how much the aluminium was to blame. Where people are aware that they have drunk water contaminated with potentially dangerous chemicals, several different mechanisms come into play. There can be the actual ill health from the substance, there can be ill health caused by the stress and worry following the incident and there can be claims of ill health brought about by a wish to punish the perpetrators of the incident or the feeling that a claim for compensation may be more successful if illness is associated with it. How much all or any of these factors played in the Camelford incident will prove very difficult to decide.

Arsenic

UK maximum concentration allowed: 10 µg/l.
WHO Guideline Value: 10 µg/l for excess skin cancer risk of 6×10^{-4}.

Arsenic (As) is an acute poison. The symptoms include vomiting, oesophageal and abdominal pain and bloody 'rice water stools', such as those associated with cholera. Long-term, chronic exposure to arsenic-rich drinking-water has been associated with cancer of the skin, lungs, bladder and kidneys. The skin is also affected by pigment changes and thickening. In China, arsenic in drinking-water has been shown to damage blood vessels leading to gangrene and a self-explanatory condition known as 'black foot disease'. Scientific studies have reported other effects but they are not conclusive. Arsenic is also, along with radon, one of the few contaminants of water known to cause a genuine risk of cancer at the concentrations they are normally found at (Mara and Clapham, 1997). WHO has set a value of 0.17 µg/l to reduce this risk, although they state that their data may have overstated the risk of skin cancer. The Guideline Value of 10 µg/l was set as the minimum amount of arsenic that could be found in water by most laboratories at that time.

Arsenic is a metal and is naturally found as a mineral in many places around the world. It will be found in untreated water that has passed through arsenic-rich rocks, particularly certain types of sandstone. It is usually found as trivalent arsenite – As (III) or pentavalent arsenate – As (V). Arsenate is the most commonly found of the two. Because it is sometimes used in industry, mainly as an alloying agent, but also in tanneries and in certain pesticides and wood preservatives, arsenic may contaminate water either from liquid effluent or by being washed out of atmospheric discharges by rain. It can also be found in the atmosphere as a by-product of the combustion of fossil fuels. Another source of arsenic is from food, often shellfish. They contain organic species of arsenic however and these are not considered to be as dangerous as the inorganic, waterborne kinds. This is because organic arsenic tends to be efficiently excreted by humans without being absorbed.

An outbreak of acute fatal arsenic poisoning associated with a private well occurred in the US in 1980. Nine members of the family became ill, four had encephalopathy and two died. Water from the farm's well (borehole) was found to contain 108 μg/l of arsenic. Arsenic was also found around the well and was thought to have come from an accidental spillage or inappropriate disposal of a pesticide containing the metal. There were several leaks in the well casing (Armstrong *et al.*, 1981).

There has been a lot of attention paid to naturally occurring arsenic in water following high concentrations being found in shallow tube wells (boreholes) in Bangladesh. The wells were drilled following a massive international aid programme. Before this, the population often had to drink from surface waters that were highly contaminated with faecal material. Until recently, the wells were considered completely safe and had been popular with the population. Now however, over 1 million people in the country are thought to drink arsenic-rich water and may be at risk from various forms of cancer associated with the metal. The water in these wells is contaminated by natural arsenic from the rock strata underlying the country. A report for WHO estimated that 35–77 million people in Bangladesh drank water that contained arsenic above the WHO Guideline (Smith *et al.*, 2000). Because the tube-wells are reasonably new, the population drinking the arsenic has not had time to develop the full range of symptoms and the main problem at present is skin lesions (Mara and Clapham, 1997).

Arsenic can be a significant problem in countries other than Bangladesh and it has been found in Australia, Chile, China, Hungary, Peru, Thailand and the US. The US Environmental Protection Agency estimates that some 13 million people, mostly in the western states of California, Arizona and New Mexico, drink water that has 10 μg/l of arsenic in it (Rogella, 2002). Many groundwaters in Arizona have concentrations greater than 50 μg/l and one water supply in Missouri has up to 70 μg/l.

Particular problems can be encountered where groundwater comes from mining areas. A water-works manager from the small coastal town of Ilo in Peru once told me that they had quite severe water problems in his area. He said that it had not rained in his part of southern Peru (it was in the Atacama Desert) for over four hundred years, so water was short. He also said that the groundwater he had to use came from a mining area and was so strong in naturally occurring arsenic that it would kill a horse. He knew this because he had carried out an experiment on one of the local nags and it had. Science in the raw, indeed.

Arsenic is not thought to be a major problem in the UK, although it can be found in the Midlands and the Northwest where some aquifers naturally contain it. This shortage of major problems is reflected by the lack of a statutory requirement to test for it when sampling small supplies. Authorities should find out if there are any local arsenic deposits in their area using geological information or contact with the British Geological Survey. If there are, arsenic should be included in sampling programmes.

Where arsenic has been found in drinking-water, the first option should be to identify an alternative source. Another option is to blend the water with arsenic-free supplies to lower the overall content to an acceptable level. Treatment systems to remove arsenic in small supplies are available but are not simple to use and need maintenance and care in operation. They may also fail to danger, which is not a good option. Where alternative sources are not available, it may be appropriate to separate water for drinking from water used for other purposes such as washing and flushing the toilet. Bottled water or treated rainwater harvested from the roof could then be used for drinking. Absorption through the skin is minimal and so there are no particular health risks associated with bathing and doing the laundry in arsenic-rich water (WHO, 1993a). The treatment systems that are available include coagulation and precipitation using iron salts and activated aluminium. Iron salts are more effective than aluminium coagulants as they have a larger surface area. Membrane filtration will also remove arsenic; particulate matter will be removed by micro or ultrafiltration and dissolved arsenic will be removed by nanofiltration (35–75 per cent) or reverse osmosis (greater than 95 per cent). Ion exchange units are effective at removing arsenate but not arsenite (Rogalla, 2002). A new development is the use of granular ferric hydroxide. It is possible to use naturally occurring ferric hydroxide, but in public water situations it is more efficient to use purpose-made, carefully milled granules. This enables arsenic to adsorb so strongly to the surface that it will not be leached out, even when the media becomes loaded and has to be disposed of to landfill. Granular ferric hydroxide is basically 'posh rust' (Steadman, 2002).

Since 1991, the UK maximum allowable level for arsenic in private water supplies has been $50 \mu g/l$. It is expected that this will soon be reduced to $10 \mu g/l$ as it was for public water supplies in December 2000 (HMSO, 2000). This followed the introduction of legislation after the 1998 Drinking-water Directive became operative (European Union, 1998). The US Environmental Protection Agency reduced their limit from $50 \mu g/l$ to $10 \mu g/l$ in October 2001, although water utilities have until 2006 to comply with this requirement. This is expected to reduce the number of lung and bladder cancer cases by up to 56 per year and between 21 and 30 deaths. The cost will be about US\$21 billion or US\$38 million per case. The US Environmental Protection Agency also expects the reduction to prevent many cases of diabetes and heart disease (Anon, 2001b). It is interesting to compare the cost of each case saved, with the number of deaths it would prevent if spent on basic clean water and sanitation provision in a developing country.

It is also interesting to note, as Whitehead (1992) pointed out, that until the introduction of modern organic pesticides, many inorganic pesticides contained arsenic. Arsenic-based pesticides were eventually considered too dangerous for British farmers and gardeners, and its use was phased out. The pesticides that replaced it were deemed much safer to humans and other biota. As an individual pesticide, arsenic had a maximum allowable concentration of 0.1 μg/l. When it was no longer a pesticide then you were, until recently, allowed to drink five hundred times as much of it in your water, that is 50 μg/l. Even with the new lower level you are still allowed to drink 100 times as much.

Asbestos

UK maximum concentration allowed: No standard set in legislation.
WHO Guideline Value: None set – not of health significance at concentrations normally found in drinking-water.

Asbestos is a naturally occurring group of fibrous minerals consisting of amosite, crocidolite (blue asbestos), tremolite, anthophyllite, actinolite and chrysotile. It has had a lot of commercial uses because of its outstanding resistance to heat, chemical inertness and high electrical resistance. The fibres may be used for brake linings and woven into fireproof cloth for use in protective clothing, curtains, etc. Asbestos is very dangerous when small fibres of it are inhaled and chronic lung problems can arise after occupational exposure. The particles lodge in the lungs and start to cause reactions of the lung tissue. It can cause the respiratory disease asbestosis, which is a form of lung cancer. Crocidolite and chrysotile are considered the most dangerous but the others are also extremely hazardous. Some people consider that there is no known safe level of exposure to airborne asbestos.

There has recently been some concern over whether asbestos in drinking-water could cause similar problems in the intestinal tract. The fibres enter the water either as atmospheric pollution, leached asbestos waste in landfill or because the water has passed through asbestos cement tanks or pipes where asbestos rope has been used. There does not appear to have been a lot of research carried out on potential problems in water but fibres have been discovered in the range of 'not found' to 1.5 million fibres per litre in the UK and US (FWR, 1982). WHO reported that there seems to be little evidence that asbestos in water causes health problems. Extensive animal studies have shown no clear incidence of tumours in the gut, and epidemiological studies, on populations with drinking-water containing naturally high numbers of fibres, show no increased illness. WHO therefore felt no need to set a Guideline Value for asbestos (WHO, 1993a). There remains some concern however and the International Agency for Research on Cancer reported some evidence of an association between asbestos in drinking-water and stomach and pancreatic cancer (Mara and Clapham, 1997).

The US Environmental Protection Agency has regulated asbestos in drinking-water for some time and there is a maximum contaminant goal level of 7 million fibres (less than 10 μm long) per litre. Fibres longer than 10 μm are considered to be of no human health significance when breathed in because they will irritate the lung tissue and be coughed up or removed by the natural protective action of the small hairs in the bronchioles. Presumably it is considered that a similar action will occur in the gut and they will be naturally removed.

Although there appears to have been no major study of asbestos in private water supplies, there should not normally be a problem unless old and damaged asbestos cement storage tanks are used or the supply source is situated close to a landfill site where asbestos has been disposed of.

Copper

UK maximum concentration allowed: 2 mg/l.
WHO Guideline Value: 2 mg/l.

Copper (Cu) is one of the essential minerals and is normally consumed in foods such as nuts, fruit and chocolate. It is required by the body to ensure that several of its enzyme systems work properly. Insufficient copper can lead to anaemia and bone problems in malnourished children. Where excess copper is ingested it is normally excreted in the bile. According to WHO (WHO, 2002b) long-term intake of copper in the diet in the range of 1.3–3 mg/day has no ill-effect. In water, once levels of copper are above 1 mg/l problems start to occur with staining of clothes during washing and unsightly marks on sanitary fittings. When the level goes above 5 mg/l, the water will start to taste bitter and show signs of colour.

Excessive levels of copper are known to cause a variety of health effects and there are several epidemiological studies in progress that may alter the Guideline Value. Excess copper can trigger stomach problems and high levels have been found to cause damage to the liver. Indian Childhood Cirrhosis appears to stem from eating food that has been in contact with copper-based cooking utensils, usually brass. Following a brief period of jaundice this can be fatal. Another copper-related condition is called Wilson's disease. This is a rare genetic disorder caused by an inability to metabolize copper. Some concerns have been raised about levels of copper in water in Europe but an extensive study by Fewtrell *et al.* (1996) found that all the children with liver problems in a major UK paediatric hospital had only consumed water with levels of copper well below the European standard (3 mg/l). A similar report from Nebraska in the US in 1994 found no association between illness and copper levels in drinking-water that were above the US Environmental Protection Agency action level of 1.3 mg/l (Buchanan *et al.*, 1995). An outbreak of drinking-water related copper poisoning was reported in 1984, with a family in rural Vermont suffering from repeated gastrointestinal illness. Levels of 7.8 mg/l were found in their water, which was supplied by a long run of copper piping. The illnesses ceased when the family stopped drinking tap water (Spitalny *et al.*, 1984).

As with iron and manganese, copper is a metal that is more likely to be found in water where the pH is low and the water is soft. Copper can be found naturally in certain rocks and groundwater coming into contact with it can dissolve it, but because many plumbing systems are made of copper, it is much more likely that any of the metal found in water will have been dissolved from pipework. For this reason water that has stood for some time will contain higher amounts of copper in it than normal and should be allowed to run to waste for several minutes before being used. Water in some small supplies can be so aggressive that the plumbing has to be regularly replaced because the water has stripped so much copper from the piping that it begins to leak – see Plate VII.

Public water supplies will often have their pH levels balanced. This reduces the aggressive effect of the water as it passes through copper-based plumbing. This was confirmed as part of the Fewtrell *et al.* (1996) study; they found that over a three-year period, copper excedences in public water supplies in England and Wales amounted to only fourteen out of the thousands of samples taken. The highest was 7.31 mg/l and that particular sampling zone only supplied a single property. In Scotland, seven public water samples failed the standard, the maximum amount being 8.26 mg/l. In contrast, private water supply sampling during the same period found failures in 46 local authorities (out of 307 authorities with domestic private water supplies). These failures ranged from 3.08 mg/l to a massive 26 mg/l. An American

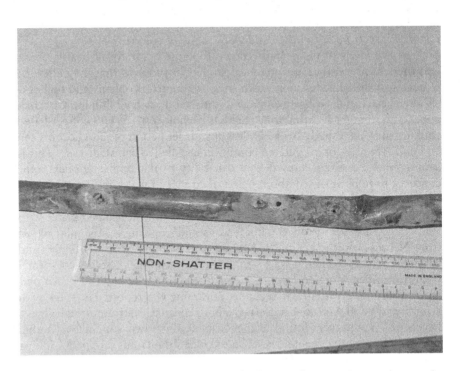

Plate VII Damage to copper piping caused by low pH from a private water supply, Yorkshire, England. (See also Colour Plate VII.)

report (Bremer *et al.*, 2001) discussed how the build-up of biofilms in plumbing leads to increased levels of copper in water and reduces the durability of pipework because of biocorrosion by the microbes. The factors that contributed to the build-up were low pH, high suspended solids and organic carbon levels with low or non-existent chlorine levels. Just the conditions found in many private water supplies. It is important therefore that sampling programmes should consider this problem and advice given to the owners of small supplies where failures of the standard occur.

One extreme example of high levels of copper in a private water supply concerns a woman in the north of England who arrived at the local authority offices to complain that her hair had turned blue. She used a well that had been no trouble for many years, until a significant drought occurred in the 1990s. Upon investigation, it was found that the well was normally fed by two different aquifers from separate rock strata and the drought had caused one of them to dry up. The remaining water supply had a much lower pH than the other one. Because it no longer had this neutralizing influence, the highly acidic water had started to strip substantial amounts of copper from the plumbing system. When the woman washed her hair in shampoo containing stearates, this reacted with the copper in the water to form copper stearate, which is naturally blue. Thus her hair took on a lovely azure hue. According to the clerk who received the complaint, the woman was not impressed by being told that most people have to pay for a blue rinse and she was getting hers for free. Once the drought had finished, the water, and eventually her hair, returned to normal.

Iron

UK maximum concentration allowed: 200 µg/l.
WHO Guideline Value: None set – complaints will be expected at 300 µg/l.

Iron (Fe) is the fourth most common element in the earth's crust. It is the third commonest constituent of shales after aluminium and silicon, and the fifth commonest element in sandstones (Mason, 1966). Iron is so common that its oxides are used to identify different types of soils by the colours they produce. Iron is found in a variety of forms, particularly oxides and hydroxides but its chemistry is relatively simple and it only has two valence states. These make either iron II (ferrous) or iron III (ferric) compounds. Iron is found in natural fresh water at levels between 0.5 and 50 mg/l (WHO, 1993a).

The solubility of iron, as with many other contaminants of water, is bound up with the pH of water. In natural waters, iron is present not only as Fe(II) but also as Fe(III) and FeOH(II) (ferrous hydroxide) ions. The concentration of ferric iron, Fe(III) in solution is proportional to the cube of the hydrogen ion concentration. Ferrous iron, Fe(II), is even more soluble. The solubility of all forms of iron at pH 6 is about 10^5 times greater than at pH 8.5 (Mason, 1966). This explains why iron is a regular constituent of water from private supplies with a low pH. This property however provides a useful way to remove iron from water. Raising the pH of water will cause excessive levels of iron to precipitate out as insoluble ferric hydroxide. This can then be physically removed by a filter or sedimentation. A similar effect will be produced if the water is aerated and the dissolved iron is oxidized to ferric iron oxide. Where this happens naturally the iron oxide sometimes causes an unpleasant brown slick along the bed and sides of small streams. This is the substance that some visitors to the countryside think is an example of man-made pollution as, in some lights and particularly when mixed with normally occurring oils in peat and soil, iron in the water also looks as if oil or petrol has been spilt in it. This is perfectly natural however and nothing to worry about.

Iron is classed as an aesthetic contaminant with no particularly harmful effects at the concentrations normally found in small water systems. It is a well-known essential trace element and forms important functions in blood chemistry. Lack of iron is known as anaemia and can be a serious problem particularly amongst the malnourished. In water, an excess of iron can cause a bitter taste and will stain clothes and sanitary fittings brown. WHO considers that levels up to 2 mg/l of iron in water will not present a health hazard. The standard for iron in private water supplies is 200 µg/l but guidance provided by the UK Drinking Water Inspectorate (HMSO, 1991b) suggests that levels of up to 2.8 mg/l are acceptable from a health point of view.

One problem with iron in water from boreholes and wells is the smell and odour problems caused by the die-off of iron bacteria that use it to metabolize. The section on wells in Chapter 2 discusses this further.

A householder who had been having trouble with iron in his private supply once told me that his drinking-water had developed an odour of a faecal nature. In order to allow me to see clearly how bad his water was, he poured some into an empty milk bottle and for some reason added a few drops of bleach. We then waited to see what happened. Fairly quickly, brown sediment appeared that gradually settled to the bottom of the bottle. He waved this at me and said that it showed that the water was contaminated with sewage. I tried to explain that rather than this being the reason for the brown sludge in the bottle, he had carried out a rather interesting chemical experiment. He had raised the pH of the water with the bleach and this had caused the soluble ferrous iron to become solid ferric precipitate and it was this that had settled out. He found this to be a somewhat less interesting example of science in the field than I did and he felt no reluctance in pointing this out. He also said that he could recognize faecal material when he saw it. Well, he did not actually use the term 'faecal material', but the more popular and shorter word of Anglo-Saxon origin.

Lead

UK maximum concentration allowed: 50 µg/l in private water supplies, 25 µg/l in public supplies reducing to 10 µg/l from 25 December 2013.

WHO Guideline Value: 10 μg/l (other measures to reduce total lead exposure should also be taken).

The health problems associated with lead (Pb) are well known and the importance of keeping it out of drinking-water is one of the few areas where everyone appears to agree. Lead has been known to be toxic to humans since the Roman times of Vitruvius the Architect. It has no biological function (Alexander and Heaven, 1992) and is not a vital trace element. It is undesirable for any lead to be in drinking-water and any that is found must be a matter of concern. A statement by WHO said: 'Lead is toxic to both central and peripheral nervous systems. Scientific evidence clearly shows prolonged exposure may lead to serious neurological damage, especially among infants, children and pregnant women' (Toft, 1992). Lead has been associated with ischaemic heart disease, hypertension, renal insufficiency, gout, premature birth, mental retardation, impaired cognitive ability and behaviour problems. The effects are permanent and cannot be negated by removing lead from drinking-water at a later date (Alexander and Heaven, 1992). It has been suggested that the association between cardiovascular disease and the hardness of water (see Chapter 3), which is one of the more well-known epidemiological relationships, may be caused by the additional lead (and other trace metals) in soft water, rather than the absence of hardness itself. It is however, the effect of lead on the mental development of children that causes most concern. It has been calculated that lead can cause a reduction of between 5 and 15 per cent of a child's intelligence depending on the amount found in the water. An average population loses two to three IQ score points when the average blood level of lead increases by 10 μg/dl (decilitre) (WHO, 1995). The reason for the additional risk to foetuses, babies and children is because they put on weight very quickly as they grow and they take up lead from their diet at approximately five times the rate of adults (Craig and Craig, 1989).

Although human beings can come into contact with lead from a variety of sources, such as paint, seals in metal food containers and car exhausts, it is ingestion in water that is seen as the main source of lead in blood (Pearce, 1992). The problem is not normally that the water picks up lead from geological deposits but that it dissolves it from the inside of old pipes, particularly when the water has stood for some time. Lead pipes are no longer fitted because of the dangers associated with them, but they were once so ubiquitous that the very word plumbing comes from the Latin, *plumbo* meaning lead. The ability of water to remove lead from pipes is known as plumbo-solvency. Plumbo-solvency is mainly associated with soft water areas because of the aggressive nature of the water (Craig and Craig, 1989) but lead dissolved from pipework can also be found in hard-water areas. The amount of lead in water depends on its physico-chemical qualities, particularly the pH, how much lead piping it has been in contact with and how long it has stood in the pipe prior to being drawn from the tap. Most water utilities in the developed world have largely removed lead from their delivery networks but water companies in the UK have calculated that the cost of removing the rest would be in the region of €3 billion. The major problem nowadays however, is old lead piping in individual houses and the cost of removing this in the UK is estimated to be at least €7 billion (WHO, 1995).

The treatment processes used in public water systems will produce water that is virtually lead-free. Many water providers also add inhibitors to the water to reduce its plumbo-solvency where problems could occur after it has left the treatment works. In the UK for example, water utilities are statutorily required by the water quality regulations to reduce the plumbo-solvency of any water in areas with known lead problems. In the USA, there is a similar requirement under the Environmental Protection Agency's Lead and Copper Rule. The process of reducing plumbo-solvency involves raising the pH to 8.5 and adding orthophosphates to the water. The effect is to reduce the uptake of soluble lead and in most cases this reduces the amount taken up in the water by 70 per cent (Edwards and McNeill, 2002). Private supplies on the other hand, often have much more acidic water than is provided in public systems. They lack basic pH balance and any treatment for potential lead problems (in many instances, they lack treatment for any type of problem). This will result in significantly higher levels of lead in their water. In the UK, this is particularly true of moorland areas of Scotland, Wales and northern England, where the water is very acidic. In addition, most houses served by small and private water systems are older and more likely to still have lead plumbing.

Although the problems in water are mainly from lead piping, there are certain geographical areas that contain lead-rich geological formations. Some parts of the UK (West Yorkshire, Derbyshire and Cornwall) have significant lead deposits and therefore it is possible for private water supplies to contain lead from these sources. The problems will normally, though not exclusively, be in areas where there have been lead mines. The water may either pass through lead deposits or run out of the mines themselves and disappear underground, to reappear later as spring water or in wells or boreholes. This, again will not be an issue with public supplies because water utilities will normally not abstract water from sources that are naturally high in lead, it would be too much of a risk.

Until recently, regulations in the UK allowed a maximum of 50 µg/l of lead in public drinking-water. This was designed to ensure that the blood of infants and children contained no more than 25 µg/dl (DoE, 1989). The European Union's Directive on Drinking-water in 1998 reduced the limit to 10 µg/l. This will bring it closer to the American US Environmental Protection Agency standard of 15 µg/l. The US standard however, does not apply to the 18 per cent of the American population who are on private supplies (Mills, 1990). Because of the cost of removing all remaining lead piping, European regulators have allowed a period of grace until 2013 to achieve the standard. Nevertheless, a 'half way' standard of 25 µg/l is now in force. Because the new standard only refers to lead as it leaves the water utilities' pipes, the main problems will still not be addressed, that is, the old pipes belonging to individual householders and those on private water supplies.

The regulations in the UK for private water supplies recognize these potential problems by including lead as one of the basic parameters to be sampled. They state, however, that monitoring is unnecessary where there are no lead pipes or solder used. This does not take account of any naturally occurring lead and it is expected that very few sampling officers will be able to state categorically that a house with copper plumbing has no lead solder whatsoever.

Where lead is a problem in small water systems, the obvious remedy is to remove all the lead piping associated with the supply. Where alkathene or similar plastic piping has replaced old lead plumbing in private supplies, the results of sampling have shown that the lead can be reduced to insignificant levels (less than 0.005 mg/l). Installing a water treatment system will remove lead but it is a secondary choice and if it fails to danger, the lead will return. Obviously, if the source of the lead is the groundwater itself, then apart from finding an alternative supply, treatment is the only option. In the short-term, householders should be encouraged to drink bottled water until the treatment or an alternative supply is connected. This is particularly important if there are young children in the house.

It has been suggested that if there are no young children in the house and the householders have drunk lead-contaminated water for some time, then requiring a costly alternative supply or treatment may not be justified. There is an argument that if people are given sufficient information about lead in their drinking-water and an explanation of the risks involved, they should be allowed to choose for themselves whether to remove any lead piping. It could equally be suggested that, where someone has drunk water with a significant amount of lead in it for a long time, expecting to get a sensible decision may be a fairly pointless exercise. Come to think of it, the effect that lead has been shown to have on mental retardation and cognition may go a long way to explain some of the conversations I have had with people about their private water supplies over the years.

For private water supplies, lead is predominantly a rural problem and with the future of the rural economy in such poor shape, it is unlikely that hill farmers and others who make their living from the land will be able to invest in new plumbing systems for some time to come. The Government is unlikely to provide sufficient financial encouragement to get the pipes removed, nor does it seem that it will introduce legislation forcing people to have their lead pipes taken out. Children will therefore continue to suffer from this problem, which we have only known about for 2,000 years.

Manganese

UK maximum concentration allowed: 50 μg/l.
WHO Guideline Value: None set – complaints will be expected at 100 μg/l.

Manganese (Mn) is a hard, grey, brittle metal that resembles iron. It is an essential trace element and humans need between 30 and 50 µg/kg of body weight per day (WHO, 1993a). Manganese is not regarded as a health-related contaminant of water and it is often said that you would not be able to drink water that is strong enough in manganese to make you ill, it would taste too unpleasant. There have been toxic effects associated with manganese but these have followed long-term occupational exposure to ores and dusts. These problems manifest themselves as weakness, anorexia, muscle pain, apathy and slowness. It is highly unlikely that humans will ever encounter sufficient manganese in water to cause these toxic effects. Manganese is therefore an aesthetic contaminant with levels above 100 µg/l causing staining of household utensils and sanitary fittings and imparting a metallic, bitter, astringent or medicinal taste to the water.

Manganese is the tenth most abundant element on the earth, but occurs naturally only in mineral form and not as the pure metal. More than 250 minerals contain manganese as an essential ingredient. The majority of the manganese present in rock structures occurs as a minor substitution for other minerals (Graham *et al.*, 1988). Most rocks are composed of various combinations of relatively few mineral types and these are the primary source of minerals in soils. Manganese is mainly associated with ferromagnesian silicates and other iron-containing minerals and often occurs as a substitute for iron II (ferrous). This helps to explain why elevated levels of manganese in water supplies often accompany high levels of iron. As well as ferromagnesian silicates, manganese is present in many igneous, metamorphic and sedimentary rocks. Mafic volcanic rocks (named after *magnesium* and *Fe* rich igneous rock) and iron-rich shales are more likely to contain it than granite or sandstone (Graham *et al.*, 1988). Concentrations in rock vary between 10 and 6,000 mg/kg with manganese (II) being the dominant form.

When exposed to the air, manganese (II) in solution is oxidized to form a solid black manganese (IV) precipitate. As with most metals, manganese solubility is strongly influenced by pH. Concentrations in natural waters therefore increase as the pH decreases (Graham *et al.*, 1988). As with iron, higher manganese levels are also associated with the group of naturally occurring organic compounds known as humics or humic acids. They are found in soils, especially peat and cause soluble manganese to be released. Manganese will often be found in water from small systems because many lack a treatment system that would remove it. Householders may call in officials because the water tastes bitter or clothes are being stained when washed. If a treatment system is fitted that raises the pH, or oxidizes the water, the dissolved manganese will form a solid precipitate of insoluble manganese dioxide that can then be removed by physical filtering or allowed to settle out. Advice about treatment and reassurance that it will not cause ill health are all that is usually needed.

Nitrates

UK maximum concentration allowed: 50 mg/l as NO_3.
WHO Guideline Value: 50 mg/l as NO_3.

Nitrates (NO_3) and Nitrites (NO_2) are naturally occurring negatively charged ions that are readily soluble in water. They form part of the nitrogen cycle. Nitrogen is a stable, colourless, odourless gas. It was discovered by Ernest Rutherford in 1872 and makes up 78 per cent of the earth's atmosphere by volume. It is ionized by storms, dissolved in rain and fixed by bacteria in the roots of certain vegetables (legumes). Nitrates are vital for normal growth in plants and farmers and gardeners apply nitrogen as a natural or artificial fertilizer. Another major source of nitrate in water is the ploughing up of grassland. This is because a lot of nitrates are naturally stored in the topsoil. Whereas arable soils may receive an annual 150–400 kg of nitrogen per hectare as fertilizer, under grassland, levels of up to 9,000 kg per hectare are common (HOLSCOTEC, 1989). Any excess not taken up by the plants can be leached through the soil and into water supplies. See Table 3 indicating the main sources of nitrogen.

Nitrates have attracted a great deal of attention. Several years ago books were published with titles such as, *The Nitrates Story, No End in Sight* (Vogtmann and Biederman, 1989) and *Nitrates: The Threat To Food and Water* (Dudley, 1990). This sort of book appeared under the banner of the environmental movement and used lurid quotes to try to persuade the public that 'nitrates are the single most important water pollution problem in Britain' (Dudley, 1990). As part of this excitement, Britain was taken to the European Court in 1992 over failures of nitrate standards in the first Drinking-water Directive (80/778/EEC). The Secretary of State for the

Table 3 The relative importance of different nitrate sources in Britain[a]

Input source	Total UK input %
Artificial fertilizer	43
Livestock manure	38
Pollution dissolved in rain	10
Biological nitrogen fixation	6
Sewage effluent	1
Seeds	1
Straw	0.5
Silage effluent	0.5

Source: Adapted from HOLSCOTEC (1989).

Note
a Ploughing as a source of nitrate is not included as it is already in the soil.

Environment at the time was quoted as saying that the UK water industry was spending £28 billion to improve water quality. He added that 'no country has unlimited funds to spend on the environment and no industry can cope with an unending series of new environmental expenditure commitments' (Heseltine, 1992). The Royal Agriculture Society also considered that European thinking on nitrates in water was misplaced. They stated that if farmers had to reduce nitrate fertilizer usage to protect water supplies, there would be 'serious, if not catastrophic effects on the UK, UK agriculture and the UK farmer' (HOLSCOTEC, 1989).

Nitrates are a problem for water utilities as concentrations in water from boreholes rise (Tompkins, 2003). As well as causing problems because simple treatment options such as blending are running out, sewage undertakers in the UK are concerned because complicated new regulations are reducing their options for spreading nitrate-rich sewage sludge to farmland. One recent report caustically pointed out that the Lords Prayer and the Ten Commandments used 56 and 297 words each to make their point, whilst the 'Notes for Guidance of Officers in the Environment Agency when looking at Applications for Exemptions from the Waste Management Licensing Regulations' (to put small compost heaps on farms) used 35,709 (Butterworth, 2002). There are therefore, two distinctly separate schools of thought over the problem of nitrates in water.

In 1986, 2.5 million people in Britain occasionally received drinking-water that exceeded the maximum allowable concentration of 50 mg/l (DoE, 1986). In the past, some environmental organizations put this figure even higher at 4 million (Lees, 1991). In 1999, 99.94 per cent of public water samples in the UK met the standard (DWI, 2003a). This equates to 32,000 people, which is a substantial improvement. The majority of small water supplies where nitrates may be a problem are in arable areas and regions with large farm animal populations. An adult dairy cow produces 16.5 tonnes of manure per annum. This contains 50 kg of nitrates (Dudley, 1990). Areas with low rainfall will also have higher levels in groundwater as dilution is less and lower rainfall will reduce the amount of overland flow that flushes nitrates into streams and rivers. This leaves more on the surface to percolate down. This phenomenon will be offset to a certain extent because the lower the rainfall, the greater will be the distance between the water table and the soil surface. This greater distance will allow more bacteria to fix the nitrates, reducing the amount that enters the aquifer (HOLSCOTEC, 1989). Because of both these factors, East Anglia and the south of England are particularly prone to high concentrations of nitrates in water. The use of nitrate fertilizer has gradually risen since the 1940s and there appear to be a consequent increase of nitrates in groundwater. Because nitrates are particularly stable they tend to remain in the ground for a long time, and it has been calculated that it takes about twenty years for applications of nitrates to reach water supplies (Oakes, 1992). It will therefore be some time before the present high rates of application start to show up in groundwater.

The risks associated with excessive consumption of nitrates are twofold. The first and most well known is methaemoglobinaemia or 'Blue Baby Syndrome'. This affects infants under the age of six months when the pH of their gastric juices is typically 4.5 as opposed to pH 2 in an adult. Nitrates in the baby's stomach are reduced (oxygen is taken away) to form nitrites by bacteria that can survive at pH 4.5 but not pH 2. The nitrites react with the chemical in blood that transports oxygen – haemoglobin and turn it into methaemoglobin. This substance does not transport oxygen and the blood therefore cannot absorb oxygen from the lungs. This causes the baby to look blue (cyanosis) and may result in death (HOLSCOTEC, 1989). To avoid the possibility of any cases in the UK, water utilities are required to provide bottled water to mothers where nitrate levels exceed 100 mg/l (although there are none that have to do this at present). However, the last case of methaemoglobinaemia in the UK was over thirty years ago, in 1972 and the water causing the problem was not from a public supply but a highly polluted, sewage-contaminated private well (Gray, 1994).

Private water supplies are much more likely to present a risk of methaemoglobinaemia from high nitrate levels than mains water. In fact, an old name for the condition is 'well water disease'. Of all the UK reported cases since the 1920s, eight were from private supplies and six from public supplies. Of those six, four were suspected of having artificially high levels because of repeated boiling of the water (HOLSCOTEC, 1989). I am unsure whether this was because the kettle was kept constantly on the Aga, or whether the people thought that repeated boiling of the water would have better disease-removing effects. Studies in the US have indicated that the only water supplies where there have been serious cases of methaemoglobinaemia and subsequent deaths have been private supplies (Hunter, 1997). Virtually every case of methaemoglobinaemia is therefore associated with *poor quality* private water supplies (Packham, 1990; Allot, 1991). Why is this? The reason is that the private supplies are much more likely to have the combination of high concentrations of nitrates and the large numbers of the bacteria necessary to cause the condition. Except where there are outstandingly high concentrations of nitrates, the bacteria may be the limiting factor as to whether the condition occurs (Dudley, 1990). As 98 per cent of the population benefit from piped water that is bacteriologically sound, this hypothesis could be the explanation for the 'almost total' absence of reported cases of methaemoglobinaemia from public supplies (HOLSCOTEC, 1989). French studies have shown that foodstuffs with high nitrate content are also partially responsible for methaemoglobinaemia and enteritis alone could cause the condition (Richard, 1989). Hunter (1997) wisely points out that the best advice is that babies should be breast-fed whenever possible and infant feed should not be made up from water in private water supplies unless it is boiled and the nitrates are below the current standards.

The second health problem concerns a suggested association between nitrates and stomach cancer. The hypothesis is that under certain circumstances, nitrates are converted in the stomach to nitrites, these react with amines (compounds produced by the decomposition of organic matter) to form nitrosamines. In laboratory studies, nitrosamines have been linked to cancer (Stupples, 1991). Pressure groups admit that the evidence is contradictory and Dudley (1990) acknowledged that epidemiological studies in Britain, Denmark, Chile and Columbia had failed to find any clear link between nitrates and stomach cancer. There has also been a continuing decline of stomach cancer in Britain throughout the period when nitrate levels in water have been increasing. In addition, the incidence of stomach cancers is lowest in eastern England where nitrate concentrations in water are highest (HOLSCOTEC, 1989). Does this mean that rather than causing stomach cancer, nitrates protect against it? This may be so, but there could be other reasons for the decrease in stomach cancers. As well as nitrates being consumed via drinking-water they are often used as a food preservative, particularly in cured meats. Regulations now limit the amount of nitrates and nitrites in certain foodstuffs. Lower stomach cancer rates are also linked to the reduction in mould on stored food, due to refrigeration, improved food handling techniques, etc. This reduction may have masked the effects of increased nitrates in water. Although 75 per cent of nitrosamines can cause cancer in laboratory animals (Richard, 1989), it seems that the carcinogenic effects found in animal studies do not present a hazard at the levels of nitrates encountered in mains water (HOLSCOTEC, 1989). A WHO working party concluded in 1985 that there was no convincing evidence for a relationship between gastric cancer and consumption of water containing nitrate at or below the guideline level (Packham, 1990). Information from WHO continues to assert that there is 'no evidence for an association between nitrite and nitrate exposure in humans and the risk of cancer' (WHO, 2002c). Richard, (1989) went further and stated that it was impossible to provide any final proof of the carcinogenic effects of nitrosamines in man.

It has also been suggested that childhood-onset insulin-dependent diabetes is associated with nitrates in drinking-water. A study in 1999, on behalf of the UK Drinking Water Inspectorate, investigated this claim and looked at cases of the disease comparing them with nitrate levels in drinking-water. The study found no evidence of any consistent or plausible association between nitrates and childhood diabetes (DWI, 1999). A second study in the Netherlands in 2000, also found no evidence of a relationship between nitrates at current exposure levels and this disease (van Maanen *et al.*, 2000).

One area of potential confusion is that there appears to be two different maximum limits for nitrates in drinking-water, depending upon where you are. In Europe, the standard is 50 mg/l but in the US it is 10 mg/l. In fact these limits are roughly the same as the standards are just expressed in different ways. In Europe, the limit of 50 mg/l is based on the total amount of nitrates in the water and not just the amount of nitrogen. The atomic weight of nitrogen is seven and oxygen is eight, NO_3 therefore has a total atomic weight of 31. If the standard for nitrates as a whole is 50 mg/l, then the contribution of the nitrogen itself is 11.29 mg/l. In other countries, the standard is expressed only as the amount of nitrogen. In 1974, WHO recommended a maximum guideline limit of 10 mg/l for the nitrogen that is present in nitrates and this is the standard that has been adopted in the US. This way of representing the standard can be written as mg NO_3 –N/l. The WHO standard is equivalent to 45 mg/l of nitrates and these units can be written as mg NO_3/l to prevent, or at least reduce, misunderstanding.

The WHO Guideline Value was set at its present level because of the potential problems with methaemoglobinaemia. It should be remembered that WHO recommendations are for all the countries of the world, which includes many that have a great deal of microbiologically poor water and where methaemoglobinaemia is therefore much more likely to occur. It is perfectly acceptable for countries that have microbiologically acceptable water to have a maximum allowable concentration above this level. The UK Department of Health however considers that in view of the proven carcinogenic properties of nitrosamines to laboratory animals, there is no reason to believe that humans would not be susceptible at certain levels. They maintain that, using the precautionary principle, nitrate levels should be prevented from rising above the present 50 mg/l standard. These levels will give a more than adequate safety margin to prevent future cases of methaemoglobinaemia.

Because nitrates are so important for agriculture, it is important that the maximum allowable concentrations are accurately set and cautiousness regarding safety is balanced against unnecessarily low and consequently economically damaging maximum levels. Many people in the water industry feel that the scientific basis for the nitrate standard is suspect. As levels in groundwater rise because of changes in farming practices, they question whether it is sensible to spend quite so much on achieving progressively smaller increments of improvement (Bellak in Allott, 1991). They point to the fact that in the 1950s, 1960s and 1970s, the recommended level in Britain was 100 mg/l, which has not been proved to cause any particular health problems (Allot, 1991). The 1980 European Community Drinking-water Directive set a new maximum level of 50 mg/l, with a guideline level of 25 mg/l. A House of Lords Select Committee at the time, which carried out a widespread review of the problem, concluded that the UK maximum should remain at 50 mg/l, although it conceded that this was a prudent limit (HOLSCOTEC, 1989).

In the USA, nitrates are considered to be a very useful indicator of local pollution, particularly in rural areas. This is the main reason why small water supplies are checked for both coliforms and nitrates on a regular basis (Brown, 2003). Using both these parameters can give an added safety factor above that of relying on faecal coliforms alone. Coliforms have a tendency to die-off if the pollution event is remote from the water source, whereas nitrates will continue to be present in the water for a long time. In the UK, local authorities will not always monitor for nitrates, if previous sampling has revealed that levels do not exceed 25 mg/l. This is to prevent money being spent in areas where farming does not cause a nitrate problem.

To keep levels of nitrates low in groundwater, local authorities in conjunction with water utilities and environment agencies should look towards limiting the use of nitrates in particularly sensitive agricultural areas. This is encouraged by European legislation. The EU Nitrates Directive (91/676) (European Union, 1991) required the introduction of measures to protect water against nitrate pollution. In 1995, the Department of the Environment designated sixty-eight 'nitrate vulnerable zones' in England and Wales. These now cover 55 per cent of England. The zones limit the amount of nitrate fertilizer used, encourage sensible disposal of farm wastes and discourage unnecessary ploughing of grassland. This is much better than relying on expensive drinking-water treatment to remove the nitrates, although it is still expensive as grants have to be given to encourage this reduction in nitrate usage (DETR, 2002). Reducing fertilizer and manure use in the zones will not only reduce nitrates in drinking-water but also limit the run-off into surface waters and thus help prevent rivers and lakes becoming eutrophic.

In common with many other contaminants, nitrates may be found in higher concentrations in private supplies than public supplies. There are several reasons for this, they include the following:

- Private supplies are particularly prone to point source pollution. Leaking drainage, washings from animal housings and excessive levels of fertilizer use can therefore easily contaminate them.
- Illegal or inadequate slurry disposal and leaks from slurry or silage tanks can produce serious problems to groundwater. Individual private supplies on farmland can be close to these sites and because they are often from shallow sources, they will be more immediately affected by pollution incidents.
- The removal of nitrates from water is not an easy process, particularly for small treatment systems. Private water supplies moreover, frequently do not have any such treatment.
- A common public utility solution to water with high nitrates is to blend it with water from a low nitrate source. Apart from supplies that incorporate nitrate removal, this option is not normally available for private supplies.
- Smaller categories of supply are rarely sampled. Consequently, elevated levels may be drunk for a long time without anyone knowing. Public water supplies would be much more regularly checked for nitrates and thus any pollution problem would be identified more quickly.
- As mentioned earlier, methaemoglobinaemia is associated with gross bacterial contamination, this is much more likely to occur in private supplies.

A sanitary survey can often identify potential sources of nitrogen even if sampling fails to find a problem. Examples of things to look for are; old or leaking septic tanks uphill from the source, excessive build-up of animal waste within the catchment area, contaminated streams running close to the source and defective silage storage. Nitrates will typically move down through the ground at the rate of a metre a year. This will continue to be the case even if the cause of the problem has ceased. Nothing, other than installing nitrate removal treatment or relocating the water supply, can be done to correct the situation in the short term. Where nitrate treatment is necessary, an ion-exchange treatment system must be installed. However a sensible approach to treatment should always be taken. The standard for nitrates in drinking-water is set with regard to the incidence of methaemoglobinaemia, but it is only a potential problem for bottle-fed babies. If there are no babies in the house, what is the problem? Householders may wish to drink bottled water as an alternative, but problems with very high levels of nitrates (unlike microbiological ones) will not go away if the water is boiled, and very few people use bottled water to make a cup of tea.

Minor inorganic contaminants

Ammonia

UK maximum concentration allowed: 0.5 mg/l of NH_4.
WHO Guideline Value: None set – complaints will be expected at 1.5 mg/l.

Although ammonia (NH_3 and NH_4) is not in itself a health-related parameter, its presence indicates possible contamination from sewage or animal wastes, particularly intensive farming. If ammonia is found in a private water supply, the owner should be alerted and the source of the contamination found as quickly as possible. Where ammonia is present there may also be pathogenic microorganisms. Ammonia is naturally produced in the body and any that is consumed in water will be relatively minor in comparison.

Boron

UK maximum concentration allowed: 1 mg/l.
WHO Guideline Value: 0.5 mg/l.

Boron is usually found as boric acid in water and has been associated with tumours of the male reproductive tract in laboratory animals. Boron occurs naturally in seawater at an average concentration of 5 mg/l. The element is also found in fresh water lakes, rivers, streams, and groundwater at concentrations well below 1 mg/l. It is rarely found in drinking-water but may be leached from rocks or waste following glass, soaps or detergents manufacture. Amounts vary widely depending on geology and any wastewater discharges. Concentrations in drinking-water have ranged from 0.01 to 15.0 mg/l (WHO, 1998b). It can be removed using ion exchange or reverse osmosis; although in areas with high natural levels blending with low-boron water might be a cheaper possibility. The Guideline Value is provisional because it will be difficult to achieve in areas with high natural boron levels.

Cadmium

UK maximum concentration allowed: 5 µg/l.
WHO Guideline Value: 3 µg/l.

Cadmium (Cd) is one of the heavy metals and its presence in water may come from industrial pollution, fertilizer use or from plumbing fixtures. It is considered to be carcinogenic when breathed in. There is no firm evidence that it causes problems when consumed in water but it may cause damage to the kidneys, where it is deposited with a half-life of decades. A WHO Guideline Value of 3 µg/l has been set.

Calcium

UK maximum concentration allowed: No standard set in legislation.
WHO Guideline Value: None set.

Calcium (Ca) is a natural trace element and is needed for bone and tooth development, particularly in children, and for blood clotting. Although it is important for health, both in water and food, calcium is a component of water hardness and is removed when water softening is used to reduce damage to plumbing systems. Where potable water is softened, people should be advised to drink the unsoftened water.

Chlorides

UK maximum concentration allowed: 250 mg/l.
WHO Guideline Value: None set – complaints will be expected at 250 mg/l.

Chlorides (Cl) are usually present in small amounts in drinking-water and are not considered to be a potential health problem. They can increase the aggressiveness of the water and cause pitting of stainless steel. As well as naturally occurring chlorides dissolved from rocks, they can be found in road salts, fertilizer, industrial wastes and sewage. They are important however as an indicator of saline intrusion (seawater getting into the drinking-water aquifer). The chloride comes from dissolved common salt (sodium chloride) in the seawater and saline intrusion can be a serious problem in coastal areas where too much groundwater has been abstracted. Once saline intrusion has occurred, the groundwater will be unusable.

Surface waters normally contains up to 50 mg/l of chloride but seawater has 35,000 mg/l. Levels up to 1,500 mg/l are considered safe but may be too salty to drink. WHO has set no Guideline Value but once the amount goes above 250 mg/l, the salty taste will become noticeable. If it is also used for irrigation purposes, water with chloride levels above 900 mg/l will be unsuitable for most plants.

Chromium

UK maximum concentration allowed: 50 μg/l.
WHO Guideline Value: 50 μg/l.

Chromium (Cr) is naturally found as a constituent of certain rock formations throughout the world. It can also be a pollutant in the air and as liquid effluent from under-regulated industries and mining operations. WHO (1993a) reported that normal concentrations in water are 2 μg/l but concentrations up to 250 μg/l have been found. It is another heavy metal and like cadmium is known to cause damage to human health particularly when inhaled. Although there is no firm evidence of damage from consumption in water, WHO has set a safety level, but this is provisional until firmer evidence is available.

Fluorides (fluorine compounds)

UK maximum concentration allowed: 1.5 mg/l.
WHO Guideline Value: 1.5 mg/l.

Fluorides are chemical compounds containing fluorine (F), a halogen in the same chemical group as chlorine. Fluorides are natural constituents of the rocks in the earth's crust and in some places are found in high concentrations. They are found in phosphate fertilizers and can be ingested in various foods including fish and tea (WHO, 1993a). There is great controversy over fluoride in drinking-water, with one camp (dentists and public health doctors) wanting it to be added to water and the other camp (health activists, organic campaigners and generally worried citizens) determined not to let this happen. Fluoridation was first introduced in the 1940s in the US (Gray, 1994) and 250 million people worldwide have it added to their water. Fluoride in suitable doses (1 mg/l) is thought to protect teeth against dental carries, and in those places where children do not have access to fluoride toothpaste or they do not brush their teeth sufficiently often, it can prevent them from getting rotten teeth. Fluoride in excess (greater than 1.5 mg/l) can cause mottling of teeth. Above 5 mg/l pitting of teeth occurs and in extreme cases, brittleness and damage to bones (fluorosis). Where it is added to water supplies it is never allowed in such high doses. The anti-fluoride campaigners claim there is no real reduction in tooth decay following its introduction, that it is a toxin, that some people may be allergic to it in the doses supplied and water should not be used to dispense medicine.

WHO has found no convincing evidence that fluoride is carcinogenic (WHO, 1993a) and considers that their Guideline Value will prevent any ill-health problems. Removal of naturally occurring fluoride has been carried out with activated carbon or activated aluminium. At a village scale, in those developing countries that have a particular problem, filtering water through charred bone meal may be effective.

Magnesium

UK maximum concentration allowed: No standard set in legislation.
WHO Guideline Value: None set.

Magnesium (Mg) is a naturally occurring constituent of hard water. It is a natural trace element that is necessary for energy production and conducting nerve impulses. It has a slight laxative effect when combined as magnesium sulphate but other than that, it appears to have no ill effects.

Nickel

UK maximum concentration allowed: 20 μg/l.
WHO Guideline Value: 20 μg/l.

Nickel (Ni) is naturally found in both food and water. It can also be found in water that has passed through nickel tap fittings. It is not normally found in drinking-water in quantities sufficiently high to cause any health problems. If consumed in sufficient amounts it may cause dermatitis, similar to the sensitized effects that have been found with nickel-based jewellery. Concerns have recently been raised over nickel in water boiled in certain electric kettles (DWI, 2003b). Research by the UK Drinking Water Inspectorate showed that water picks up nickel when boiled in kettles with elements not made of stainless steel. The amounts found when they were used were about 1 mg/l. The report also found that this could be exacerbated if the water was first passed through a jug filter. The US has a guideline advisory level of 100 μg/l. WHO's Guideline Value, will probably be tightened in the light of recent research (Fawell, 2003).

Nitrites

UK maximum concentration allowed – 0.5 mg/l.
WHO Guideline Value: 3 mg/l (acute provisional level) 0.2 mg/l (chronic provisional level).

Nitrites (NO_2) are not normally contaminants of water but are the reduced state of nitrates. Therefore some of the information relating to nitrates in drinking-water will apply to nitrites. Because it is the nitrites that cause methaemoglobinaemia, the European standard allowed in water is lower than that for nitrates and is 0.1 mg/l or 0.03 mg/l of nitrates as nitrogen. In the US, the standard is 1 mg/l of nitrate as nitrogen. WHO has set a Guideline Value of 3 mg/l as nitrites are considered to be about ten times more potent than nitrates.

Phosphorus

UK maximum concentration allowed: No standard set in legislation.
WHO Guideline Value: None set.

Phosphorus (P) can be naturally found in water but is also an indicator of sewage pollution or agricultural run-off. In surface waters, it may lead to algal blooms. In drinking-water it does not have a WHO guideline concentration, or in the USA, a Guideline Value.

Potassium

UK maximum concentration allowed: No standard set in legislation.
WHO Guideline Value: None set.

Potassium (K) is an essential trace element but too much in the diet may cause problems with kidney functions. It is not considered to be toxic in the amounts found in water.

Sodium

UK maximum concentration allowed: 200 mg/l.
WHO Guideline Value: None set – complaints will be expected at 200 mg/l.

Sodium (Na) is a constituent of common salt (NaCl), so will be found in waters subject to saline intrusion and road-salt run-off. It is also found in industrial and sewage pollution. Sodium is added to water as part of water softening and other ion exchange processes. When the ion exchange medium is exhausted, it is regenerated with common salt and once it is put back into operation, the sodium ions are preferentially replaced by the calcium or magnesium ones that cause the hardness.

Sodium is an essential trace element but too much can be a problem for people with heart problems or hypertension. Therefore, people who have been put on low sodium diets should check the amount they get in their drinking-water. If the water has more than 20 mg/l a doctor should be consulted. Sodium in excessive amounts can cause symptoms of illness in children and the kidneys of premature babies are not efficient at removing high levels. The US Guideline Value is 20 mg/l. The taste threshold for sodium is around 200 mg/l and it makes water taste salty above 250–300 µg/l. Levels above 400 mg/l may prove unsuitable for irrigation as water that is high in sodium causes burning to the leaves of plants. This is because when it dries on the surface, the crystals pull water from within the leaf via osmosis.

Sulphates (Sulfates in the US)

UK maximum concentration allowed: 250 mg/l of SO_4.
WHO Guideline Value: None set – complaints will be expected at 250 mg/l.

Sulphates (SO_4) are mainly of natural origin, dissolved from rocks by groundwater. They are also found in food and the amount consumed in water is fairly insignificant in comparison. Some sulphates however find their way into water from industrial discharges. They are not considered to be a particular health problem and are a constituent of Epsom salts (magnesium sulphate), which is taken for its laxative properties. Because of this, WHO only recommends that levels above 500 mg/l are drawn to the attention of health authorities. WHO has also stated that taste thresholds range from 250 mg/l for sodium sulphate to 1,000 mg/l for calcium sulphate (WHO, 1993a).

Tin

UK maximum concentration allowed: No standard set in legislation.
WHO Guideline Value: None set – not of health significance at concentrations normally found in drinking-water.

Inorganic tin (Sn) is not considered to be a health problem in water and is unlikely to be found in any significant quantity. It is efficiently excreted by the body and WHO have set no Guideline Value for its presence in water. This is because it is not toxic up to three orders of magnitude greater than the amounts normally found (WHO, 1993a).

Zinc

UK maximum concentration allowed: No standard set in legislation.
WHO Guideline Value: None set – complaints will be expected at 3 mg/l.

Zinc (Zn) is an essential trace element found in many different foods and occasionally in water. It occurs naturally but can also come from industrial pollution and the corrosion of copper–zinc alloys in plumbing fixtures. WHO does not consider that it is sufficiently a problem to give it a Guideline Value but suggests that levels above 3 mg/l may not be acceptable to consumers. The US Environmental Protection Agency has set a Guideline Value of 5 mg/l. Where levels are encountered above 4 mg/l, an astringent tang may be present and the water can take on a chalky appearance.

16 Organic contaminants

The significant contaminants of small water systems have traditionally been microbiological, followed by inorganic chemicals and physico-chemical parameters such as pH and turbidity. For the foreseeable future microbiological contaminants will continue to be the most important ones, but there are certain organic chemicals that are attracting attention because of their potential to cause chronic ill health. The main ones are disinfectant by-products (DPBs), trihalomethanes (THMs), organic solvents, pesticides and polynuclear (or polycyclic) aromatic hydrocarbons (PAHs).

Many of the Guideline Values and maximum concentrations for these chemicals use the 'linear no-threshold dose response' model. This is used where very low doses of chemicals will be consumed and animal studies at these levels do not produce any ill effects within the lifespan of the experiment. This model uses higher amounts of the chemical to elicit a response but assumes that there is no safe level for these chemicals. Information about the damage they cause at various concentrations during the experiments is put in a graph and the line is projected backwards until the amounts reach zero. Any amount of the substance above this can be read against the figure on the graph for the amount of cancer or ill health it can theoretically produce. It should be remembered however, that the no-threshold dose response curve is an assumption and is not proven. If the theory were wrong then most fears about low levels of chemical consumption or radiation would disappear. Many people question this basic assumption and consider that the human body may be quite efficient at dealing with low levels of these chemicals.

Over six hundred organic chemicals have been found in water (WHO, 1984), many are chemically similar and they are sometimes grouped together for analytical purposes. It is also appropriate to group them together for discussion here.

Disinfection by-products and trihalomethanes

UK maximum concentration allowed: Trihalomethanes 100 μg/l.
WHO Guideline Value: Values are set for individual disinfection by-products, these range from 200 μg/l (chloroform) to 1 μg/l (trichloroacetonitrile).

Disinfection by-products were first identified in water following the revolution in detection techniques in the 1960s and 1970s. The discovery led to a great deal of furore and research into their health effects. They are a group of organic chemicals sometimes formed following the disinfection (mainly with chlorine) of drinking-water; the main group attracting interest is the trihalomethanes. Some people think the terms are synonymous, but there are some disinfection by-products that are not trihalomethanes.

Disinfection by-products are normally produced as a result of a reaction between chlorine and/or bromine and organic material in the water. Recently however, they have also been found to be a potential problem where disinfection has been carried out using ultraviolet radiation or ozone. The chlorination of water has probably saved more lives than any other single public health measure in history but many people are now looking for alternatives to chlorination due to worries about disinfection by-products. It seems that once we have removed one problem, we have to find another to be anxious about. For example, a French company experimented with the replacement of chlorination by nanofiltration, in a small area of Paris. The initial cost was £3.5 million, with a subsequent rise in water bills to customers of 10 per cent (Patel, 1992).

Although some disinfection by-products have been found to cause cancer in laboratory animals, in most instances the epidemiological evidence is insufficient to prove that the amounts found in drinking-water are enough to cause disease in humans. The evidence was often based on short-term experiments using high levels of individual substances. Bromate for example, had been found to be carcinogenic and mutagenic in animal experiments. In 1987, WHO had no definite proof that normal levels would actually affect human populations (WHO, 1987). By 1993, more research had been done and WHO were a little more circumspect. They therefore introduced wide safety-margins due to the unknown nature of the risks (WHO, 1993a).

The term trihalomethanes covers several separate substances. For the technically minded, WHO (1993a) describes them as 'halogen-substituted single carbon compounds with the general formula CHX where X may be fluorine, chlorine, bromine or iodine or a combination thereof' – so now you know. It was originally proposed that trihalomethanes are formed as a result of pollution of the water supply but this has been found to be untrue (Fawell *et al.*, 1986). The nature and extent of the reaction is controlled by several factors including pH and the ratio of free chlorine to organic products. Many of these organic compounds will be naturally occurring in the water, particularly in small systems.

The main trihalomethane of interest is chloroform. Chloroform is best known for being an anaesthetic and thus has saved many people from untold misery. John Snow, of cholera fame, persuaded Queen Victoria to use chloroform when giving birth to one of her many children. This caused it to be widely accepted by the masses, who up till that point were somewhat sceptical of its benefits. Unfortunately, it has also been found to have carcinogenic properties in animals, but only at extremely high doses (WHO, 1984). The three other key compounds in the trihalomethane group are bromodichloromethane, which with chloroform is classed by WHO as being 'probably carcinogenic to humans' (this means there is limited evidence of their carcinogenicity) and bromoform and dibromochloromethane, which are defined as 'not being classifiable as to their carcinogenicity to humans'. This appears to mean that there is inadequate evidence to prove anything, but no-one is prepared to admit it.

WHO, when first considering Guideline Values for trihalomethanes, concentrated on chloroform as the only trihalomethane with proven problems (Packham, 1990). The original European Drinking-water Directive (80/778/EEC) did not include limits for trihalomethanes or chloroform but the Private Water Supplies Regulations 1991 followed the US Environmental Protection Agency's lead with a maximum limit for total trihalomethanes of 100 µg/l. When WHO published the second edition of its water quality guidelines (1993a), rather than having a limit for trihalomethanes as a whole, it suggested individual values for bromoform (100 µg/l), dibromochloromethane (100 µg/l), bromodichloromethane (60 µg/l), and chloroform (200 µg/l). These Guideline Values incorporated a safety factor of 1,000, that is the amount that has been found to cause illness is 1,000 times the amount allowed in the drinking-water. It is worthy of note that chloroform, which was the most studied and the one considered to be in greatest need of a Guideline Value, had a limit at least twice that of the others. The precautionary principal has obviously been generously applied until there is definite proof either way.

Because they often occur as a group, some experts consider that it is the total amount of risk that needs to be determined. WHO (1993a) have therefore produced a formula for calculating a suitable upper limit in water. For each parameter, the amount in the water is divided by its Guideline Value. These are then added together and the final total should be less than or equal to one. The costs involved in removing trihalomethanes from water are high and the process is difficult, so the risks involved should be commensurate with the benefits that could be accrued by other uses of the money. Unfortunately this may not be the case.

The other groups of disinfection by-products, apart from trihalomethanes, are as listed here:

- Chlorinated acetic acids, with three separate chemicals identified in the group. Two have been given Guideline Values – trichloroacetic acid (100 µg/l which includes a 10,000 safety factor) and dichloroacetic acid (50 µg/l – a 1,000 safety factor).
- Halogenated acetonitriles – four substances, three of which have provisional Guideline Values; dichloroacetonitrile (90 µg/l – a 1,000 safety factor), dibromoacetonitrile (100 µg/l – a 1,000 safety factor), and trichloroacetonitrile (1 µg/l – a 5,000 safety factor).

- Bromates – these have a European and thus a UK limit of 25 μg/l falling to 10 μg/l by the end of 2008. The 1998 Drinking-water Directive introduced these standards. WHO has given a Guideline Value of 25 μg/l, which assumes an excess risk of one extra case of cancer in every 700,000 people drinking the water at that concentration.
- Individual chemicals that are classed as disinfectant by-products and have provisional Guideline Values are chloral hydrate (10 μg/l – a 10,000 safety factor) and cyanogen chloride (70 μg/l, based on the toxicity of cyanide).

In all, 38 products have been identified in drinking-water following chlorination (Fawell *et al.*, 1986). Those that have not been mentioned here do not have sufficient data for Guideline Values to be established.

The media attention over the potential carcinogenic or mutagenic effects of disinfectant by-products has caused some people to assume that chlorination is inherently dangerous. Users of small water systems, when considering which form of treatment to install, have echoed this worry. When I asked one owner of a consistently faecally contaminated private supply what form of treatment he intended to use to resolve the situation, he replied that he was considering membrane filtration, because using chlorine 'would give you cancer'. He had read this was the case in one of the Sunday papers, so there was absolutely no doubt in his mind. This could be a bit of a worry to the 98 per cent of the population drinking chlorinated water – if a single human being had ever been proved to die from drinking it. A UK government committee on the Medical Aspects of Air, Soil and Water discussed the conclusions of much of the research into the possible ill health effects of trihalomethanes (DoE, 1989). They stated that there was no sound reason to conclude that the consumption of the by-products of chlorination, according to current practices, increased the risk of cancer. They went on to conclude that although individual by-products of chlorination cause cancers in laboratory animals when administered in large doses over long periods, modification in the use of chlorine was unnecessary. There is some concern that statements about the safety of trihalomethanes have been exaggerated and that these could lead to the undermining of confidence in the vital process of disinfection (Packham, 1990). It remains the official view in most of the world that the unproved risks involved with minute quantities of trihalomethanes do not warrant any changes that could stop the use of the well-proven health benefits of chlorination.

Is the risk from either disinfection by-products or trihalomethanes greater to private water supply users than to people on public water supplies? The easy answer in the UK would be no, because the vast majority of small water supplies do not use chlorine as a treatment system. Many people in the water treatment business are wary of installing chlorination to small supplies because chlorine, as a gas or liquid, can be dangerous from a health and safety standpoint. Installers like to put systems in place that will not damage the operator if something goes wrong. Many of them have had a lifetime's experience of engineering processes and know that if something can go wrong, it probably will do so. It should be pointed out however that in the US, chlorination is standard in many small and private supplies without any significant problems. UK treatment engineers however, prefer to install something like an ultraviolet disinfection system, where the worst that can normally happen is that the light in the tube goes out. There are systems available nowadays however, that use electrolysis to automatically generate chlorine on-site from common salt. This is obviously a lot safer and should encourage more people to install chlorine disinfection for small systems.

It is technically possible that the private water supplies using chlorine may have an increased risk of disinfection by-product creation. This is because the incoming water, especially from springs and surface supplies, can contain high levels of organic material and the chlorine dosing will be less accurate than for public supplies. One of the aims of a good chlorination system is that the process should leave free chlorine in the water. But, in private water supplies, because of the propensity to overdose rather than under-dose, there could be an excess of free chlorine at the end of the treatment. This excess chlorine could react with the naturally present organic material, particularly at the low pH values often found in small supplies, to form trihalomethanes (Fawell *et al.*, 1986). In fact however, it is unlikely that they will pose a particular problem in private water supplies because disinfection by-product formation is slow, typically taking 72 hours or more of contact time (Jackson *et al.*, 1989). It is extremely improbable that the amount of time between chlorination and consumption in a small supply will be more than a day or two. Where there is a large storage tank with a long turnover time or if occupants have been away for a few days, trihalomethanes may be produced and officials must be aware of this when determining what to test for in a particular supply. In the larger independent systems in the UK, where contact times may be greater, there is a requirement in the regulations to sample for trihalomethanes four times a year.

Private boreholes in cities have been found to contain quantities of chloroform because of extensive aquifer contamination from leaking chlorinated mains water (Rivett *et al.*, 1990a). This appears to be more of an example of the quality of chemical detection in modern laboratories than an example of a research study finding a serious health problem in urban neighbourhoods. It is difficult to believe that we should spend a great deal of time worrying about cleaned, treated and disinfected water leaking from a mains water pipe into a private borehole.

Organic solvents

UK maximum concentration allowed: Values are set for individual chemicals.
WHO Guideline Value: Values are set for individual chemicals.

The term 'organic solvents' covers a range of chemicals that have been widely used as dry-cleaning agents or industrial cleaners, solvents and degreasers. They are a group of chemically similar organic molecules but have been given individual guidelines by WHO because they have different health effects (WHO, 1993a). Many organic solvents are used in industry but the following list contains the ones most likely to be found in groundwater.

- Carbon tetrachloride, used in the production of refrigerants (Guideline Value 2 μg/l).
- Dichloromethane, used to remove caffeine from coffee and for paint stripping (Guideline Value 20 μg/l).
- Two forms of dichloroethane. These are used to make vinyl chloride (Guideline Value 10 μg/l for the 1,1-form and 30 μg/l for the 1,2-form, which also has a UK standard of 3 μg/l).
- Trichloroethane (Guideline Value 70 μg/l), and tetrachloroethane (Guideline Value 40 μg/l). Both are used as dry-cleaning agents and degreasers. The UK standard of 10 μg/l is for a combined total amount of both substances.

Because of their chemical composition, organic solvents are readily adsorbed through the skin, gastrointestinal tract and lungs. They have been found to cause acute and chronic damage, with carcinogenic properties in some laboratory animals. The 'linear no-threshold dose response' model has been assumed when calculating their maximum allowable concentrations. WHO has indicated that the way it calculates the guidelines for some organic chemicals is fundamentally different to those for inorganic ones because of the lack of suitable data. They have stated that the most important influence on the quality of drinking-water for some organic compounds is aesthetic and organoleptic (smell and taste), rather than their health effects, 'Such substances can often be completely undrinkable when present in concentrations well below those that cause concern for health reasons' (WHO, 1984). Since the publication of the original WHO Guidelines, a lot of experiments have been carried out to find acceptable daily intake values for these chemicals. The uncertainty factors in setting guidelines remain high however. For carbon tetrachloride and tetrachloroethane, this uncertainty factor is 1,000 and for trichloroethane it is 3,000. This indicates that the data are still limited and health officials have erred on the side of caution when setting standards.

Water is not the only route whereby these chemicals will come into contact with the human body. They can be found in food and inhaled from industrial airborne effluent streams. The justification for these organic solvents being considered a problem in water is that despite being only slightly soluble, they are very persistent. They consist of small, uncharged molecules that theoretically can travel great distances underground because they are not readily adsorbed to clay particles, etc. This could lead to water supplies

Table 4 Organic solvent contamination of drinking-water wells[a] in the US

Compound	Number of wells sampled	Wells with detectable concentrations (in %)
Tetrachloromethane	1,659	18
Trichloroethane	2,894	14
Tetrachloroethane	1,586	13

Source: Adapted from Craun (1984).

Note
a This term is used to cover all groundwater sources.

being polluted far from the point of contamination. Pollution problems are usually confined to groundwater sources because these organic chemicals are highly volatile and levels quickly reduce in surface waters. Studies in the US have found up to 79 per cent of groundwater sources are contaminated in certain states (Craun, 1984). A larger study of 18 states provided a wider-based illustration of the problem, as Table 4 shows.

In the UK, there have not been extensive studies of organic contamination in private supplies and the Private Water Supplies Regulations 1991 only require the largest of supplies to be monitored for organic solvents. Most private supplies are located in rural areas and will not be within range of the major users of these chemicals. In the cities, most of the older domestic wells and boreholes have been connected to the public supply, therefore the only remaining supplies tend to be industrial or commercial boreholes with specialist uses for the water. A study of the Birmingham and Coventry aquifers found extensive contamination by chlorinated solvents (Rivett *et al.*, 1990b): 40 per cent of the boreholes sampled were above the UK limits, including all the boreholes located in food-processing sites. Some food businesses will still use water from city boreholes, and breweries are likely to be the biggest group in this category. Many of the bigger breweries have had to connect to the public supply, because their borehole could not produce enough water, but many smaller ones have not (Ball, 1993). They are located near reliable boreholes and, because water is so important to their product, may have resisted moving to the public supply to prevent changes to the quality or taste of their beer. The boreholes of breweries located in the older parts of towns may, however, also be close to businesses that have used industrial cleaners and solvents. The question then arises – is it possible for any breweries to be using water from a private borehole and not be monitored for the organic solvents under the regulations? That is, could they be using water contaminated by organics solvents but be unaware of it? A medium to large brewery can use about 110 m³/day. In the UK, the wording of the regulations state that the average volume of water supplied for *food production purposes* should be used when calculating what to test for. This figure should include all the water used for cleaning, which is typically five or six times the amount used for making the beer (Drury, 1993). This will ensure that nearly all breweries and food processes will be class one or two supplies and therefore regularly tested for organic solvents. It needs pointing out as well that, because of the importance of water to the brewing industry, they take great pains to ensure the quality of their water. In-house testing will usually be for more substances and carried out more regularly than required by any governmental standards. This question may also be academic as the water will be boiled and therefore most solvents will be evaporated.

When organizing drinking-water monitoring programmes, people should be aware of the potential problems of groundwater contamination in urban areas. Despite their ability to travel large distances, most organic solvent contamination is, in fact, local (Lerner and Telham, 1992). A sanitary survey should therefore take note of surrounding sites and their current and previous usage. Shallow sources may be even more at risk than the sampling boreholes used in the Birmingham and Coventry study as they were deep bores and so had higher dilution rates from uncontaminated water. A sanitary survey in towns may therefore be equally as useful as one locating sources of microbiological contamination in remote country areas.

Pesticides

UK maximum concentration allowed – individual pesticides 0.1 µg/l, total for all pesticides 0.5 µg/l. An individual lower limit has also been set for pesticides considered to be particularly toxic. This is 0.03 µg/l and has been set for aldrin, dieldrin, heptachlor and heptachlor epoxide.
WHO Guideline Value: Values are set for individual pesticides, these range from 'unnecessary' because they are not hazardous to human health at concentrations normally found in drinking-water (glyphosate) to 0.03 µg/l (aldrin and dieldrin).

Pesticides have become a very emotive issue. The reason for this is partially (but only partially) because they are all designed to kill or injure some form of life. Hurst *et al.* (1991) stated that all pesticide compounds potentially pose an environmental health hazard as they are chemically tailored to be toxic. There is a vast amount of material available on pesticides in newspapers, magazines, books and scientific papers and not only is media reporting sensational but many of the books on the subject fail to be as objective as they should be. Experts are seen to be at odds with one another and research results are sometimes contradictory.

What is it that causes so much alarm about pesticides? The answer is that pesticides are in the forefront of the 'chemicals' problem. Many people today do not like 'chemicals', despite the fact that everything, including people, is made of them. There has been a rash of popular press and TV articles about agricultural chemicals and the green movement has spear-headed a move away from their use with an almost religious fervour. Lack of toxicological proof of safety at low levels, particularly to non-target species, especially humans, is the main worry. It is easy to observe the effects of acute exposure to chemicals but the chronic results from low-level intake are more difficult to assess. As well as the consequences to healthy adults, hazards to health can include additional problems for children, the undernourished, the aged, the infirm and to both pregnant women and the foetus. There may be risks to genetic material or the nervous and immune systems. Safety factors for chemicals in water usually have to be at least two orders of magnitude where the toxicological data are inadequate or inconclusive. Several organizations question whether even this is enough and complain that the figures are taken in isolation without consideration of the confounding and compounding effects of other potentially dangerous chemicals taken into the body at the same time.

The risks are not merely associated with direct intake by humans. There are worries about low concentrations of pesticides at the bottom of food chains building up through bioconcentration to levels potentially in excess of 80,000 times the original (Morgan, 1992). Big mammals may therefore consume unacceptably large amounts of pesticides originally intended for other species. There are also the environmental problems associated with pesticide usage including the build-up of resistance in pests or, where a pest is removed, a previously non-problematic species suddenly becomes a nuisance because of the disappearance of its natural competitor. Critics point out that while there may be a global need for pesticides, there is a surplus of food in Europe and America and therefore no need to use such efficient agricultural methods where alternative organic methods are available.

Pesticides therefore attract an interesting collection of protestors. Animal rights activists see the use of animal experiments as immoral, particularly for things they consider unnecessary, such as cosmetics and (in some people's opinion) pesticides. Single-issue political groups with a mission to protect the world, object to the misuse of what they see as dangerous chemicals. They see the lack of detailed toxicological data as worrying and build up a case for stopping the use of pesticides until they are proved safe. This is the precautionary principle. It is interesting to note that only certain things are singled out for the precautionary principle and it does not apply to everything that is known to cause damage to human health. This is illustrated by the Home and Leisure Accident Surveillance System, a report from Britain's Department of Trade and Industry. It gives figures for accidents reported by people admitted to a group of hospitals and then produces extrapolated estimates for the country as a whole. A report from *New Scientist* (Anon, 2001c) details the worries: 'The toll of accidents caused by tea cosies is up again, with a national estimate of thirty-seven tea cosy injuries, compared with twenty the previous year. Equally alarming, the number of accidents caused by place mats – a menace we have paid too little attention to in the past – is up from 157 to 165 across the country as a whole. These worrying figures are somewhat balanced by a welcome decline in another area of concern – sponge and loofah accidents. The shocking previous total of 996 nationwide is now down to 787.' There appears to be no major campaigning groups using the precautionary principle to demand the removal of all tea cosies or loofahs until they can be proved safe.

Pesticide enthusiasts (apologists, defenders – even the names have value judgments attached to them) counter the arguments against pesticides with the view that the opposite of surplus is shortage and that organic farming would never produce sufficient food for the world's needs. They also insist that as well as increasing cropping, pesticide applications take away much of the time consuming, back-breaking menial labour associated with traditional farming methods (Whitehead, 1992). Pesticide manufacturers take great pains to counteract what they see as misunderstanding and over-reaction against their product. Organizations such as the Crop Protection Association spend a lot of time arguing against claims of problems with pesticides. It is interesting however that they have decided not to include the word 'pesticide' in their title.

Crop production since the Second World War (when the modern generation of organic pesticides such as DDT were introduced), has kept up with the increase in population, even though the area of land under agricultural production has decreased (Graham-Bryce, 1983). It has been estimated that if pesticide usage stopped, cereal crop yields would be reduced by 24 per cent in the first year due to pest damage and 45 per cent in the third year due to competition from weeds (Morgan, 1992). There is still an estimated 20–40 per cent loss of crops worldwide due to pests and diseases. This percentage is even higher in developing countries. There is also a subsequent 10–20 per cent additional loss due to harvesting, storage, drying and milling (Morgan, 1992). These figures should be looked at alongside estimates of global population increase of 250,000 per day and the consequent reduction in available cropping area that these additional people will require for living space. Many therefore see pesticides as a global necessity.

The pesticide industry also points out the public health benefits of pesticide usage, not only in the reduction in the number of insect and rodent-borne diseases, but also in the increased nutrition of the population and the resulting greater ability to withstand disease. In addition, they maintain that the reduced costs of food and non-food crops such as cotton, makes them available to an ever-widening section of the community (Whitehead, 1992). Manufacturers' organizations also attempt to put pesticides into perspective with other chemicals in the environment. They point out that a once popular herbicide such as atrazine is half as toxic as aspirin. In addition, common household bleach (which has an acute toxicity at least twice that of a weed killer) has few of the restrictions that pesticides are subject to.

What is the answer to this dichotomy? Both sides would agree to the need for further toxicological data, more accurate application of pesticides, alternative methods of pest control, protection being given to non-pest species and the replacement of unduly persistent chemicals. The situation is complex, in the UK the use of atrazine and simazine, two well-known contaminants of water supplies was restricted about ten years ago. They were the most commonly used agricultural herbicides and local authorities and British Rail employed them for weed control on pavements and tracks. They were so commonly found above the maximum allowable concentration (689 samples in 1989) that water utilities were required by the government to look for them in all their water supply zones. In 1993, their use was banned for local authority and British Rail use (MAFF, 1992) and they were gradually phased out for agricultural use, beginning with a bar on aerial applications. It has been suggested that outright bans on substances used by farmers is counter-productive as the unscrupulous ones then tip the outlawed substance straight into the nearest watercourse (Oakes, 1992). It is considered preferable to allow the substances to be used up gradually. The result of the ban has been to force local authorities to use the less hazardous, but more expensive and less persistent herbicides such as glyphosate. Because of this lower persistence, several applications are necessary rather than the single one used previously. The government report at the time (DWI, 1992) also stated that adequate limits of detection were not available for glyphosate. Although all newly approved pesticides must have an analytical detection method, this may not fit in with the suites of pesticides set up in laboratories or may be too expensive to be economically viable (Pound, 1992).

The limits in the 1998 European Drinking-water Directive are largely the same as those in the 1980 version. These are thought to have been set at the 1970s detection levels for organochlorides (King, 1991; DWI, 1992). The message in 1980 seems to have been that if a pesticide was detected, the concentration was too high. It has also been suggested that the reason the limits are as low as they are is because the Europe Union accepted that pesticide contamination should be kept to an absolute minimum, known as the 'surrogate zero' level. Another proposal is that it was politically prudent to show that Europe wanted no pesticides in drinking-water (Fawell, 1991) and others consider that the setting of the limits was merely based on political pressure from environmental groups.

The maximum allowable concentrations for individual and collective amounts of pesticides in the European and UK legislation bear no relationship to actual risk (Fawell, 1991). The UN Food and Agricultural Organization, WHO and individual countries, including the UK, have produced figures for various pesticide concentrations in drinking-water based on toxicological data. The data are often confusing; simazine for example, had a WHO limit of $17 \mu g/l$, a US limit of $1 \mu g/l$ and a UK Department of the Environment safety limit of $10 \mu g/l$. These different figures result from differing safety margins due to the lack of definite information but they all show that the European standard has little to do with health. Even the most cautious is ten times that of the 0.1 μg/l legal limit. It is also not clear why European regulators originally took only one average figure for all pesticides instead of separate limits for individual substances. This was especially germane in the case of aldrin and dieldrin, which have a WHO advisory level of 0.03 μg/l based on toxicological data. Fortunately, the latest European Drinking-water Directive has incorporated the WHO figure for these two chemicals.

To illustrate the problems associated with keeping within the bounds of the European maximum allowable concentration, an often quoted figure is that 2 kg of any pesticide (about the same amount as two bags of sugar), if put directly into the national water supply, would cause all the drinking-water in the UK to fail the standard. The chair of the Department of the Environment's Pesticide Panel, at the time the Private Water Supplies Regulations were introduced, considered the standards to be 'potty', because they were set 'with no regard to toxicity' (Wilkins, 1992). The expenditure necessary to achieve this standard is expensive, not only the cost of treatment plant at the water works but also monitoring. Installing pesticide analysis equipment for the Severn Trent area alone cost £1.5 million and their annual monitoring of 6,000 samples cost over £1.2 million over a decade ago (Wilkins, 1992). The present day cost of the removal of pesticides in UK public water supplies is over £1 billion pounds a year (Taylor, 2003). If the limits are not based on toxicity, would some of the money being used to achieve them be better spent on something that shows a clear health benefit? More than likely – yes, but it should be pointed out that there are some hidden benefits that accrue with the installation of suitable pesticide removal equipment. This is usually granulated activated carbon, ozonation or increasingly as costs are reduced, membrane filtration. This technology will efficiently remove other contaminants of water such as colour and microbiological agents.

In the UK, the body that controls pesticide approval is the government-appointed Advisory Committee on Pesticides. The Advisory Committee and its scientific sub-committees review toxicological data and other information on individual pesticides and decide whether to recommend to the government to allow or ban their use. It is an independent body with various expert advisory panels, such as for medical and toxicological assessment and pesticide monitoring. Twenty-two new pesticides were reviewed in 2001. It is important to note that safety levels for many older pesticides have not been afforded the full attention of the Advisory Committee and it therefore carries out reappraisals of long-established pesticides. Popular interest in a certain pesticide can cause them to be moved up the waiting list, as happened with Maneb, Mancozeb and Zineb, when the US government banned them in 1989. The UK Advisory Committee subsequently scrutinized these products, which had been in use for over forty years and concluded that there was no risk to consumers from these fungicides or their metabolites (breakdown products from chemical reactions within living organisms, such as digestion, involving the substance in question)

(Lee, 1992). It should be noted that the time involved in these assessments either caused a possibly useful pesticide product to have to wait an even greater time before its appraisal or prevented the re-evaluation of an existing pesticide considered to be potentially dangerous.

Following the setting of water quality standards, it is necessary to calculate how much of a specific pesticide can be applied. Researchers try to calculate the application rates at the surface that will produce pesticide levels in water supplies less than the 0.1 µg/l limit. To achieve this, the vulnerability of aquifers and surface waters to pesticide contamination has been mapped using mathematical models and computer databases. The calculations are also the subject of much research by universities and manufacturers. This covers general factors influencing degradation and leaching as well as information on specific pesticides. Studies are carried out either *in situ* or under laboratory conditions using large soil samples (or lysimeters).

In order to understand the behaviour of a pesticide, various properties of the soil must be taken into account. These are as follows:

- *Adsorption.* The factors found to be important to adsorption are particle size, organic content, calcium carbonate content, pore size and the distribution of fissures. The smaller the particle size and the greater the organic content, the more likely pesticides are to be adsorbed. Adsorption is more likely to happen in the top layers of soil because of the greater organic content, although when the pesticide reaches clay levels, which consist of very small particle sizes (typically less than 2 µm), adsorption can increase.

- *Degradation.* Usually identified by using the half-life of the pesticide in the ground. This is also mainly dependent on organic content, due to the increased number of microbiological organisms that can metabolize the pesticide. This action is also conditional on temperature and the availability of oxygen and water, which are in turn affected by the pore size of the soil. The larger the pore space, the more water and air is available to the microbial organisms to carry out degradation. One interesting aspect of the microbial breakdown of pesticides is that some subsoils with a low organic content show a marked lag-phase of between 3 and 6 days before they reach optimal activity. This indicates that microbial populations can rapidly adapt to breaking-down new compounds (Campbell *et al.*, 1991).

- *Hydrolysis* (chemical decomposition or ionic dissociation caused by water). This is affected by the availability of water, its pH and the salts within the water.

- *Retention.* Again, this is dependent on organic content plus the amount of water in the soil. The more organic matter in the soil, the more water it will hold and the less likely precipitation will wash the pesticides down to the underlying aquifer. Leaching is additionally affected by pore size, particle size and the nature and size of fissures.

When a particular mixture of chemicals is first disposed of to land, there will be some natural removal, but it will not be very efficient. If the mixture is regularly disposed of in the same place, the microorganisms that can use any of those particular chemicals to metabolize, will thrive, multiply and they will start to form an efficient natural treatment system. The chemicals may also kill any rivals or natural predators, thus increasing the numbers of bacteria feeding off it. In lysimeter tests, it has been found that indigenous microbial populations can rapidly adapt to degrade new compounds. Once the pesticide has reached the aquifer its transport through rock structure is known to occur at rates between 0.5 and 1.5 m per annum (Oakes, 1992). However, because of the fractured nature of aquifers, they often permit more rapid movement and less subsequent opportunity for retardation by adsorption, chemical reaction and degradation (Foster and Chilton, 1991). Some pesticides are very long-lived and in deep aquifers they may take many years before they eventually emerge.

Polynuclear or polycyclic aromatic hydrocarbons

UK maximum concentration allowed: 0.1 μg/l.
WHO Guideline Value: Values are set for individual chemicals. Benzo [a] pyrene – 0.7 μg/l, bromoform and dibromochloromethane – 100 μg/l, bromodichloromethane – 60 μg/l, etc.

Polynuclear aromatic hydrocarbons (also referred to as PAHs) are widely found in the environment and human exposure is via food and air as well as water. Some have been found to be carcinogenic to laboratory animals. The relative contribution to average human intake from water is in the order of 0.1–0.3 per cent, with 0.9 per cent from the air. We therefore obtain nearly all our polynuclear aromatic hydrocarbons from the food we eat (WHO, 1993a) and this point should be borne in mind when considering the relative risks from water consumption. Polynuclear aromatic hydrocarbons in water are usually associated with the degradation of coal tar pitch linings in water pipes (Bridges *et al.*, 1991). Many cast iron water mains laid between the beginning of the last century and the 1970s were lined with coal tar pitch and up to half of these linings consist of this particular form of hydrocarbon. The chemicals have subsequently leached into water supplies and, where the lining is corroding, have also been found in particulate form (Healey, 1991). It is not known however, whether the health effects are different for these two types of contamination. Bitumen-lined pipes are now used and these have considerably lower amounts of polynuclear aromatic hydrocarbons; so they are not such a problem (DoE, 1989).

Private water supplies that are over thirty years old and have long runs of iron pipework may therefore contain polynuclear aromatic hydrocarbons. Larger supplies in the UK are automatically monitored for them and local authorities that have smaller supplies may include them in their monitoring programmes, if problems are suspected. Where a problem is found and the maximum allowable concentration is being consistently exceeded, the pipes should be replaced.

The carcinogen potential of different polynuclear aromatic hydrocarbons varies and research is continuing into their health effects. The one that has caused the most interest, due to its carcinogenic and mutagenic properties is, benzo [a] pyrene. In animal studies, it has been found to cause tumours in several different sites in the body. In the UK, the maximum allowable concentration was set at 10 ng/l (0.01 μg/l), which is the lowest limit for any parameter in the regulations. Benzo [a] pyrene was not part of the first European Drinking-water Directive, but was included in regulations by the UK government following the publication of WHO's 1984 Guideline Value (DoE, 1989). The 1993 WHO Guideline Value took account of further research and corresponds to a lifetime cancer risk of 1 in 100,000.

Other individual polynuclear aromatic hydrocarbons have had less research carried out on them but are generally thought to be less dangerous. The European maximum allowable concentration of 0.1 μg/l for polynuclear aromatic hydrocarbons is based on a combination of chemicals and is half the 1980 standard of 0.2 μg/l. However, the most commonly found polynuclear aromatic hydrocarbon – fluoranthene – has been taken out of the calculations. This is because it is no longer considered a health risk, so the latest standard is not a particular tightening of the legislation.

Minor organic contaminants

From time to time other groups of organic chemicals are associated with water contamination or possible health problems. They are then subject to media attention and scientific interest. This will either lead to concentration guidelines being set for them and inclusion in the general sampling programme or the interest will die down. Four groups of chemicals are currently attracting attention. These are endocrine disrupters, Methyl Tertiary Butyl Ether (MTBE), personal care products, pharmaceutical and veterinary medicines and phenols. There have been no standards set in UK legislation for any of these substances in private water supplies.

Endocrine disrupters

Animals use their hormone (or endocrine) systems to regulate development, growth, reproduction and behaviour. Some chemicals may adversely affect these systems and are collectively known as endocrine disrupters. These have been attracting attention for over three decades and are generally thought to consist of natural or manmade versions of the oestrogen hormone or substances that resemble it. Worries first began with the introduction of the female birth control pill; it was thought that oestrogens excreted by women using the pill passed through sewage and water treatment works unaffected. This led to lurid headlines about the number of times Londoners' water passed through other Londoners, but I suspect this was more to do with the media having fun rather than any actual problems with endocrine disruption. It has now been found that many chemicals can disrupt the hormone systems of animals, and concern has often involved worry over changes or damage to the human male reproductive system (although as the research tends to be carried out by male scientists, this may not represent the complete set of problems). The evidence is still far from conclusive and other reasons, including smoking and lifestyle changes, have been suggested for the increasing levels of problems such as testicular cancer.

The publicity surrounding endocrine disrupters has also involved the feminization of male fish. Some species of fish (e.g. roach) naturally show signs of feminization of males, so it is not quite such a shocking revelation as it might be first thought (although I would not want to be seen to be encouraging such behaviour). The substances principally considered responsible are natural human oestrogens. A nationwide survey found evidence of changes to male fish in many waterways, especially downstream from sewage treatment works. The River Aire in Yorkshire has nearly forty sewage treatment works on its banks, and despite a lot of cleaning and improved treatment processes, was found to have a problem from a particular treatment works serving a large traditional mill town (Morris, 1997). The source was eventually traced to organic chemicals being washed off imported wool. The problem was identified following research where trout were put into the river in cages, just below the treated sewage outlet. They were left for three weeks and then examined for signs of sex changes. On the one hand, this showed that changes were indeed taking place and led to the situation being investigated and eventually put right. On the other hand, it also showed how clean the river was getting. It was not too long ago that trout put in the River Aire would be spluttering and gasping within minutes and would not have survived three days, never mind three weeks.

It is now thought that most of the endocrine substances found in surface waters are a natural constituent of female urine and do not present a particular problem. A report in 1995 (MAFF) found no evidence of risk to health from the amount of oestrogens present in drinking-water at the time. The worries that they were unaffected by water treatment processes have also proved unfounded. Both sedimentation and chlorination have been found to be effective against oestrogens and modern treatment plants using ozonation or activated carbon are effective in removing them and a whole range of other organic substances causing similar concerns over effects on the hormone system (DWI, 2002). These other potentially problematic substances found in trace amounts in water include alkyphenols (from some detergents), phthalates (used as plasticizers in plastics such as PVC) and phyto-oestrogens (these are naturally occurring in plants – especially soya beans apparently) (Anon, 1997). Both the UK Drinking-Water Inspectorate and the European Union have carried out research work on potential endocrine disrupting chemicals, both in rivers and in drinking-water. The DWI are confident that they are only present in such small amounts that they are undetectable in public water supplies (DWI, 2002).

MTBE (methyl tertiary butyl ether)

This is a petrol additive that has caused concern in the US, particularly in California. It can be toxic but not until the chemical has passed the concentrations at which it becomes unacceptable to drink because of organoleptic problems. It is unlikely to be a problem in the UK because it is not added to petrol in such quantities as in the US. There has only been one problem with MTBE in groundwater in the UK. This was close to a US airbase in East Anglia where fuel had been spilt (DWI, 2001b). Despite this lack of problems, it is always important to ensure that petrol is not disposed of where it may contaminate groundwater.

Personal care products

Personal care products such as shampoo, deodorant and soap are a potential problem, similar to, though less worrying than, pharmaceutical chemicals. The range of chemicals used on the hair and body will be washed into the sewage system after someone takes a bath or a shower. Unless removed by sewage or water treatment, they may then find themselves in drinking-water supplies. Because we use them on our skin we expect them to be safe, particularly in the amounts that would be expected to get into the water of private supplies. The products all contain a variety of chemicals and even some that claim to be natural appear to contain a great number of substances that people would not like to consume in their drinking-water. Glycol distearate, phenooxyethanol, sodium benzoate, astrocryum, murumuru, tetrasodium EDTA, sorbitol and isobutylparaben are just some of the chemicals listed as ingredients of one natural bath cream in my house. Murumuru in particular seems worrying – no doubt it is perfectly harmless but it sounds just like one of the exotic substances found on the end of poisoned blow-pipe darts. I cannot imagine that these substances, natural or not, would ever accumulate in the water of small and private supplies in sufficient concentrations to cause any degree of realistic concern without the smell or taste of the water becoming unpleasant.

Pharmaceutical and veterinary medicines

Once consumed, small amounts of pharmaceutical drugs can be excreted via the urine or faeces. They will then enter the sewage disposal system and, unless removed by sewage treatment, will be deposited in surface waters or onto agricultural land in sewage sludge. Animals can also deposit veterinary medicines directly onto agricultural land. These medicines can then go into drinking-water supplies. As Fawell (2000), points out, the substances most likely to be found in water are those used widely and for a long time. Whether they enter the water system will depend on their physico-chemical properties and degradability. Antibiotics, steroids and anti-inflammatory drugs are amongst those of most concern. The interest in pharmaceutical and veterinary medicine residues in water started in the 1960s and has naturally followed on from the ability of laboratories to isolate minute traces of these chemicals.

Water company treatment systems, particularly those taking water directly from rivers downstream from sewage works, are very sophisticated. Water companies will be aware of potential problems and should look for these chemicals when sampling the water. Existing treatment systems such as activated carbon, reverse osmosis and ozonation are likely to remove any minute traces found. Conversely, private water supplies take water directly from agricultural areas and often the water has had little time for chemicals to degrade or be removed by the natural action of the subsurface soil and rock structures. Private water supplies are also regularly found to contain evidence of animal excrement. Any animal medicines defecated onto land could therefore quickly make their way into a small water supply. Once there it is unlikely to be removed, as treatment is often absent, not working or inefficient. Regular sampling of the water will not be carried out for these types of products so it is unlikely to be considered a problem in private water supplies – whether it is or not.

Drinking-water has been found to contain some pharmaceutical products but because of the large dilution factors involved in public supplies, only at exceedingly low levels, usually below 100 ng/l. Because of these low levels they are unlikely to be of any health significance. However as data are limited, research will continue to try and identify whether these particular products are a problem or not.

Phenols

Phenol is a white crystalline solid with the chemical composition C_6H_5OH. There is also a group of chemicals known as phenols, which are organic compounds that contain a hydroxyl group, bound directly to a carbon atom in a benzene ring. Phenol and phenols have occasionally been found in water supplies following accidental spillage, or in the case of an outbreak of illness at a UK private water supply – cross-contamination of the supply with sewers from a laboratory. Several rural wells were found to be contaminated by large amounts of phenol (also known as carbolic acid) in July 1974 in Wisconsin. Symptoms of illness caused by this contamination included dark urine, mouth sores, burning sensations and diarrhoea. Geological testing indicated that the contamination would take many years to clear (Baker *et al.*, 1978). In 1984, the UK River Dee was contaminated by phenol from an accidental spillage, which entered a drinking-water treatment works. This contaminated the water of 2 million households and was thought to have caused some gastrointestinal illness amongst those drinking the water (Jarvis *et al.*, 1985). A third accidental spillage, this time into the River Nakdong, near Teagu in Korea, contaminated the water of 2 million people in March 1991. Gastrointestinal illnesses including nausea, vomiting, diarrhoea and abdominal pain were reported (Kim *et al.*, 1994).

It is unlikely that private water supplies will be subject to pollution problems from phenols considering the rural location of most of them. There may however, be occasional accidental spillages around the source or upstream of surface water intakes and in such cases, it is likely that the changes in taste and smell will alert the people drinking the water to the contamination. There is no UK maximum concentration.

17 Radioactive contamination

UK maximum concentrations allowed – no individual standard set in legislation, but there is a total indicative dose for radioactivity (including gross alpha and gross beta activity) of 0.1 mSv/year. Radon and its decay products however, are specifically excluded from this standard along with tritium and potassium-40. There is a separate standard for the radioactive element tritium (100 Bq/l) but not others.
WHO Guideline Value: Gross alpha activity 0.1 Bq/l, gross beta activity 1 Bq/l.

There is a general degree of public concern about risks from radioactivity and there are regular calls for increased safety within the nuclear industry. However, the most significant risk from radioactivity to human health comes from natural radionuclides. In fact, the nuclear industry's contribution to average human annual exposure is trivial (O'Riordan and O'Riordan, 1991). The United Nations Scientific Committee on the Effects of Atomic Radiation has estimated that exposure to natural sources contributes more than 98 per cent of the radiation dose of the population, excluding medical exposure. Thus less than 2 per cent of radiation exposure comes from nuclear activities. When medical exposure is included (which is about 11 per cent) natural radiation provides 87 per cent of the total contribution, with radon's contribution alone being 32 per cent (Castle, 1988).

The effects of radiation are categorized in two ways. One is a direct effect caused by the energy of the radiation on human tissue. This is known as a 'deterministic' effect. It is considered that there is a limit below which this does not occur and radiation above the limit is not normally encountered, particularly in drinking-water. The other effect is called 'stochastic' where defects may or may not occur because of the effect of radiation on the individual cells of the body. There is usually considered to be no level below which these problems do not occur (the linear no-threshold dose response theory) although, as with low levels of carcinogenic chemicals, this theory has not been proven. The main problem found with radiation is cancer, although other damage occurs with different radionuclides on different tissue.

Levels of radioactivity are measured in becquerels (Bq) named after Antoine Henri Becquerel, a Nobel Prize-winning French physicist, who discovered radioactivity in 1896 after finding that invisible rays from uranium salts could affect a photographic plate even through a lightproof wrapper. One becquerel is equal to an average of one radioactive disintegration per second. WHO has suggested safety limits for radioactivity (1993a) in water and because different radionuclides are found in water and their effects will combine, WHO decided that the best way to limit radiation exposure would be to measure the gross alpha and beta activity (0.1 Bq/l and 1 Bq/l, respectively). The WHO limits for water are considered to be very low and precautionary in intent. The latest European Drinking-water Directive has a total annual dose limit for radioactivity of 0.1 mSv, which is in line with the WHO Guidelines. A sievert (Sv) is the unit of dose equivalent and is equal to joules per kilogram of body weight (the becquerel is a unit of activity rather than dose received). This compares to an individual's total annual dose from all sources of around 2.5 mSv. The US measure of radioactivity is the pCi. A pCi is a picocurie, which is 10^{-12} curies. A curie, named after Pierre and not Marie Curie, is equal to 3.7×10^{10} becquerels, therefore 1 pCi is equal to 0.037 Bq and 1 Bq equals 27 pCi).

Although many substances can be radioactive, either naturally or following anthropogenic practices such as bomb making and nuclear power production, there are two main elements that cause problems in water supplies and they are both naturally occurring. The most important by far is radon. The other is uranium, which also causes problems in water because of its chemical toxicity. Others, such as caesium and strontium are very rarely found in water and are not considered here. Nevertheless, if they were present in water in any significant amounts, they would raise the levels of gross alpha and beta activity and this would initiate further specific radionuclide studies (WHO, 1993a).

Radon

No standard set for UK concentration or WHO Guideline Value.
Draft European action level – 1,000 Bq/l.

The element radon (Rn) is a colourless, odourless gas that was discovered by Ernest Rutherford in 1899 (although a German called Friedrich Dorn also claimed to discover it in 1900). It was originally known as nitens, but its name was changed to radon in 1923. It is heavier than air and is in fact the densest gas known. It is radioactive and emits alpha particles with a half-life of 3.82 days. It is also know as ^{222}Rn because of its molecular weight when radioactive, although there are other, less stable forms. It is one of the naturally occurring series of products that follow the decay of uranium.

The UK National Radiological Protection Board (NRPB, 1990) estimated that radon in dwellings throughout England causes around 2,000 deaths per annum. Radon is the single most significant source of indoor pollution and is the second most important cause of lung cancer after smoking

(Green *et al.*, 1992). Radon is also the radionuclide of greatest interest in water (Hesketh, 1982). Crawford-Brown (1992) has stated that 'in considering the induction of fatal cancers by contaminants in water, the calculated incidence from radon may be shown to be larger than the sum of all the others. Clearly, radon has the potential (and the word 'potential' is chosen deliberately to avoid claims of certainty) to represent the primary (chemical) health risk in water supplies'. Milvy and Cothern (1990) calculated that 17,000 fatal cancers in the US occurred over a seventy-year period from exposure to radon originally present in drinking-water. The combined effects of airborne and waterborne radon compound the problem. Together they contribute about half the global dose of radiation, although the contribution from water is the significantly lower of the two.

The gas is normally a problem because it rises up from particular geological formations and enters dwellings, where it accumulates in cellars and under-floor spaces. The efforts of the UK's National Radiological Protection Board have very largely been focused on dealing with this problem by encouraging people to fit physical barriers to stop the gas getting in or to install pumps to get rid of it.

Less generally well known is the fact that radon can be a problem in some drinking-waters. Radon is highly soluble in water and easily given off if aerated. Groundwater amasses radon as it passes upwards through small spaces between the rocks where the gas is produced. The groundwater is then pumped from wells and boreholes into domestic plumbing systems where it comes out of solution due to aeration from taps, showerheads or kitchen appliances.

Radon in water is not a problem with larger, treated supplies because of the amount of aeration caused by the treatment they receive, but it can be a problem in small groundwater supplies. Boreholes will normally carry a higher load of radon than spring or surface supplies because the groundwater is from deep underground, with more recent contact with radon-containing rock and will have had very little aeration. The results of a survey of boreholes in the South-west of England (Heath, 1991) first attracted British attention to this problem. The survey found high levels of radon, up to five times the recommended limit. Studies in the US show that the greatest problems normally occur in homes that are located in areas with high levels of radon in the groundwater and are served by an individual borehole or small community water system (Lao, 1990). Because radon increases in supplies that serve less than one hundred people (Crawford-Brown, 1992), there is a disproportionate risk to those on private supplies. This has been calculated by one study as two hundred times that of people on public surface supplies (Mills, 1990). An earlier study put this figure at only ten times higher than local mains supplies (Hess *et al.*, 1985) and this difference can be explained by the later study comparing only surface supplies.

The actual reason for this higher level of radon in private supplies has been explained in two ways. The first is that because many private supplies have little or no treatment and a short time between pumping and consumption, the radon will not have had the aeration necessary to dissipate or the time to decay. The second explanation concerns the relationship between system size and rock type. Crystalline rocks, such as granite, which are high in radon, do not produce enough water to satisfy large users, but they do provide enough for smaller supplies (Michel, 1990) so in these areas public supplies will be absent but private supplies may be present. Private supplies also tend to come from small aquifers that in general have a larger granular surface area, which encourages the take-up of radon (Milvy and Cothern, 1990).

Radon normally comes from two different types of rock. The main one is granite where the rocks naturally contain uranium; the other is karstic limestone. Problems occur in limestone areas such as the Mendip Hills, due to large-scale natural underground drainage that allows water from deep aquifers to surface quickly and easily. Concentrations in water of over 10,000 Bq/l have been recorded in limestone areas (Castle, 1988).

As with many contaminants, radon concentration in soils varies with meteorological conditions. Work by the British Geological Survey has shown that during heavy rain there is an increase in radon followed by a return to normal. If the ground becomes waterlogged there is a dramatic but temporary fall, which appears to be caused by the water sealing the gas in the soil. Radon concentration varies directly with barometric pressure as would be expected and to a lesser extent, indirectly with wind speed (Brown, 1992).

If water is drunk quickly from the tap, radon will be ingested rather than released into the air. Health problems associated with the ingestion of radon are less well understood than those from inhalation. From the stomach, water migrates to the intestines and the bloodstream where radon lodges preferentially in fatty tissue. Although 95 per cent of the radon is excreted after passing through the body, the rest undergoes radioactive decay in the organs. This leads to the possibility of cancer. The National Radiological Protection Board considers that radon in drinking-water represents only 1 per cent of the total risks from all sources of radon (Green, 1992). These risks have however been calculated for a population predominantly on public supplies. For people who rely on private boreholes, water may be a much more important source of indoor radon (Lao, 1990).

Because of the short half-life however, very few decays will take place while drinking-water is in the body. The real concerns are with airborne radon because it attaches to airborne particles that then lodge in the lungs. Here the alpha particles from radon decay, and its daughters come into close contact with human tissue where they can cause damage (O'Riordan and O'Riordan, 1991). The National Radiological Protection Board has set an action limit for airborne radon in homes at $200\,Bq/m^3$. They have estimated that 100,000 houses in the UK are above this action level. It has also been found that isolated communities often have proportionally higher levels (Green *et al.*, 1991). This, of course, includes those with a greater possibility of being on private supplies.

Boiling releases nearly all the radon contained in water because the solubility of water decreases rapidly with increasing temperature. Aeration will also quickly release radon from water and typical removal rates from household appliances are: taps – 30–70 per cent; showers – 66 per cent; dishwashers – 95 per cent; and toilets – 30 per cent (Nazaroff *et al.*, 1987). The amount of radon being released from household appliances will quickly raise levels of the gas inside dwellings. A Swedish study for example, found that levels in bathrooms were forty times higher than those in living rooms (Castle, 1988). A British survey found that levels of radon were raised in an otherwise protected house every time the tap was turned on (Green, 1992). Clearly, radon may be a problem in kitchens and bathrooms of houses using private boreholes but generalized British studies for radon have only located detection devices in living rooms.

The reason why waterborne radon levels are so important is because, where existing airborne levels are already high, the additional load supplied by the water may bring the house within the action level. It seems germane for properties on private water supplies, in areas of naturally high radon levels, to be involved in an integrated surveillance programme and the dangers from the water supply to be included in the factors used to assess the safety of the dwelling as a whole.

In the UK, there appears to have been very little work carried out on radon in private water supplies. Until recently, very few UK local authorities were aware of any particular problem with radon in private water supplies (6.1 per cent) and the vast majority had not carried out any sampling to see whether they had any problems (96 per cent) (Clapham, 1993). Their efforts were much more likely to be concentrated on airborne levels of radon in houses. One study in a granitic area of the North-west of England found levels between 1.5 and 44.5 Bq/l in drinking-water supplies. It concluded that there was little problem of overall risk to health from these levels. Nevertheless, the researchers queried the accuracy of their dose calculation methods (Smith and Welham, 1990). Heath (1991), also estimated that the normal risk associated with radon in drinking-water was not great, despite the high concentrations sometimes found (1 MBq/l is not uncommon). Assuming a daily intake of 2 l, he calculated that the maximum level in water could be up to 30 MBq/l without exceeding the recommended annual oral intake of 20 MBq. This calculation however does not take into account the added risks from the additional airborne radiation being released from the water or soil. The latest study was by West Devon Borough Council (2001) which showed that there were some problems in private water supplies in its area: 116 different supplies were tested and 8 per cent were found to be above the National Radiological Protection Board's advisory level of 1,000 Bq/l.

Many people consider that radon should be included in the list of parameters that contribute to a water supply being considered unfit. WHO have specifically excluded radon from their Guideline Values for radioactivity because it cannot be detected by standard methods. One of the problems is collecting the water and getting it to the laboratory without loosing some of the gas on the way, so special precautions have to be taken. WHO points out that water should be carefully sampled if the gross alpha level is exceeded and this activity is from radon.

The US now has a standard of 15 pCi/l (0.55 Bq/l) for alpha particles but no set standard for radon. In February 2005, the US Environmental Protection Agency is proposing to have a maximum contaminant level in water of 4,000 pCi/l (148 Bq/l) where states have airborne radon programmes and 300 pCi/l (11.1 Bq/l) where they do not (EPA, 1999). The National Radiological Protection Board's action level is 1,000 Bq/l in private water supplies as is the European Union's proposed guideline (NRPB, 2003). It must be remembered however that in small water systems, the actual risk will be low even if the levels are high. A typical small system that treats water to achieve a lifetime risk of 1 in 100 (at least two orders of magnitude greater than the minimum recommended), only one case of disease is avoided every 680 years. A similar level in a system serving 63,000 people would save one life every seven months (Schnare, 1990).

Smoking has been found to increase the risks of cancer from airborne radon and risks can be reduced by a factor of five by not smoking (Crawford-Brown, 1992). There may be an even more complicated relationship between radon and smoking as it has been found that radon levels in houses of non-smokers are 10 per cent higher than in smokers' houses; this may be because smokers open their windows more often to reduce tobacco odours and unwittingly let the radon out (Cohen, 1990). The average concentration of radon in all US public water supplies is 3.9 Bq/l. This produces a lifetime risk of 10^{-6} for smokers and less than 10^{-7} for non-smokers (Mills, 1990).

Correlations between average radon exposure and lung cancer rates show that counties with higher radon levels have lower lung cancer rates. This does not imply that individuals with higher exposures have a lower risk, that is what epidemiologists call an 'ecological fallacy' (Cohen, 1990). The group whose exposure is measured (the whole county) is not the group at risk; it is actually a small subgroup with individually high levels. An average study thus cannot determine whether radon causes lung cancer or not. Wolff has stated (1993) that it will be impossible to determine radon risk through epidemiology alone because there are too many factors to be taken into account. As an example, he points to the correlation between radon levels and wealth. This is because the richer you are, the more likely you are to have a house with a bigger average floor area and this will allow more radon to enter. However, the better off you are, the less likely you will be to die of lung cancer, because of other environmental and personal factors.

I recently stayed in a house in Connecticut in the US that had a borehole supply with a radon problem. The problem only came to light when the owner, who was employed in a nuclear power station, went to work after taking a shower and set off all the alarms as he entered the plant. At first it was thought that he had been contaminated while at work and so he was scrubbed down and cleaned up. When this happened a couple more times, he put two and two together and decided to test his water supply for radioactivity. He was shocked to find high levels of radon in the family's borehole water supply. Because he lived in a rural area where everyone had private water supplies, he kindly told all his neighbours about the problem and explained the dangers. One neighbour said that she was not worried because radon was natural and therefore it would be safe. She was only going to worry about man-made radiation because that was not natural and therefore must be dangerous (Altvater, 2001).

Uranium

UK maximum concentration allowed – no standard set in legislation, but there is a total indicative dose for radioactivity (including gross alpha and gross beta activity) of 0.1 mSv/year.
WHO Guideline Value: 2 µg/l provisional value set for acute chemical effects. No value set for radioactivity.

Uranium (U) is interesting in that it can be a health problem as a mineral, causing damage mainly to the kidneys in excessive amounts and as a radioactive element with carcinogenic properties. Radioactive uranium is an alpha particle emitter and can be found as two radionuclides, ^{234}U and ^{238}U. Both will provide an annual dose of 0.1 mSv/year if drunk in concentrations of 4 Bq/l, according to WHO data. It is found naturally in the earth's crust and can also leach into water from nuclear plants and mine tailings (WHO, 1993a). Small amounts can also be found in coal and phosphate fertilizers.

  The US Environmental Protection Agency has set a maximum contamination limit of 30 µg/l rather than an individual radiological standard. WHO has not set a limit for chemical toxicity of uranium, due to a shortage of available data but has recently set a provisional Guideline Value as more research data have been received. It recommends that individual countries set standards with regard to the guidelines for its radioactive properties. Like radon there is no individual standard in UK legislation for uranium.

18   Treatment options

If a private water supply is found to contain pathogens or the indicators of such contamination, the first consideration should be whether to change the supply. If the supply is easily contaminated or under the direct influence of surface water, it may be safer, cheaper or more effective to abandon it and arrange for the property to be connected to a public supply or to have a new borehole drilled. If a new source or borehole is not feasible, checks should be made to ensure that the supply is adequately protected. Earlier in this book, ways of identifying problems and protecting the source have been outlined, whatever the type of supply. As has been said before, it is much better to stop pollution getting into a supply than to attempt to remove it once it is there. Once protective measures have been fitted, repaired or improved, the water should be re-examined under a variety of meteorological conditions to see if there is anything in it that could still cause harm or unacceptable aesthetic problems. Unless the supply is proven to be totally free from any contamination, consideration should then be given to installing a treatment package. The use of technology to treat drinking-water is not a modern invention. Greek and Sanskrit texts dating back 6,000 years contain guidance on the use of charcoal (activated carbon filters), boiling, straining and exposure to sunlight (ultraviolet irradiation) to improve the aesthetic quality of drinking-water and over 4,000 years ago, the Egyptians used coagulants to reduce the turbidity of water (WHO, 2003). If the quality of the water is permanently satisfactory, treatment should be avoided as it may provide an opportunity for introducing possibly pathogenic material into what would otherwise be a sealed system. Treatment systems nowadays are effective at removing chemical contamination and dealing with physical problems such as low pH and turbidity. For most problems, a suitable form of treatment can be found that is designed to deal with the low flows associated with private water supplies.

Once the water has been improved to an acceptable standard from a health and an aesthetic point of view, some form of disinfection system should be considered. The safety of drinking-water must always rely on more than one protective measure and a 'multiple-barrier approach' should always be taken. It cannot be stressed too strongly or too often that disinfection of all water supplies is very important and relying on good luck and mechanical processes to stop pathogenic organisms getting into drinking-water is imprudent. Ideally, the disinfectant should be chlorine or another chemical that will provide a residual disinfectant to carry on working up to and including the tap. Unfortunately, some people do not like chlorine but there are other effective disinfection methods, such as ultraviolet irradiation, that, if fitted close to the tap, will adequately treat drinking-water without the residual effect.

When considering a treatment system for small water supplies, a reputable installer should be contacted. Deciding which form of treatment and disinfection are the most suitable for any particular water supply is a skilled task and needs a thorough understanding of the problems of the water and the effect the treatment processes will have. Some installers are linked to specific manufacturers and will recommend their products; others will use a variety of different treatment systems dependent on the problems they find with the water. Many installers are members of an association called British Water, which can be contacted at 1 Queen Anne's Gate London SW1H 9BT. They have a list of current treatment installers and contact details are on their website (British Water, 2003). Local authorities may also have lists of local contractors but being public bodies will not be able to officially recommend anyone.

Before addressing treatment methods in detail, the question of whether it is better to have a point-of-entry or point-of-use system needs to be considered. As is implicit in their names, a point-of-entry system treats all the water entering a property and the point-of-use normally treats just one tap, which should be the only one used for drinking. This unit is normally attached to the kitchen tap and is either fitted next to or under the sink. The US Environmental Protection Agency has for some time considered that for small community systems, neither method is better than central treatment because of the lack of control over maintenance and water quality (Clark, 1990). It will however accept point-of-entry as a suitable system for small private supplies because the only people who may be affected by lack of maintenance are the people responsible for the supply. The potential problem with point-of-use devices is that not all the water in the house is treated and exposure to microbiological health risks could result from using other taps. If the water is disinfected or bacteriologically safe, as happens with some well-protected boreholes, there may be some advantage in using a point-of-use system for nitrates or organics removal. This is because there will be no health risk from the water other than by ingestion over long periods. If there are bacteriological problems however, whole-house treatment should be insisted upon wherever possible and will ensure that contaminated water is not ingested while teeth are being brushed or accidentally drunk by babies in the bath. It can be argued that whole-house water treatment is a luxury and that it is wasteful of resources to treat water that is flushed down the toilet. It will however, guarantee microbiologically safe water to everyone in the house at all times, if maintained properly. Because faecal contamination in private water supplies is so widespread, it is easy to disregard other contamination problems. Care should be taken to consider all potential problems and ensure that where health is concerned it is better to err on the side of safety.

There are also questions about the safety of various treatment systems. This has been discussed previously with regard to disinfection by-products but it is pertinent to most systems. Questions have been being asked about the safety of granulated activated carbon (Regunathan *et al.*, 1990), reverse osmosis (Payment *et al.*, 1991), ultraviolet irradiation (Bryant *et al.*, 1992), ion exchange (HOLSCOTEC, 1989), distillation units (Health and Welfare Canada, 1977), ozonation (Rice, 1990) and water conditioners (DoE, 1992). It should be borne in mind that many of the discussions on the safety of treatment systems have very little direct evidence of illness and rely on epidemiological or laboratory experiments to justify the claims. It is important that people responsible for the safety of water in private supplies are not dissuaded from fitting a suitable type of treatment because they have become aware of vague reports concerning potential health problems. Fawell and Miller (1992) have pointed out that it is unlikely that the general public have any better understanding of water quality parameters and their significance today than they ever had, but they have been extensively informed by the media of problems with water, couched in terms of threats to health.

A final point that needs consideration is the use of health claims to persuade people to install particular treatment systems. This problem first came to light in the US and Canada in the early 1980s. Producers of treatment devices claimed that mains water was unsafe and that their system would provide health-giving properties to the water. Legislation has now been introduced in both countries regarding false advertising claims. In the UK, any false claims made about water treatment filters need to be investigated by the Trading Standards Department of the local authority. There are two issues of concern: the first is one of claims that tap water is intrinsically unsafe. Information about water filters has claimed that 'contamination is rampant' and that 'lead, copper, pesticides, fertilizers and bacteria are all present in our water supplies'. One company even states that tap water 'can cause kidney failure' and 'chlorinated water destroys much of the intestinal flora'. They have also turned their attention to the dangers from showers and state – 'Surprisingly, the water contaminants a person absorbs in one shower is far greater than if they were drinking the same water. This is because hot water opens the pores so chlorine, radon and other contaminants in water are absorbed through the skin...This is cause for concern because it is well established scientifically and medically that chlorine and its by "products" Trihalo methane (*sic*) are carcinogenic' (Paragon, 2003). It is surprising that the health benefits of unchlorinated water and not washing are not more widely known. Another advertisement more subtlely affirmed that the company's system provided 'water the way it used to be' (National

Safety Associates, 1991); presumably this does not mean teeming with *Vibrio cholerae*. Pressure groups are also responsible for some of this problem. It is startling that claims of 'poison' in public water supplies have become commonplace, while in excess of 25,000 people worldwide die each day from a lack of an adequate and safe supply of water (Fawell and Miller, 1992). The second issue of concern is claims that water filters will actually improve health. These advertisements are more muted, but many contain hidden messages to this effect, such as claiming to 'give peace of mind knowing you are doing something for your family'. It is important that such claims are provable.

Disinfection

There are several ways to disinfect drinking-water supplies by the removal or inactivation of microbiological organisms. They are listed here in alphabetical order. Their effectiveness is usually measured in log (logarithmic) reductions. A simple explanation of log reduction is that numbers are reduced by a factor of ten. If there were 1 million (10^6) bacteria per litre in a sample of water, a 1-log reduction would divide that number by 10 to give 100,000. A 2-log reduction would reduce the number by a factor of 100, which would be 10,000. A 7-log reduction would therefore mean that the 1 million bacteria would be reduced to 0.1 per litre. Another way of describing this is to translate it as a percentage decrease, for example, 2 logs is equal to 99 per cent, 3 logs – 99.9 per cent, 4 logs – 99.99 per cent and so on.

It is also important to realize that microorganisms are often found in huge numbers and unless there is a substantial reduction, the treatment process may make little difference to the chance of infection. For instance, a 3-log reduction, killing 99.9 per cent of all household germs may be impressive, but if there were 10^9 in a litre of raw, surface water to start with (remember that infected young animals are known to excrete up to 10^{10} *Cryptosporidium* oocysts a day) this means that there are still 10^6 left, which is a million, more than enough to upset a tummy or two.

Chlorination

Chlorination is the addition of chlorine (Cl) to water and this simple act has saved millions of lives by destroying nearly all the pathogenic organisms found in drinking-water. *Time Magazine*, when celebrating the Millennium, picked the introduction of chlorination and water treatment as the forty-seventh most important event of the last thousand years. It has been used as the disinfectant of choice in the UK since it was first widely adopted in 1936 following a major outbreak of waterborne typhoid in the Croydon area. Chlorine is a powerful oxidant that inactivates pathogenic bacteria and viruses. *Salmonella*, *Campylobacter* and *E. coli* O157 for example, are all easily destroyed by chlorine. Chlorine can be added to water in a variety of ways: as a solid (calcium hypochlorite tablets or powder), as a liquid (sodium hypochlorite solution) or as chlorine gas. It can also be electrically generated on-site from common salt. Because chlorination has been the most important factor in reducing the incidence of waterborne disease, it would appear to be the natural choice where there is any risk that a small water supply may be bacteriologically contaminated.

One of the most important aspects of chlorination is that the chlorine carries on working after being added to the water; this is known as its residual effect. This is because as well as inactivating any post-treatment contamination, it reduces biofilm and microbiological aftergrowth in the distribution system. The amount of chlorine that needs to be added will depend on the level of contamination in the water. In water, chlorine quickly forms hydrochloric acid (HCl) and hypochlorous acid (HOCl) – this takes less than a second. The hypochlorous acid then breaks down to form a hypochlorite ion (OCl^-). The water will contain a mixture of the ion and the acids, depending on its pH. The more acidic the water, the more efficient the chlorine is at disinfecting. The amount of chlorine needed to deal with the contamination is known as the chlorine demand of the water. Chlorine reacts with most things and the demand can be from microorganisms or organic and inorganic chemicals. The organic chemicals absorb chlorine and the inorganic chemicals, such as iron or manganese, use chlorine to form chlorides.

Where treatment is fitted, it is recommended that 0.5 mg/l of chlorine is added to the water and that a contact time of 30 minutes be allowed for the chlorine to disinfect all the susceptible pathogens. The amount of chlorine and the time it is in contact with the water is sometimes known as the 'Ct' value. The figure is arrived at by multiplying the chlorine concentration in mg/l by the time in minutes, thus the units are mg/l·min, and the larger the Ct value the more effective will be the disinfection.

It is better to remove as much of the contamination as possible before chlorinating water because it not only reduces the chlorine demand but also reduces fluctuations in water quality. This is important because it will ensure that a regular dose of chlorine per litre will satisfactorily deal with the pathogens and not be used up on other material. To find out what the chlorine demand is involves a sample of water having chlorine added to it in measured small amounts and the water being regularly tested for available chlorine. As it is added, the chlorine will be used up until all the possible reactions have taken place. This point is known as the breakpoint and the process is known as breakpoint chlorination. Once this point is passed, any surplus chlorine is available to disinfect bacteria or viruses found within the distribution system and at the tap. This extra chlorine is known as the free chlorine residual. Another way of measuring the necessary dosing levels is to test the final tap water for free residual chlorine by trial and error whilst the treatment unit is being installed and adjusted. The amount of chlorine that should theoretically be in the water when it is drunk is supposed to be nil, that is, it has all been used up just prior to being drunk. In practice, however,

0.1–0.2 mg/l is a reasonable amount at the tap. Because private water supplies are subject to variations in water quality, the dosage calculations need to take account of the water when it is most contaminated. The amount of residual chlorine has to be carefully controlled because chlorine imparts a taste to the water that should be minimized. One of the reasons why people say they drink bottled water or have an activated carbon filter unit next to their tap is because of perceived problems associated with the taste and odour of chlorine. Some people may also be allergic to it. Levels of 0.6 mg/l and above of chlorine will begin to be noticed by consumers.

Adding chlorine to water containing ammonia is slightly more complicated. The nitrogen in the ammonia reacts with the chlorine to form chloramines. Three chloramines can be produced and are known as monochloramine (NH_2Cl), dichloramine ($NHCl_2$) and unsurprisingly, trichloramine (NCl_3). The chloramines have some disinfecting qualities themselves. Where ammonia is present, as chlorine is added the chloramines build up until a point is reached where the major one present, monochloramine, is oxidized and nitrogen gas is produced. This process reduces the residual chlorine, but as more chlorine is added, the residual starts to build up again (Gray, 1994). Some water treatment plants use chloramines as a disinfectant. They are less reactive than chlorine and therefore have a longer residual effect. Other treatment plants add ammonia to the water so that the process itself produces the chloramines.

Where odours and tastes are a problem in raw water, excess amounts of chlorine can be added to oxidize the organic contaminants and break them down into less offensive chemicals. The chlorine is then taken out by the addition of sulphur dioxide. The process is called super-chlorination and a similarly strong oxidizing effect is produced by ozonation. It is unlikely that super-chlorination would be encountered in small water supplies. High level application may however be used when the supply is first being used or problems in the water such as iron bacteria need shock treatment.

Many environmental health officers and other professionals have concerns about the use of chlorine for private water supplies. As well as the health and safety aspects, they consider that ensuring accurate dosing and regularly testing for free residual chlorine is a job for specialists. Despite the advantages of residual chlorine, many people feel that systems such as ultraviolet disinfection, which can be installed and serviced periodically by trained personnel, are far more likely to prove effective. In the US, where chlorine is much more commonly used, this worry does not seem to be justified, particularly nowadays when chlorination systems can use electrolytic conversion.

An interesting method of chlorination that may sometimes be found (but unfortunately is no longer available to buy) uses an automatic system where a ball-valve control device tips an accurately measured amount of chlorine bleach into a tank when the water level reaches a set point. This tank then empties itself into a larger storage container using siphonic action as the water level continues to rise. The two-tank system ensures good mixing of the chlorine and does not rely on electricity or complicated electronics.

In private water supplies, as water quality varies, chlorine demand changes plus people may adjust application rates downwards to save money or upwards 'just to make sure'. It is unfortunate that a foolproof system is not in universal use that provides the undoubted benefits of chlorination, without the attendant dangers or reliance on untrained personnel.

Recently, chlorination has been the subject of two concerns. The first is that, at normal dosing levels, it cannot completely deactivate *Cryptosporidium*. Chlorine has been used for many years and it is obviously unfortunate, but probably inevitable, that an organism has eventually come along that is not eradicated by it. It is interesting to note however, that most tests for the efficacy of a disinfectant against *Cryptosporidium* are done in laboratories using fresh oocysts. Where oocysts have been in the environment for some time they have been found to be more susceptible to chlorine, as the outer protective wall has become damaged or aged. In addition, with two chlorine applications, the dual action has been found to be more effective than a single dose. In most cases, chlorine should be recommended because of its residual effect however, where there is a particular risk from *Cryptosporidium*, additional treatment such as ultraviolet irradiation or ozonation should be considered. Source protection to prevent contamination will nevertheless always remain the best and most effective option and wherever possible owners of small supplies should try to ensure that pathogenic protozoa do not enter the water. The second potential problem with chlorine is from the disinfection by-products formed in water with a high organic carbon content. The mechanisms and implications for private water supplies are covered in the section on disinfection by-products in Chapter 16.

Other chemicals with similar oxidizing properties can be used as alternatives to chlorine, particularly the other halogens – bromine and iodine. Iodine has been found to cause problems with thyroid abnormalities when used in small ceramic purification units that are used for drinking-water but only when used over a long time period (Khan *et al.*, 1998). Because chlorine is used much more frequently than these alternatives, potential problems with its use are better known, but that does not mean that the alternatives are necessarily risk-free.

So many lives have been and continue to be saved by chlorination that it is vital it continues to be used in water treatment. Much more illness will result from stopping the use of chlorine than will be caused by the potential carcinogens in disinfectant by-products. If there ever proves to be a real risk from chlorination by-products at the levels found in private water supplies, they can of course be removed by point-of-use filters.

Ozonation

Ozone (O_3) has been known for over one hundred and thirty years. It is a colourless, acrid gas that is a very strong oxidizing agent – it is over one and a half times more powerful than chlorine. Because it is so reactive it cannot be transported any distance and therefore has to be generated on site.

This involves passing very dry air or oxygen between two high frequency, high voltage electrodes. The voltage causes some of the O_2 molecule to split and each of these oxygen atoms then combines with other O_2 molecules to form ozone. The ozone 'wants' to revert to its original state and so it tries to loose the extra oxygen atom at the earliest opportunity. This gives it its strong oxidizing potential. As with chlorine, ozone needs a contact time to work. At low concentrations it has taken nearly nine minutes to inactivate some *Pseudomonas* species (Lezcano *et al.*, 1999), but the reaction is usually much quicker.

Ozone is often installed to remove difficult organisms such as *Cryptosporidium* that are resistant to chlorination. The ozone is bubbled through the water being treated from diffusers at the bottom of a contact tank. After treatment it has to be vented off and any remaining ozone left in the waste stream turned back into oxygen. This is achieved by heating the gas, passing it over a special catalyst or through activated carbon. Most ozonation plant has been installed in large-scale commercial or industrial settings and is therefore unlikely to be found in small private water supplies. Large community systems however, may choose to use it where they have particular problems that are not adequately dealt with by other treatment systems.

Ozone can be used for other types of water treatment and will oxidize iron and manganese in solution, so that they form a solid precipitate that can be physically filtered out of the water. Additionally it will remove colour caused by humic acid in soils and taste or odour problems from anaerobic decomposition of organic material. The process is also particularly useful as a final polish for treating water that has a problem with trace organic chemicals, particularly pesticides or toxins from cyanobacteria. Its strong oxidizing potential will tear apart most organic chemicals and break them down to their basic constituents. However, this has been known to raise the organic carbon content in the water and may lead to increased biofilm formation in the distribution system (WRc-NSF Ltd, 2002). As concerns about pesticides, *Cryptosporidium* and trihalomethanes grow, there will be pressure on ozonation equipment manufacturers to scale down their product for use in larger private water supplies.

Unlike chlorine, ozone does not have any residual effect and so some manufacturers recommend that after ozonation, chlorine is added to drinking-water before it leaves the treatment plant (Ozonia Ltd, 2000). Because of the effects of the ozone however, a much smaller amount of chlorine will be needed.

Ultraviolet irradiation

Ultraviolet (UV) units consist of a long metal canister with a single lamp or series of mercury vapour lamps, which look like fluorescent tubes, running along its center. Water passes over and around the lamps and any microorganisms in it absorb the radiation they emit. At wavelengths of around 254 nm, the irradiation will denature the double helix of chromosomal DNA (Denny *et al.*, 1991). What actually happens is that microorganisms absorb the ultraviolet radiation until the adenine and thymine molecules within the DNA are fused and this prevents the organism from reproducing. This property was discovered in 1847 and water purification units using ultraviolet have been commercially available since 1901 (Foust, 1990). Properly fitted they can reduce high levels of pathogenic bacteria to a virtual zero. Ultraviolet filters are normally added at the end of a treatment system to provide final disinfection. They are electrically powered and are reasonably cheap and easy to install. Because of this, ultraviolet is a very popular means of treating private water supplies.

One of the problems that has been associated with trying to get people on small water supplies to install disinfection is that they say they do not like the taste and smell of chlorine and are worried about the potential ill health from disinfectant by-products. Ultraviolet irradiation will not impart any taste or smell to the water and does not usually cause disinfection by-products to be formed. There are questions being asked about the possibility of ultraviolet irradiation encouraging trihalomethane production in chlorinated supplies but presumably, it cannot be a problem where there is no chlorine available.

Ultraviolet disinfection systems consist of two distinct types. Low-pressure filters, with quartz or Plexiglas mercury vapour lamps, deliver a single wavelength output (245 nm) and are known as monochromatic units. Medium-pressure filters deliver a range of outputs between 240 and 310 nm and are called polychromatic units. These larger filters swirl the water around a number of tubes to ensure that the water has been in close contact with them for a sufficient length of time. Light travelling through water is subject to the inverse-square rule, that is, the power of the light on a microbiological pathogen is proportional to the reciprocal of the square of the distance from the source. Therefore it is important for the microorganisms to pass as closely to the lamp surface as possible. Like other forms of disinfection, a contact time is necessary for the filters to work effectively. Dosage is measured by the intensity of the ultraviolet light, multiplied by the exposure time. The units are milliwatts seconds per square centimetre $(mW \cdot s/cm^2)$ and, as a joule is a watt second, the units are sometimes known as millijoules per centimetre squared (mJ/cm^2).

If the water is turbid and there are particles in the water or there is excessive contamination or colour, the microorganisms in the water can be physically protected from the ultraviolet rays. Suspended solids will also tend to scatter or absorb the light and make the unit less effective. Another problem is that bacteria sometimes clump together and provide additional protection for each other. For the units to be effective, turbidity and colour must therefore be low, but they are not detrimentally affected if the turbidity and colour are within the maximum allowable concentrations in UK regulations (Denny *et al.*, 1991). It is important to test the water prior to fitting ultraviolet treatment to find out what is in it and to install appropriate pre-filters to reduce particulate contaminants (Figure 8). When testing the water, the same rules apply as to normal sampling to ascertain fitness. It

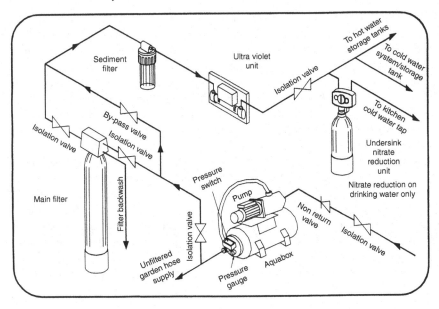

Figure 8 An arrangement for a treatment system incorporating ultraviolet irradiation and nitrate reduction.

should take place several times, particularly after heavy rainfall, so that any increased chemical and microbiological contamination is taken into account.

A common problem in areas where there are high concentrations of iron or manganese in the water is that they can coat the tube with a brown deposit and reduce the amount of light it will emit (Aquafine, 1982). This has caused several outbreaks of gastrointestinal disease. During a recent survey, 12 per cent of treated private water supplies were found to contain viable faecal coliforms (Rutter *et al.*, 2000) and one of the reasons for this will be iron build-up. If iron in particular is found to be a potential problem it should either be removed from the water at the pre-disinfection stage or the lamp should be regularly checked. Where it is necessary, a reputable installer will recommend iron and manganese removal as part of a complete package.

Ultraviolet treatment of water does not provide residual disinfection and microorganisms in the distribution system will not be dealt with. In some countries, for instance Holland, it is not considered necessary to provide residual disinfection if the water has been adequately treated (Kool, 1990). Nevertheless, in the UK and the US, the presence of residual disinfectant remains important. Some ultraviolet units have been found to increase the amount of organic carbon in the water and because of the lack of residual effect, this may encourage biofilm production (Shaw *et al.*, 2000). Small water supplies do not normally have extensive pipework between treatment system and the tap and so aftergrowth may not be a problem. Even so, to reduce this possibility, ultraviolet filtration units should be fitted as close to the final tap as possible.

Some viruses are more resistant to chlorination than indicator bacteria and it has been suggested that they are also more resistant to ultraviolet light (Karanis *et al.*, 1992). However, tests carried out on the ultraviolet energy needed to destroy viruses have shown that it is within the normal operating range of the filters and is similar to the $20\,mW \cdot s/cm^2$ needed for a 6-log reduction of *E. coli*. Correctly fitted ultraviolet filters operate within wide safety margins and viruses should not therefore pose a particular problem (Combs and McGuire, 1988; Denny *et al.*, 1991). Early research on ultraviolet inactivation of *Cryptosporidium* led people to believe that it was not particularly effective at normal operating levels but recent research is challenging this. Medium pressure ultraviolet lamps at $3\,mW \cdot s/cm^2$ have been found to inactivate oocysts by 3.4 logs and low-pressure units at the same rating by 3 logs (Clancy *et al.*, 2000). In the UK, private water supply-sized units rated at $40\,mW \cdot s/cm^2$ have been shown to produce a 4.7-log reduction of *Cryptosporidium* in clean raw water (Murphy, 2003). Because ultraviolet treatment does not actually remove the oocysts from the water, when the systems are being evaluated they have to be subject to a viability analysis to ensure that the water is safe to drink. Different countries use different method to test for viability and this may have been the reason why early studies indicated that ultraviolet was less effective against *Cryptosporidium*. One newly developed system, at present only available for large-scale treatment, uses ultraviolet in concert with a fine metal grid that temporarily holds the oocysts in front of the lamp. This ensures retention time is lengthened to ensure inactivation.

One problem with some ultraviolet systems is that if the lamp breaks, the water from the unit will not be treated. It is possible however, to fit an alarm that alerts the users to this failure. Efficiency will also diminish with time and the lamps have to be periodically replaced. The frequency is determined by the manufacturers and typically is once a year or when the lamps are operating at 70 per cent of their original power. This replacement frequency allows for a safety factor to be incorporated, as manufacturers expect most lamps to remain perfectly satisfactory for at least two years. The unit should also have a removable cover or small window to allow periodic inspection to check that the lamp is still operational and not covered by an iron coating.

Contaminant removal

As well as disinfection, water may need treating to remove unwanted contaminants that could cause harm or aesthetic problems. There is a range of devices that can provide this but the treatment they provide is often specific to certain problems, so no one system will remove all the problems that are encountered.

Activated carbon filters

Activated carbon filters remove contaminants from water by adsorption and to a lesser extent entrapment (Figure 9). The removal is therefore physical rather than chemical. Activated carbon has a large surface area and under a microscope looks like a lump of coke with holes all over the surface. The chemicals stick to this large surface area because of electrostatic attraction. The carbon in the filters is usually derived from wood or bone charcoal and can be supplied in a variety of sizes. Activated carbon units have been found to remove suspended solids, lead in both soluble and insoluble forms, turbidity, colour, taste, odours, chlorine, organic contaminants, including trihalomethanes, petroleum products such as benzene, volatile organic carbons (VOCs), organic solvents and pesticides (Jackson, 1992; Kuennen *et al.*, 1992). Activated carbon is not a generic commodity however. The source of the carbon and how it is prepared provides a range of capacities and selectivities.

Aside from ancient designs (see Treatment options discussed earlier) the modern type of filter was originally designed for public supplies but they have now been introduced into private supplies because of their removal efficiencies. Sometimes they have been fitted without strict adherence to manufacturers' instructions as some companies state that their product should not be used for water that may contain microorganisms (WRc-NSF, 2002). This will apply to many private supplies, so disinfection is therefore usually necessary when an activated carbon filter is installed. If chlorine disinfection is fitted, it should come after the activated carbon unit because of their ability to remove chlorine. Activated carbon for water treatment comes in two forms – powdered activated carbon (PAC) and granulated activated carbon (GAC). PAC is usually only used for public water supply systems and is added to the water during the treatment process. After it has adsorbed the contaminants, it is physically removed from the water by

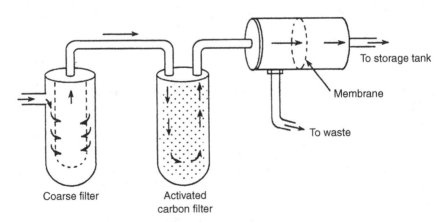

Figure 9 Activated carbon filter.

filtration, taking the pollutants with it. GAC systems normally consist of a cartridge packed with the media within a sealed filter unit. Water goes in at one end, is passed over the granules, loses the contaminants to the activated carbon on the way and then exits via a small pipe or tap for use or storage.

GAC is normally found as a point-of-use filter and the cartridges can often be seen attached to the tap in a kitchen or fixed in-line underneath the sink. These cartridges are also the ones normally used by householders on a mains supply. They are fitted if there is a chlorine or odour or taste problem and if people want nasty chemicals like pesticides removed from their water (if they are there, which they probably are not, but it is a good advertising ploy). GAC is also the treatment system used in jug filters. The process is the same, with the water being poured over a GAC medium. There is no reason why GAC could not be used as a final polishing element to a point-of-entry or larger treatment system but it may be pointless to remove organic chemicals, etc. other than at the kitchen tap.

For many small water systems, pre-treatment may need to be installed so that it removes as much contamination as possible before the GAC filter. This will ensure that the cartridges last longer and, because they have a more uniform quality of water passing over them, will work more efficiently. There is competition for available sites on the carbon and more strongly attracted molecules will displace the weakly retained ones (Van Dyke and Kuennen, 1990). Consequently when a GAC device is fitted, it is important that aesthetic parameters are not strongly adsorbed to the detriment of the health ones.

GAC units are relatively cheap and because of this, owners of private water supplies may fit one when told they need some form of treatment. It is important for them to realize that not only are most filters not guaranteed to remove all pathogens but also that the filters have a limited life. Manufacturers may assume that their filters are being fitted to public supplies; therefore the length of time they expect the filter to work is calculated using water of reasonable quality. If they are fitted to a small water system, with varying quality or uniformly poor quality, the available sites may be used up very quickly. Some people have been known to fit a unit for much longer than the manufacturer's recommended time limit and the reappearance of an odour or taste problem is often the first sign of deterioration of the filter. If the reason for installing a GAC filter was not due to organoleptic problems, the failure of the unit will not be so readily apparent. When a GAC filter is first put on a private system, it should be thoroughly tested to see whether (a) the filtration system can cope with heavy loads following rain; and (b) how long it lasts before needing changing. This is true for both GAC-only systems and where a full treatment system has been installed. When the filter has stopped working, the carbon needs reactivating. This is done by carefully heating the carbon to drive off all the organic chemicals. This has to be carried out at a specialized factory, normally the filter manufacturer's. Obviously with some smaller cartridges and jug filters they are just thrown away and a new one fitted.

Some GAC filters include silver as it is thought to be bacteriostatic, that is, does not kill bacteria but inhibits their growth. Silver will not inactivate coliforms completely unless there is extended contact times but some individual studies on silver have occasionally reported significant reductions in bacteria, viruses and protozoa (Regunathan *et al.*, 1990). Its use has also been granted a registration by the US Environmental Protection Agency because of its potential to limit the growth of bacteria. There is some confusion about its efficacy however, as data from some studies have shown no significant difference between silver-containing and non-silver devices (Bell, 1991).

The ability of GAC to adsorb organic material ensures a perfect setting for adventitious bacteria to grow on. These are heterotrophic organisms and they obtain their carbon from the organic material trapped in the filter. Problems concerning microbial growth on GAC filters will increase if the manufacturer's recommended cartridge replacement periods are not adhered to (DoE, 1992). Harold *et al.* (1990) reported that water from GAC point-of-use devices, used to remove lead from water in the Glasgow area, quickly began to develop an unusual taste. When the filters were examined they were discovered to have a thin coating of green algae on the surface of the granules. Without some form of bacteriostatic action, filters may eventually produce more bacteria in the treated water than are found in the influent water. The build-up of the heterotrophic bacteria also requires that any first flush of water be run to waste after standing overnight (Reasoner, 1990).

There is some speculation about whether the bacteria that colonize GAC filters are harmful. A large epidemiological study carried out by Yale University showed no significant differences in illness rates between healthy people drinking water from a GAC unit and a control group (Dufour, 1990). When considering heterotrophic bacteria in water filters, research has shown the following:

- Conventional experiments underestimate the numbers of bacteria growing on activated carbon. Heterotrophic bacteria and coliforms will colonize GAC filters and be released into the water on small particles of carbon, even in well-maintained filters (McFeters *et al.*, 1990).
- The heterotrophic bacteria may suppress the detection of coliforms (Health and Welfare Canada, 1977).
- Bacteria attached to carbon particles are much more resistant to chlorine than free bacteria (McFeters *et al.*, 1990).
- Pathogenic bacterial colonization of GAC will depend on the other bacteria present (McFeters *et al.*, 1990).

As with most types of individual treatment or disinfection systems, GAC filters are not guaranteed to be the only answer and many of the filters are not designed for private supplies. However, they can work well, particularly if they are regularly maintained and will remove chemicals that other systems have problems with.

Ion exchange units

When salts dissolve in water, the two parts separate to form charged ions. Calcium nitrate ($CaNO_3$) for example, separates into a positively charged calcium ion (Ca^+) and a negatively charged nitrate ion (NO_3^-). Positively charged ions are called cations and negatively charged ones are anions. Ion exchange units are filled with a media, called a resin and, as the water passes through the unit, 'sacrificial' ions on the resin will swap places with similarly charged, but different ions in the water. Water containing excessive nitrates, for example, will lose nitrate ions but will be replaced with chloride ions. The *modus operandi* of ion exchange units is therefore, to replace potentially dangerous or aesthetically unwanted ions with others that are safer or more acceptable. The resin can exchange either cations or anions, but within a single unit only one type of ion is removed. Naturally occurring minerals were originally used in the units but synthetic resins are now more common. When fitting an ion exchange unit, the water needs to be tested when it is likely to be most polluted to ensure that the unit is sufficiently robust to cope with peak demand levels. Ion exchange will not remove pathogens but can be used on its own as a point-of-use device to remove a single problem chemical or small group of chemicals. When used on small systems with variable microbiological problems, ion exchange is normally only part of a complete treatment and disinfection package.

Cation exchange units are usually used as water softeners. This involves calcium or magnesium ions being replaced by sodium ions. It should be remembered that overly softened water should not normally be used for drinking, due to the health-protecting properties of hard water. People on a low-sodium diet should also be aware of the sodium ions added to the water. As well as softening water, cation exchange units can be used to remove a number of undesirable chemicals including barium, beryllium, cadmium, chromium, lead, mercury, radium and thallium. As with activated carbon, cation exchange resins are incapable of distinguishing between beneficial and undesirable substances. For example, zinc and copper are both essential elements and are as equally likely to be removed by the process as the highly toxic elements cadmium and mercury. If calcium and magnesium are present at high concentrations they will dominate the resin exchange capacity and this will restrict the ability of the resin to extract more harmful substances. Because cationic treatment units remove hardness, they need to provide enough water to supply washing machines, boilers, dishwashers, etc. They are therefore normally larger than anionic units.

Anion exchange is usually used to remove nitrates. It can also be a treatment option for a mercury compound ($HgCl_3^-$), nitrites, some forms of arsenic and sulphates. As the problem of nitrates continue to rise, anion exchange units on small water systems may become more common. Nitrates are fairly difficult to remove by other treatment methods on a small scale; they are stable, highly soluble in water and there is no chemical compound that precipitates nitrates or has a specific affinity for them. The resins used in anion exchange display selective preferences for different ions and normally they prefer sulphates to nitrates. Units treating water containing high concentrations of both can become loaded with sulphate ions and therefore stop removing nitrates before the manufacture's recommended regeneration time is reached. To overcome this problem, nitrate selective resins have been developed. Unfortunately they have a lower overall capacity and are more difficult to regenerate (Solt, 1991). There is also some concern that microbial growth will occur on the resin and because of the high nitrate levels on some of the filters, there are additional worries that these bacteria will reduce the nitrates to nitrites. It has also been pointed out that because the nitrates are usually replaced by chloride, areas of naturally high chloride may find that a salty taste is given to the water.

When the units have lost most of their sacrificial ions they have to be regenerated or backwashed. Many units, water softeners included, regenerate with a strong solution of common salt (NaCl). The captured ions are driven off to waste and replaced by either sodium ions (cationic) or chloride ions (anionic). Regeneration can be automated and the units consequently need a minimum of skilled attention. For private supplies, units can be purchased with replaceable cartridges and this removes the necessity for regeneration. Regeneration with common salt is to some extent disinfecting and as long as it is carried out every five days or so, microbial growth should not be a problem (DoE, 1992).

Membrane filtration units

Membrane filtration uses a thin film or semi-permeable membrane to remove unwanted particles, large molecules and microbiota from water. There is an argument that membrane filtration should have been included in the section on disinfection, as some membranes remove pathogenic microorganisms. I have included it here nevertheless, as not all of them will do this. The term covers a variety of different processes such as ultrafiltration, nanofiltration and reverse osmosis. Osmosis is the process where water moves from a less concentrated to a more concentrated solution through a semi-permeable membrane and has been known about since 1748. Reverse osmosis is the process where a pressure is applied so that water passes from the more concentrated solution. The membranes are made of a variety of materials particularly thin film polyamides and cellulose acetate or triacetate and are about the thickness of a human hair. Ceramic membranes are nowadays also used in a wide range of pore sizes (usually ultrafiltration and microfiltration) and can be found in the chemical, pharmaceutical and food industrial sector as well as wastewater treatment. Where hostile environments are encountered normal membranes are not viable and ceramic filters may be the only viable option, despite being more costly (PCI-Memtech, 2003).

Electro-dialysis, used for kidney patients, is also a membrane filtration process. Water is passed through a tube that has two electrodes attached to it. Charged ions dissolved in the water are caused to migrate across the membrane by the applied electric field. This is normally a highly sophisticated operation as it is vital that the water used for dialysis is of the very purest quality.

The various filtration types are differentiated by the size of particle they allow through the membrane. Reverse osmosis filters have the finest pore sizes and will remove dissolved chemicals with molecules above 10^{-10} m, which is not much bigger than the size of a hydrogen molecule. In fact, each pore constantly changes size from 0 to 5×10^{-10} m as the molecules in the membrane vibrate and one molecule at a time diffuses through holes in the structure of the membrane material (Gray, 1994). Nanofiltration will remove molecules above 10^{-9} m. Next in size is ultrafiltration; this is used to stop larger molecules and particles above 10^{-7} m. Finally, microfiltration stops particles bigger than 10^{-6} m. Other filters, known as microstrainers, have a pore size of 25 or 35×10^{-6} m. These are not used as a treatment process in themselves but as an initial filter to clean debris from raw water supplies.

Membrane filtration is described more accurately as a liquid separation technology. For microfiltration and microstrainers, all the water is usually forced through the membrane under pressure but below 10^{-6} m the process is separation rather than filtration. In reverse osmosis and nanofiltration, for example, the water runs alongside the filter membrane with some of the water passing through, and about 80–90 per cent running to waste. This process is normally self-cleaning, as the unfiltered water washes away any build-up of detritus on the membrane, but to work efficiently the water has to be reasonably clean. Pre-filters are therefore often added for the treatment of small supplies. The most common type of membrane filter found in private water supplies is reverse osmosis. The filters are capable of removing almost every contaminant of water both chemical and microbiological. The removal efficiency is also known as the rejection rate and for individual chemicals will be between 85 and 99 per cent of the amount found in the water. The slight electrostatic charge of water molecules attracts ions of most of the inorganic salts dissolved in water. These clump together and the particle sizes become greater than the filter pore size. Non-ionic organic molecules will not form these clumps and are not filtered as efficiently, although removal efficiencies of 85–90 per cent have been claimed by some manufacturers (Gray, 1994). Application of particular membranes can also result in considerable selectivity, for example for nitrates, which are otherwise quite difficult to remove (Burke, 1991). Recent microbiological testing has shown that reverse osmosis is very effective at removing protozoa (4-log removal), viruses (6.5-log removal) and bacteria (7-log removal). Reverse osmosis or other fine-scale filtration can therefore efficiently remove most of the contaminants of a private water supply but because power is needed for the units, electricity must be available.

Problems associated with reverse osmosis:

- If the membrane becomes worn out or damaged, contamination may pass through into the drinking-water side of the unit. In some instances, people will not be aware that the membrane has failed and may inadvertently drink raw water. To counteract this possibility, many units are fitted with a switch that is activated when the pressure of the influent water drops. This sets off an alarm that lets the operator know there is a leak.

- Because of the nature of the reverse osmosis process there is a large reduction in water pressure and a lot of water is run to waste. This can be a problem when there are droughts or water shortages. However, where water is not restricted this is less of a problem and the water can be stored in a pressure vessel to replace head-loss.

- Some years ago, variable quality filter membranes tended to be dependent on a narrow band of temperatures and pH in order for them to work properly. If they were operated outside these ranges they did not work as efficiently (Health and Welfare Canada, 1977).

- Cellulose acetate membranes can be subject to biofilm build up with adventitious bacteria colonizing the filter itself. This reduces efficiency and the bacterial build-up can slough off and enter the distribution system. For this reason, some people consider that they should not be recommended for immuno-compromised people who use a private supply, as usually there is no disinfectant added to the water (Bray, 1990; DoE, 1992). Even the thin film polyamides that are less prone to bacterial growth require carbon pre-filters to remove the organic contamination the bacteria feed on (Bray, 1990).

- Because of their highly efficient nature, reverse osmosis units can remove beneficial chemicals from the water. In addition, many people do not like the taste of this highly purified water, as it can be somewhat insipid. To counter these problems, the treated water is sometimes re-mixed with some of the raw water. This will keep any unwanted contaminants within acceptable limits, without causing unwanted problems with loss of taste or essential elements. This is of course unacceptable where microbiological problems occur.

- The minimum standards for alkalinity and hardness may not be achieved using reverse osmosis (DoE, 1992). For health reasons, many countries have minimum standards for hardness when treatment systems artificially soften the water. Removing hardness and alkalinity will also make the water more aggressive and can dissolve lead and copper from distribution networks.

The use of membrane technology can be a very efficient form of water treatment and it may be the only practical means of rendering potable a small water supply containing an excess of dissolved inorganic constituents.

Other treatment systems

Activated aluminium

This system is not often used for private water supplies in the UK, but consists of aluminium ore or aluminium hydroxide that has been treated, so it becomes porous and highly adsorbent. The media is kept in a cartridge, like activated carbon and similarly, when it becomes used up, needs regenerating or the cartridge replacing. Activated aluminium will remove arsenic and fluorides.

Aeration

Where groundwaters have become contaminated with volatile organic chemicals after accidents or leakage problems, they can sometimes be removed or 'stripped' by mixing the water with air. This is normally done in aeration towers with the water passing downwards over an inert media whilst compressed air is pumped upwards and ventilated at the top. The aim is to provide as big a surface area as possible between the air and the water, so that the volatile chemicals can be efficiently transferred. This type of treatment can be effective (DWI, 1987) if properly set up. Other methods of aeration include different ways of increasing the surface contact area, such as bubble diffusion in basins and spray technology.

Distillation

Distillation simply means boiling water and then collecting the distillate. The steam is passed through a coil that is cooled (usually by passing through a water bath) and the clean condensate removed. This system produces very pure drinking-water that has had all its pathogens and nearly all its chemical contaminants removed (especially inorganic salts and particulate matter). Distillation does not remove all organic chemicals and therefore it is sometimes used in conjunction with activated carbon. It should not be used for radon or benzene removal however, as harmful vapours are produced by the process. As with reverse osmosis, distillation will remove beneficial chemicals as well as dangerous ones.

Compared with the other, more usual treatment systems, distillation uses a lot of energy per litre of drinking-water produced. It is most often used where the only available source of water is seawater and therefore any other drinking-water will be expensive. Where clean water is a rare commodity, distillation may be considered an option to produce a small amount of water for drinking, personal hygiene and cooking purposes. Any other water can be treated to a lesser standard to save money – if it is only going to be flushed down the toilet or used in washing machines, it does not need to be absolutely pure. As water shortages become more acute and distillation technology improves by using alternative energy sources, it may be possible that this form of water treatment becomes more common. At the moment, however, it is rarely seen away from small islands and cities in desert coastal regions.

Sand filters

For larger small water supplies, if that is not oxymoronic, treatment systems are similar to those normally used for public water supplies. Public or mains water treatment normally consists of physical filtration by roughing filters or metal screens, followed by flocculation, clarification, filtration by sand filters, chlorination and storage. This section deals with sand filtration. Other processes in public water treatment can include super-chlorination, microfiltration, ozonation, activated carbon, water softening, adding orthophosphates to reduce plumbo-solvency, the use of alternatives to chlorine, fluoridation and pH balance. This book is not designed to cover large public water treatment systems in depth, as there are many books that already deal with this subject.

As with other systems, a larger private or community water treatment plant has to be designed to fit the specific water being treated and should cope with peak contamination loading and high demand. It should also consist of several barriers to prevent illness in the people drinking the water, so that if one of the barriers fails there are others to protect the supply. Several recent cryptosporidiosis outbreaks have been caused by traditionally low-risk public supplies relying on disinfection alone to remove pathogens. When problems have occurred, usually during a period of heavy rainfall, high levels of oocyst-containing faecal contamination have been washed into surface waters or faulty aqueducts. The disinfection system has been unable to cope and this reliance on a single barrier has proved ill advised.

There are two types of sand filter, slow and rapid, and they have both proved to be very reliable over the years. Slow sand filters, in particular, are a very simple and trusted method of treatment for the removal of microorganisms, iron, manganese, hydrogen sulphide, nutrients, etc. Slow sand filters consist of a bed of very fine quartz sand (0.5–2.0 m deep), supported by a bed of coarser sand or gravel (1.0–2.0 m deep) (Gray, 1994). Water is trickled through the sand and collected at the bottom, where it runs under gravity to the disinfection unit and thence into the distribution system. If they are properly looked after, with the depths of the beds maintained and cracks in the surface of the sand not allowed to build up and form short cuts, they will remove a great deal of contamination.

The 'secret' of the slow sand filter is that both physical filtration and biological treatment takes place because a layer of microflora and microfauna builds up on the top of the bed. This consists of a mixture of bacteria and algae and is known by the German word 'schmutzdecke'. The microbiota in the schmutzdecke feed on the various contaminants in the raw water. Its make-up will be different for each treatment works, depending on a variety of factors such as the organic content of the water, which organism is feeding on which contaminant and the natural predator/prey relationships in any complicated microbiological community. The layer will take time to build up before it is at its most efficient but once established will be very effective at removing whatever contaminants normally pass through the filter bed. The schmutzdecke will remove nitrogen and phosphorus in its top or autotrophic layer (i.e. it contains organisms that synthesize their organic requirements from inorganic chemicals) and colloidal and organic material in its lower (heterotrophic) layer. The dual treatment action will also remove tastes, odours and pathogens. Eventually this bacterial and algal layer builds up so much that it starts to block the passage of the water and the top has to be skimmed off. In large treatment works, this is done by tractor but in smaller works it often has to be done by hand using shovels. It is important not to completely remove the schmutzdecke or treatment efficiency will suffer. After two or three months, so much sand has also been removed that it too must be topped up. The slow sand filter employs natural processes, so it is important to protect the bacteria and algae from accidental spillages of bleach or other chemicals that may kill them and severely reduce the effectiveness of the filters.

Rapid sand filters work by the physical removal of contaminants and are therefore not as efficient as slow sand filters in removing bacteria, odours and tastes. They have largely replaced slow sand filters, despite their treatment efficiencies, because they have higher loading rates (about fifty times that of a slow sand filter) (Gray, 1994), they take up much less space and are less labour intensive. They are small, compact units with a coarser grade of sand than slow sand filters, in beds 0.6–1.0 m deep; the medium can also be a mixture of anthracite and sand or activated carbon and sand. There are different designs for rapid sand filters and the water can pass downwards or upwards. They are designed to follow coagulation and sedimentation and the sand traps the particles not removed by the sedimentation process.

The filter eventually starts to become blocked and so needs backwashing, using water or an air and water mix. The sand is agitated so that the trapped particles come loose and the backwash water containing these physical contaminants is run to waste. The sand in the filter is then allowed to settle back down again and the process starts over. This process is called ripening and is necessary for the filter to return to full efficiency. The first flush of water through the filter after backwashing is also run to waste in case it picks up any remaining unwanted particulate matter or pathogens. The backwashing process can be largely automated with pressure sensors or time switches initiating the process. Sometimes backwash water is partially mixed with the raw influent water at the start of the treatment plant, but this needs to be carefully watched. A major outbreak of cryptosporidiosis in the south of England was thought to have been caused by recycled backwash water containing a large number of oocysts that presented too large a challenge for the filter beds (West, 1998). Since that time several water utilities have stopped recycling backwash water, even though this may amount to 20 or 30 per cent of the partially treated water.

Coagulation, flocculation and sedimentation

These interconnected processes will only be encountered in public water supply-sized treatment works, providing clarified water to rapid sand filters. Coagulation is the process of gathering together the negatively-charged particles and microorganisms that are so small that they would never naturally settle out. Coagulants with positively charged ions such as alum (aluminium sulphate), ferric sulphate and polyelectrolytes are added to the water and their electrostatic charges attract the small contaminant particles. Coagulation is followed by flocculation, which is the growth in size of the coagulated particles, which then tend to aggregate together and form clumps. The end result of this process is a relatively dense, woolly-looking precipitate known as a floc, which will readily settle out or can be easily removed by filtration (Bassett, 1999). Sedimentation (or clarification) is achieved in deep tanks, of which there are many designs. Most designs ensure that the floc (normally in a layer called the floc blanket) remains in contact with the water stream for a period of time as it is an efficient collector of many impurities as well as physically collecting more floc particles from the water. Most tanks work on an upward flow principle, water entering the tank at its base and rising through the floc blanket. A small proportion of the floc is continuously removed as a sludge (this is sometimes known as sludge bleeding) to prevent excess amounts building up. The clarified water is removed via weirs at the top of the tank. The upward flow of the water must carefully balance the downward movement of the floc due to gravity, so that the floc blanket stays in the same place.

Ceramic candles

At the other end of the scale is the very traditional form of treating small water supplies known as the ceramic candle. The candle is a small, domed tube of low-temperature fired clay that physically removes small particles from a water supply. Clay consists of very fine particles so the spaces between them are very small and will stop a lot of contamination. Because bacteria tend to clump together and adhere to small particles in water, physically removing these particles also removes a lot of microorganisms. The water is supplied under a head of pressure on the outside of the filter and the water trickles through to the inside where it is collected for drinking. This idea has been used around the world for many years, with bowls made of permeable stone being used to physically remove contaminants of water under gravity. The bowls sometimes have a pointed bottom so that the filtered water drips into a collecting cup underneath (see Plate VIII, which shows a bowl filter still being used in a nunnery in Arequipa, Peru). Interestingly,

Plate VIII A seventeenth-century version of a ceramic candle. Water is poured into the permeable stone vessel and clean water drips out of the pointed base. Santa Catalina Monastery, Arequipa, Peru. (See Colour Plate VIII.)

modern membrane technology has recently started to use ceramic membrane filters and it could be argued that ceramic candles should strictly be placed in the membrane filtration section of this book.

Ceramic candles require regular cleaning as the outside of the filter gradually builds up with a layer of the physical contaminants. This reduces the amount of water capable of being passed through the filter and so alerts the user to the problem. The filter is then dismantled, scrubbed clean, boiled and replaced. This system is therefore a very cheap and low-tech solution for very small water supplies. Unlike some treatment methods, ceramic candles do not usually fail to danger – they just block up and the water stops. Of course the ceramic candle can become damaged, but that will not happen very often without the user being aware of it. It is more likely however that the candle is not refitted properly after cleaning and the raw water bypasses it. Although they will not remove all contamination, ceramic filters are considered to be reasonably effective if used properly, but have been criticized because of the large drop in water pressure following installation.

Apart from physically removing clumps of pathogenic bacteria and faecal contamination, most ceramic filters do not have disinfectant properties. To counter this, some candles have a thin layer of silver applied to them because of its bacteriostatic properties. Tests have been carried out on the effectiveness of this as a pathogen removal system, but results have been mixed. Research has also recently been carried out on ceramic candles for their efficiency in removing *Cryptosporidium* oocysts (DoE, 1996). Surprisingly perhaps, the research indicated that they were reasonably efficient; as long as when they became blocked they were replaced rather than cleaned. One explanation for this is that the scrubbing may have forced the oocysts through the pottery wall. Because the main selling point of ceramic candles is that they are cheap and if cleaned carefully will last a long time, it seems unlikely that they will be installed as a regular treatment system for *Cryptosporidium* removal.

pH balance

A pH-balancing filter is usually part of well-designed whole-house treatment system. It is usually used to aid the removal of various metals, mainly iron, manganese and aluminium from acidic groundwaters or upland catchment areas with little natural buffering in the soil. The pH of the water is adjusted upwards so the water becomes less acidic. As the water is raised to pH 7 or 8, the dissolved metals are oxidized in the presence of air or oxygen and become particulate. They can then be removed by physical filtration. This can be by sand, membranes or the media within the filter system itself. As with nearly every physical filtration system, the filter unit (which looks like a large plastic or fibreglass aqualung) will eventually start to block up and need backwashing; this can be arranged so that it occurs automatically. As the neutralizing granular media get used up, the particles will get smaller and tend to break through into the treated water. Before this occurs the media needs replacing. It is important therefore, that where this form of treatment is installed, a regular maintenance programme is in place to ensure consistent quality of the drinking-water. Other methods of correcting pH are the addition of solutions of sodium hydroxide, calcium hydroxide and sodium carbonate, a simple tank filled with limestone chippings or forced aeration to remove dissolved carbon dioxide (WRc-NSF, 2002).

Another use for pH balancing (or pH adjustment) is to raise the pH of acidic water to reduce its plumbo-solvency. This will stop the 'aggressive' water stripping copper or lead from plumbing systems. One public house on a private water supply in the North of England periodically replaced all its central heating system because the acidic water ate holes in it. It would of course been much cheaper to install pH balancing.

Alternative water supplies

Sometimes clean water is needed for temporary purposes, particularly emergencies. In such cases, instead of installing expensive or short-term treatment, alternative sources of water can be used. While the problem is being attended to, bottled or boiled water can be used for drinking, washing teeth and preparing food intended to be eaten raw. Boiling the water is a sensible precaution if there is a microbiological problem. As a long-term measure however, it is a wasteful, environmentally unsound and less than satisfactory means of ensuring safe water.

Whilst considered to be an expensive short-term alternative in the West, drinking bottled water or boiling drinking-water is a way of life to many people in developing countries. In rural areas where water is untreated or in urban areas where the public water system cannot be relied upon, drinking boiled or bottled water is the only choice. In most megacities (those greater than 10 million inhabitants) and in many other large conurbations throughout the world, tap water is regularly contaminated. This is either because the local water treatment works need repair or is overloaded, or pressure losses have caused contaminated groundwater to be sucked into the distribution system via damaged or broken pipes. A normal morning practice for millions of people is to boil pans of water and leave it to cool down for later use. This of course can become re-contaminated, as it is often stored without a close-fitting lid. Of course, many millions more do not even have a tap or any money to spare on fuel for boiling.

In emergencies local authorities or water utilities may provide drinking-water bowsers for people on private water supplies. These are large wheeled containers filled elsewhere and driven to the point of use. The emergency can either be a contamination incident or the small water supply may have run dry during a drought. As with other alternatives, a bowser is a short-term measure and an expensive system of water provision. They are also difficult to clean properly and usually rely on rinsing with highly chlorinated water.

In some developing countries, drinking-water is delivered in tankers to remote towns and villages. They deliver either to storage tanks in or outside people's homes or to a large central storage container. The storage ranges from the good (specially constructed concrete or metal tanks) to the very poor (badly made and leaking brick cisterns or old oil drums). The water is supplied from a central treatment works or borehole where the water is usually disinfected.

Unfortunately, although it is relatively expensive, the water is not always guaranteed pure. I have seen tankers being filled with relatively clean borehole water and then subject to a disinfection process that consisted of dosing the inside of the tanker with chlorine first thing in the morning and hoping (totally illogically) that sufficient chlorine would remain in it to disinfect the tanker after it was refilled several times during the day. In the Brazilian Amazon, huge canvas bags, about half the size of a basketball court, are filled with water and the whole village is encouraged to use them. Unfortunately, many villagers do not really like the taste of the water and often go back to drinking river water. Of course they soon build up an interesting collection of worms and other creatures in their intestines. Fortunately (perhaps), the bark of one of the trees in the rain forest has amazingly purgative properties and if swallowed, all internal parasites are thought to be either killed or flushed away in a very unpleasant and explosive bout of diarrhoea. If you survive this treatment you resume drinking the water and start a new crop of parasites. Of course, when the villagers purge their intestines of exotic worms they probably deposit them back in the river for other villagers to drink further downstream.

Storage

If water is carefully stored for any length of time there will be a natural reduction in the amount of contamination. Most pathogenic organisms will start to die-off, due to variety of factors including natural ultraviolet irradiation, the lack of suitable nutrients and temperature stress. In addition, solid particles will begin to settle out and this will not only remove the particles themselves but many microorganisms attach themselves to particulate matter and will also be removed. In fact, many large treatment works, where river or surface water is used, have a certain amount of 'bankside storage' to let the water stand before the main treatment processes begin. (See also the subsection on storage tanks in Chapter 2.)

Radon treatment

Although not a particularly common problem, where radon is present in a water supply it can be very worrying. Fortunately, if private water supplies are found to contain high levels of radon, the removal of the gas is relatively simple. Various methods using aeration will generally solve the problem. These can include a compressor to bubble air through the water, using a series of specially designed steps like a waterfall to ensure that the water is mixed with air or spraying the water through a fine nozzle like a shower. The radon-filled air is then vented at high level to be dispersed into the atmosphere. Aeration is recognized by the US Environmental Protection Agency as the best available technique to remove radon for larger community supplies (Dixon *et al.*, 1991). A combination of these approaches can be also used. A US system for single property supplies consists of a large, dustbin-sized unit kept in a garage or room away from the main living area. The water is sprayed downwards though a very fine nozzle over an inert media with a large surface area. Compressed air is pumped up through the unit and taken to the top of the building and vented.

Radon removal can also be achieved by using activated carbon, as the radon will adsorb to the surface of the charcoal. Experts in the US have suggested that GAC removal is the best technique for smaller private supplies due to its superior efficiency and cost-effectiveness (Lowry and Lowry, 1988). It has also been claimed that GAC units, with small sediment pre-filters attached, would be expected to remove between 82 and 99.8 per cent of the gas, depending on the size of the unit. The units would also be expected to last for up to a decade, if the water was otherwise of good quality and other contaminants did not use up the GAC's adsorption properties. Backwashing is detrimental to the radon removal performance of a GAC unit because a large amount of the gas will be desorbed and released during the process. Backwashing should therefore only be carried out once a year (Lao, 1990). It must be borne in mind when using GAC as a method of radon removal that the filtering equipment should be located as far away from the dwelling as practicable, to ensure dispersal of the gas and accumulated radiation. Another factor to note with GAC removal is that after a certain time the adsorbed radon decays to ^{210}Po and ^{210}Pb. These are gamma radiation sources with half-lives of 138 days and 22 years respectively. Disposal of the GAC cartridge must therefore be undertaken very carefully.

19 Action to improve supplies

Where water from a private supply has been checked and found to fail the relevant standards or a sanitary survey has highlighted problems, what should the regulator or concerned owner do? The first question is – are the problems trivial and does anything need doing at all? To help decide, further questions need to be asked:

- Is the problem a bacteriological one, indicating faecal contamination and pathogens in the water?
- Is it a failure of a chemical contaminant and if so, is it a health-related parameter or an aesthetic one?
- If the parameter is an aesthetic one, is the amount in the water within the levels where a relaxation (or derogation) of the standard would normally be allowed?
- If it is a health-related chemical contaminant, is it one where the standard is set with a substantial safety factor and if so, has the standard been breached by a large amount or is it only just over the maximum allowable concentration?
- Is the problem a short-term infringement of a standard that has been set based on a lifetime's consumption? Will anyone be drinking this water at this contaminant level for an appreciable amount of time?
- Is the contravention a regular occurrence, intermittent or a one-off?
- Is the water likely to make someone ill in the short or long term?
- Is the contamination from the water source itself, from the tap or added during the sampling procedure because of poor technique? Minor total coliform infringements, which arise where the sanitary survey has revealed no obvious cause, should be treated cautiously.

Where standards for physical properties are exceeded, the effect is often that the water will become undrinkable before causing injury. Similarly, where aesthetic chemical parameters are involved, the water will frequently smell or taste unpleasant before making someone ill. For chemical parameters where harm may be caused, it is important to understand how the maximum allowable concentration or guideline value has been reached and to compare this with what is happening in a small water supply. Guideline values are based on a lifetime's consumption of the water and because testing for the safety of many chemicals is done on bacteria or animals, there are many safety factors built in. These safety limits can be several thousand times those causing observable effects in animals. For the better-documented dangerous chemicals, such as lead or arsenic, the safety factors are lower. Other standards, even though they are for health-related chemicals have no relationship to health. Microbiological failures remain the most important contaminants in the majority of private water supplies and if an alternative superior supply cannot be identified, treatment should always incorporate a multiple-barrier approach. Radiological problems should also be taken seriously, particularly radon.

For single property supplies there is no point in taking formal enforcement action if there is no likelihood that anyone's health will be affected. Where a reasonably large number of people are involved or the water is being supplied to a third party, a more robust approach to enforcement may be more appropriate, even if it is for minor failures of harmful parameters or problems with aesthetic contaminants. The decision regarding whether the breach is trivial and therefore not requiring any further action will depend on the parameter and the amount by which the allowable concentration has been transgressed. In the UK this decision is left to the discretion of each local authority and it is advisable, when there are known poisonous or carcinogenic parameters in the water, to err on the side of caution.

National government bodies, such as the Environment Agency in the UK or the Environmental Protection Agency in the US, are usually responsible for the quality of groundwater. Where serious problems are found in small drinking-water supplies the agency should be notified in case the problem can be rectified or there is a need to warn other users of the aquifer. It can be assumed that individual cases of faecal contamination will not require notification but, if a number of supplies in an area are contaminated with a specific chemical such as a pesticide or an organic solvent, notification should be automatic. The important question is – at what point between these two extremes will the agency want to be advised? Local authorities will need to decide for themselves but the decision-making process must involve discussions with the local office of the agency involved. In most cases anything 'out of the ordinary' should be reported. The agencies will expect that where serious contamination is found they should be informed straightaway, so that their investigations can begin immediately.

Where a decision is made to take action to correct a non-trivial problem, local authorities will usually want to re-sample the supply before proceeding further. Re-sampling to try to establish whether a particular chemical problem is transitory is sensible, as is sampling in different places to locate the source of the problem. Secondary sampling can also be useful to show that the authority is acting reasonably. However, if a second sample fails to find the problem again, this does not necessarily mean that it has gone away.

Where drinking-water is found to contain faecal indicators or chemicals in quantities that are considered a risk to health, the owners and any other people responsible for the supply should be informed as quickly as possible. It would then be expected that the water would quickly be made fit to drink; unfortunately this is not always the case. Getting people to improve their water supply may prove to be relatively easy or it may be surprisingly difficult. Many people who use private supplies are very proud and protective of their water. They may have drunk it for many years and do not associate it with being a source of danger. If they have not been ill with a gastrointestinal disease for some time they will not automatically welcome advice to improve the water or to spend money on source protection or treatment. Indicator organisms are used to show the potential for contamination of the water by pathogens and are not expected to mean that illness will occur forthwith. If this is not explained clearly, the people drinking the water may loose faith in the authority's advice. If they are merely told the water will make them ill but within a short space of time it does not, they may assume that the authority is not telling the truth or has some hidden agenda.

The responsible authority therefore has to decide what steps it intends to take to encourage or enforce the improvement of the supply. This should be based on the legislation and official guidance that is available, knowledge of the supply following a sanitary survey and the results of one or more sets of samples. It is important that the sampling authority has a clear set of measures that are proportional to the risks involved, taking into account the wishes of the people drinking the water without unduly compromising their health. A formal procedural document with guidance for staff and incorporating a clear enforcement policy is a good starting point. It should ensure fairness and consistency; there is no reason why people with the same problem should be treated differently because they have been visited by different officials or live in different villages. A formal document outlining the policy and procedures allows the authority to consider its aims and objectives with regard to water sampling and should guarantee that any requirements they have to improve supplies are clear, precise and logical. It should be drawn up after consultation with water supply owners and users, and be in keeping with any other official enforcement policies the authority has.

An important question is – at what size of supply is it appropriate for an official body to force individuals to take steps to protect their own health? Various countries have set levels at which they consider action is appropriate. In the US, supplies serving over twenty-five people or that have twelve permanent connections are classed as public supplies and are required to provide water of a high standard. Some states also require this standard for all supplies serving more than a single property (Brown, 2003). In Europe, the original Drinking-water Directive covered all supplies except those to a single property. The latest Directive has changed this by allowing countries to exempt domestic supplies serving less than fifty people. It is now up to individual European countries to decide if they will stay with this number or reduce it.

There is a reasonable argument that says it should be up to the owners of a single domestic supply to decide whether they want to drink contaminated water or not. Many owners of smaller supplies object to being required to improve their water supply and claim that it is not a decision that an external body should have power over. If they choose to drink unwholesome water should that be their prerogative? In the UK, local authorities have traditionally had powers that may be regarded as infringing personal liberty. Where people or premises are filthy or verminous for example, councils can force them to be cleaned up. The Public Health Act 1936 (HMSO, 1936) also contains powers for a court to order a water supply to be cut off, if it is prejudicial to health. Public health measures have often been subject to criticism from libertarians. A *Times* editorial of 1 August 1854, for example, claimed that 'people would rather take their chances with the cholera than be bullied into health by Edwin Chadwick and Dr Southwood Smith' (the Poor Law Commissioners). The aim of public health measures has traditionally been a macro issue where individual sensibilities have sometimes taken second place to the health and general well-being of the wider community. As has been discussed previously, where private supplies contain an excess of contaminants that are injurious to health, it is not always only the individual householder that is at risk. Other members of the family and guests should not be forced to drink potentially dangerous water and it is these people who perhaps need to be protected.

If it is not normally appropriate to take formal action against a single property supply – how many households above this number would be a sensible figure? For water supplies where there is a commercial operation involved, the situation is somewhat simpler. Where someone provides water for commercial gain whether as a separate charge or in the normal course of their business, it is proper to expect that the water will be safe to drink and should comply fully with all relevant standards. This is all the more important where the people drinking the water are visitors and will not only expect the water to be of public supply quality but will have no immunity to it. Although for domestic supplies the number will always be somewhat arbitrary, it seems that if water is supplied to five households or above, then it is likely that its quality will not be within the control of everyone on the supply and thus formal action, where necessary, may be justified. Below that number, advice and persuasion will usually be the main means of ensuring the water is made fit to drink.

Advice

Most local authorities continue to believe that serving a notice is a last resort and try to ensure improvement by persuasion. This is a laudable aim as long as the risk is put into perspective and, if people do not respond to reasonable requests for improvement, a more formal approach is seriously considered. It is interesting to note that public water suppliers would invariably issue a boil notice for accidentally contaminated water, while many private supplies are left for months with equally poor (or worse) quality water, while the local authority tries to cajole the owners into improving them. UK advice currently in force states that where non-trivial failures of the regulations occur, the first thing to do is consider a relaxation and this is discussed in more detail in the section on 'Regulation'. If a relaxation is not an appropriate way forward, the next step is usually to attempt voluntary improvement (unless there is a serious risk to health, when formal action should be taken straightaway). If the responsibility rests with a single person or company, who own the land where the source is located and the building where the water is consumed, this process is really the application of health education techniques. Once people are informed of the outcome of a sanitary survey or the test results from the laboratory, they may not need much persuading that protection and treatment are necessary. They may however need the results interpreted to enable them to understand what the health implications are.

Most sampling authorities discuss this face-to-face or send an explanatory letter or leaflet that explains the significance of each parameter and what it means if it fails the maximum allowable concentration. Unfortunately, many of these leaflets can be less than successful at persuading owners to install adequate protection or treatment. Much of the advice given by authorities is scientifically complicated and talks about very small units and bacteria with long names, with no real explanation as to what it means. It would be far better where faecal coliforms were found in the water if authorities were to simply explain that the consumers were drinking dilute sheep or cattle manure and that a large percentage of cow pats contain a germ (*E. coli* O157) that can and does kill children. Normally the language used is either too technical or couched in careful governmental language designed to protect the authority rather than inform the reader. It may also be full of words that normal human beings rarely use, for example – 'notwithstanding the above-mentioned parameters, a faecal indicator, namely *Escherichia coli*, may suggest pathogenic contamination', etc. The leaflet or letter should be clear and simple and explicitly state what the problems will be. The potential for illness should not be under- or over-stated. If you inform people who have drunk water from a suspect spring for many years that they will fall ill immediately, it goes against the logic of the situation – they are probably immune. It is more accurate to state that as they get older their immunity may start to fade or that other people visiting the premises will not have that immunity. It may also be more effective to tell them that if the cows whose dilute faeces they are regularly drinking, pick up a new disease, they will probably go down with it three days after the next downpour. Alternatively, if the farmer re-stocks with new animals they will have a different selection of bugs in their gut that the people drinking the water are not immune to.

Another problem with standard leaflets is that they often do not look particularly interesting. They seem to include every parameter the authority has ever had a result for and to save money, it has put all the information on one piece of paper. Although this is easier for the organization sending out the leaflet, many people will not appreciate a long technical list, in a small font, containing many things they have no interest in and are not problematic at their house. By the time they have reached their own particular problem they may be confused, bemused or annoyed. It is better to concentrate on specific problems, such as a failure of faecal indicators and have individual sheets for each separate problem. Leaflets or letters should be thought of as advertisements trying to sell safe water; they should be bright, interesting and keep to simple messages.

The accuracy of modern health education techniques in European or American rural communities has probably been studied less often than in many developing countries. Often the problem is that small water systems are treated on a very individual basis and the big picture missed. Examples of which education techniques work best in one area are often not shared with other sampling officers or technical staff. Neither are many scientifically robust experiments carried on with alternative approaches used in different locations and results compared to see which are the most effective. In developing countries, health interventions have to be effective and sponsoring organizations require proof that the money they are donating is being used effectively. Graduate or post-graduate scientists carry out many of these projects and there is an academic rigour to their work that can be missed in a local authority setting.

Newcomers to small or private supplies are used to drinking public or mains water and when they move to the country often expect their own supply to be of the same quality. This is a reasonable expectation and should be encouraged. They can often be easily persuaded to install treatment to make the water safe. In some cases, they may even ask for the water to be tested before they move in, so that any problems can be put right before they begin to drink it. They may also be financially better off and more able to afford the necessary treatment. On the other hand, it can often be quite difficult to obtain voluntary improvement where people have drunk the water for most of their lives. They may also not have as much spare cash to invest. It is the responsibility of the enforcing authority therefore, to ensure that any treatment they suggest will be the cheapest available alternative and that it will work. It is reasonable for owners of small water systems to expect a certain amount of expertise in the people coming to test their water. The water sampler or sanitary surveyor should know which contaminants are likely to be found and have sifted through the treatment and disinfection technologies available to identify the ones that are appropriate for the problems in the area. They should also be able to give a reasonable estimate of the cost of effective treatment. Owners of supplies will expect the advice to be sound, up to date and aware of economic restraints. It will also be expected to cure the problem. In the past, many people have recommended a disinfecting system, for example, only to find it has proved ineffective because the necessary pre-treatment had not been insisted upon.

Although it is sometimes difficult to persuade people who own their own supply to make it safe, where the owner and the consumer are different, it may be nearly impossible. The person who owns the field where the source arises may have inherited the situation and feel less than happy that someone else has the right to be supplied by them, often for free. They may also have fallen out with the person using the water and want to include a dispute over the supply in with the feud. If the owner of the field is asked to improve the supply, at his or her own expense, this will rarely be seen as a reasonable request by the enforcing authority. A more formal approach may therefore need to be taken. Of course, this should also be accompanied by a full explanation of why they have been chosen by the authority as the person who should improve the supply (which should be a reasonable decision), what other action, if any, is being required of the consumers, what the sampling results mean and what needs to be done.

The first reaction of many supply owners is to threaten to stop supplying the water. If mentioned by them, even as a possibility, they should be strongly dissuaded and if necessary, threatened with legal action either in the criminal or civil courts. Depriving someone of a water supply is never acceptable and there will usually be suitable legislation ensuring this cannot be done. In the UK, for example, this would arguably cause a statutory nuisance, as the premises would be injurious to health without water. It may be found that tenants are suddenly asked by the owner to sign agreements accepting responsibility for the water quality, sometimes under duress. It is essential that landlords do not require tenants to sign any documents that take away their rights. Landlords are required by the Rent Act 1977 to maintain a supply of water (HMSO, 1977). They also have a civil duty to maintain any facilities that were provided at the beginning of a tenancy. The people receiving the supply can therefore sue the perpetrator if it is withdrawn. A person supplying water for no cost has some justification in claiming that it is unfair to also expect them to provide treatment for free. Situations have arisen where landowners have threatened to evict tenants, rather than be forced to provide them with a safe water supply. It is best, however, to try and avoid a situation where disruption of a water supply takes place. Conversely, some people who have a free supply of water feel no compunction to pay anything towards the cost of their water supply, even when another party has to spend money to improve it.

When requiring improvements to supplies, it is important to move carefully in order not to make a potentially difficult situation worse. Sometimes deeds have to be examined or a written contract exists where the responsibilities have been formally decided. At other times, agreements can be drawn up to ensure that costs are apportioned and repaid fairly. If faecal indicators, pathogenic organisms or dangerous chemicals have been found, it is difficult to justify not ensuring a safe supply by whatever means are available; it seems appropriate that the protection of the source, by physical barriers and keeping animals and humans away from the area of recharge, is the responsibility of the person who owns the land where it arises. However, if the water is supplied free of charge or on a small-scale, the householders who use the water may be considered responsible for installing disinfection and any necessary treatment. Nevertheless, I appreciate that it is often difficult to ensure an equitable approach in practice.

Boil water advisory notices

The usual method of dealing with short-term microbiological contamination is to boil the water. Bringing water to boiling point will inactivate all pathogens, even *Cryptosporidium*. Previously, the advice given to consumers was to keep the water on a rolling boil for four minutes. Unfortunately, this was also thought to have caused many accidents, as people kept the 'on button' of their electric kettle pressed down for the four minutes, causing scolding from the escaping steam.

Following an outbreak of waterborne disease or a contamination episode in the UK, local authorities and Outbreak Management or Outbreak Control Teams under the auspices of either Consultants in Communicable Disease Control or Regional Epidemiologists can send out advice to boil water to all the households and businesses affected by the problem. These are sometimes called 'boil water advisory notices'. For public water supplies, these will be sent out by the water utility. This should have the effect of protecting the population on the supply whilst the problem is resolved and the contaminated water flushed away. Boil water advisory notices have been subject to some research lately, to find out just how effective they are. The question that always needs to be asked is – will there be more illness due to the notice than will be saved by it?

Now that scalding is not so likely, it may be that boil water advisory notices will be used more often, but they are not the complete answer. A study in the US (Angulo *et al.*, 1997), found that during one outbreak, almost a third of people drank unboiled water after the notice was issued, including fourteen who subsequently became ill. The main reason cited was 'forgetting', followed by 'not believing the order' and then by 'not understanding that ice should be made with boiled water'. Compliance improved after further advice was given detailing the reasons for the notice and 'clarifying the boiling procedure' (although any procedure for boiling water cannot have been that complicated, I would have thought). A UK study (Willocks *et al.*, 2000), found that nearly 60 per cent of people did not fully comply with a boil water advisory notice even though 88 per cent thought that they did. The main problems found were that people washed uncooked food in unboiled water (20 per cent) and continued to brush their teeth in it (57 per cent). Another study in the northwest of England (O'Donnell *et al.*, 2000) found that 81 per cent of the households surveyed failed to fully follow the guidance provided. Again the main reason was 'forgetting'. This time 58 per cent brushed their teeth in unboiled water and 17 per cent washed uncooked food such as salads in it.

Official advice given to Outbreak Management Teams says that they need to decide at an early stage what the criteria for lifting the advice to boil water will be (DETR, 2001). This is often easier said than done at the beginning of what may be a large waterborne outbreak of unknown cause. It is easy to feel the need to 'be doing something' and the press and politicians may clamour for action. However, what will the benefit of applying an order be? Did the incident causing the problem happen some time ago and would the slug of affected water now have passed through the system? What is the infective dose and would the slug of water be so diluted that there is no longer any risk? Should the notice be removed when the number of cases starts to fall or when there has been a suitable gap after the last case? This could be problematic if there is secondary spread from infected cases. Finally, how much will a boil advisory notice cause people to lose faith in the water supply unnecessarily and waste money on bottled water? These questions may lead to the conclusion that a notice is not necessary and the Outbreak Control Team will therefore need to be united in this opinion when questioned about it by politicians or the media.

It is more likely that water from a small supply will be subject to long-term contamination. If so, a boil water advisory notice may alert people drinking the water that it is unfit to drink until treatment is installed. Boil water advisory notices however, may have little effect on people who do not think there is anything wrong with the water, particularly those who have been drinking it for some time. A long-term boil water advisory notice on a private water supply may soon cease to have an effect as people forget or continue to have contact with the water through brushing their teeth or washing their salad. How a boil water notice is drawn to the attention of transitory visitors to small supplies, such as to a campsite, where the notice may put off potential campers, is also questionable. Even where signs are put up they are easily 'lost'.

There has been little scientific study on compliance with boil water advisory notices on small water systems. The results would, I suspect, be much worse than that found by the research on public supplies detailed earlier. One of the few times where this has been tested was in 1999, at a small campsite, where a routine sample of the water to the private supply was found to contain *E. coli*. This resulted in a boil water notice. Onset time for *E. coli* is usually up to four days; a week later there were six new cases (License *et al.*, 2001).

Although there is a lot of evidence that boil water advisory notices are not totally effective, this does not mean that they should never be used. Sometimes the risk to health is such that it would be a failure of duty if an Outbreak Management Team or local authority environmental health department did not issue one. Because many people do not comply completely with an advisory notice does not mean that they should not be given the choice and encouraged to treat their water with caution until the problem has been resolved. It is still important, however, that whoever is considering a boil water advisory notice makes sure it is the most appropriate step, that the advice is well presented and understandable to the general public and that the process for removing the order has been decided or at least thoroughly discussed.

Formal statutory action

Once it has been established that a private water supply must be improved and gentle persuasion has failed, authorities need to consider what formal action to take to ensure the safety of the people drinking the water. For long-term problems boil water advisory notices are not appropriate and permanent improvements are necessary. In the UK, the powers to carry out formal enforcement activities are contained in the Water Industries Act 1991 and the Private Water Supplies Regulations 1991 (or regional equivalents). Both follow on from the 1980 European Directive on drinking-water quality.

Directive on the Quality of Water Intended for Human Consumption

The basis of all European legislation on drinking-water safety is the European Union's Council Directive on the Quality of Water Intended for Human Consumption (the Drinking-water Directive, as it is known). The original Directive (80/788/EEC) was introduced in 1980 and laid down cross-European requirements for water quality for the first time (EEC, 1980). Because safe drinking-water is so essential for health, it was felt necessary to have a community-wide standard, based on a specified list of microbiological and physico-chemical parameters. Countries were then expected to translate its requirements into local statutes. The European Council approved a second Directive (98/83/EC) on 3 November 1998 (European Union, 1998). This was several years after the date intended for the review of the original, so it is probable that the latest Directive will be in force for many years to come. In fact, as the Directive does not require full implementation until five years after it came into force (on 20 December 1999), the original Directive will not be officially repealed until that time. Where member states bring in new legislation on drinking-water quality however, it must be in line with the latest Directive. The new laws and regulations necessary to ensure countries comply with the Directive should have been written, agreed and become operational by 20 December 2001 (two years after the Directive came into force). This appears to be the date when any new private water supplies regulations in the UK should have been written. The quality of all drinking-water has to comply with the Directive within five years of its coming into force, and this may be being relied on as the final date for the new regulations. If any country has a particular problem with this timescale, it can be extended by up to six years in exceptional circumstances.

The aim of the Directive, as detailed in Article One, is to protect human health from the adverse effects of the contamination of water intended for human consumption. It defines this as 'all water either in its original state or after treatment, intended for drinking, cooking, food preparation or other domestic purposes, regardless of its origins and whether it is supplied from a distribution network, from a tanker, or in bottles or containers'. The Directive then goes on to allow member states to exempt water from an individual domestic supply serving less than fifty people or less than $10 \, m^3$ per day. This therefore means that individual countries can exclude nearly every small and private domestic water supply from the requirements of the Directive. If they do so however, they must ensure that the population concerned is told and that if their health is put at risk, they are given suitable advice. It seems unfortunate that the European Union has decided to allow this exemption for such a large number of consumers and it is difficult to understand the justification for thirty or forty people being supplied by a contaminated water supply with no legal protection. However, the cost of rigorously applying this legislation no matter how small the water supply could be astronomical for the newer countries to the European Union. It may be that individual countries were given some discretion, assuming that most would continue to provide protection for all supplies except single properties, but allowing poorer countries to set a more economically viable level. By giving countries the ability to concentrate on their larger supplies, this ensures that the Directive can be complied with fully by all member states and not be ignored or fall into disrepute. The Directive does not allow countries to exempt small supplies that are part of a commercial operation or that serve the public.

An overarching requirement of the Directive is that any water, whether it meets the general requirements in the legislation or not, must be safe to drink. If it is constitutes a danger to health, its use has to be prohibited and any consumers who may have drunk it must be notified. The Directive also requires that all water should be wholesome and clean and at least to the standards set for the parameters listed in its annexes. This includes water used for food production, unless the processing renders the water wholesome. Water should be of sufficient quality that someone on the supply is able to drink it for seventy years without any ill effects. The only exemptions to this are waters used for medicinal purposes and natural mineral waters; they have their own Directive (European Union, 1996). It is accepted that mineral waters are drunk as beverages and therefore lifetime consumption will be much less than for normal drinking-water. This relaxation of standards also applies to waters taken for health-giving properties. These are allowed to contain larger concentrations of certain non-microbiological parameters. The Directive also obliges countries to establish suitable monitoring regimes with laid-down timescales for sampling frequencies, establishes standards for chemicals added during treatment and allows for water protection measures for ground and surface waters.

The Directive sets water quality standards based on firm scientific principles, mainly taken from the WHO Guidelines (WHO, 1993a). Where the science is uncertain, it uses the precautionary principle – disinfection by-products, for example, must be kept to a minimum. The standards are designed to maintain public health and are set at the minimum concentrations necessary to maintain that safety level. Countries cannot however, use the Directive as an excuse to reduce the existing quality of their water supplies and if any member states have a health concern about any contaminants not in the Directive, they have to set their own additional standards. While countries can produce legislation of their own that is more stringent than the Directive, they cannot make any where the requirements are less onerous. Even when they make them more stringent, they must notify the European Commission of this fact.

The quality standards must be met at the point where the water is made available to the user. The Directive acknowledges that once inside a domestic property the water cannot be the responsibility of the state supplier or the water company. If a satisfactory water supply enters the householder's own plumbing system and loses quality, the Directive is still complied with. If the water is affected by the domestic plumbing to such an extent that it would fail the maximum requirement for one of the parameters (usually lead), member states should take measures to try and reduce the risk, such as advising the householder of the potential dangers. However, in any building where the public drink the water, such as in schools and hospitals, the quality must be up to the standard of the Directive at the tap itself. This means that action will have to be the taken to negate any deleterious effects of plumbing systems including upgrading the pipework and fittings where necessary. This will require new regulations in many countries. The Directive also allows for the introduction of chemicals to reduce any ill effect of household plumbing. This means that orthophosphates can be added to a supply to reduce plumbo-solvency.

For water in containers, the standards apply to the water as it goes into the container or bottle. Although this ensures that the water is wholesome at that point, it does not have to be of the same quality when purchased in the shop or when drunk. As already discussed, if water is left in a warm shop for a long time, the number of adventitious bacteria, and thus the colony count, can be high (Fewtrell *et al.*, 1997).

Each country has to ensure that its drinking-water is satisfactory and that monitoring results are reliable. Where failures occur they should be investigated and action taken to put matters right as soon as possible. The Directive allows member states to grant derogations (exemptions or relaxations) where standards are not being met, but only as long as health is not being compromised. Derogations are allowed for three years at a time and the member state has to justify why the relaxation should be given and what it is doing to improve the situation. A maximum of three derogations can be given for a single problem, but even this will not happen in many cases as the third one can only be granted in exceptional circumstances.

UK Drinking-water legislation

Water Industries Act 1991

The overarching legislation for drinking-water in the UK is the Water Industries Act 1991 (HMSO, 1991a). The Act mainly deals with water undertakers and their obligations, the role of government departments, the setting of standards, reporting arrangements and allowing regulations to be made. Sections 77 to 85 cover local authority functions and private water supplies. Section 77 requires local authorities to keep themselves informed about the wholesomeness and sufficiency of the water in their area and allows the Secretary of State to make regulations about water quality in private supplies. Wholesomeness is defined in those regulations (see later). Section 80 gives powers to require the improvement of a private supply through the service of notice. It states that where a local authority is satisfied that the water from a private supply is failing or is likely to fail to provide a wholesome supply, that authority may serve a notice on one or more of the relevant persons. This is known as a Private Supply Notice and applies to supplies that are insufficient as well as defective. 'Relevant person' is defined as the owners or occupiers of the premises supplied, the owners or occupiers of the premises where the source arises or any other person who has control over the supply. Section 81 explains the appeal procedure against the notice and section 82 allows the local authority to carry out the work in default and recover the costs. The local authority can acquire land to enable the notice to be complied with under section 83 and section 84 gives them power of entry to premises where they need to test the water or check on works. Finally, section 85 gives local authorities the power to obtain information in order for them to serve a notice or find out about water quality.

Private Supply Notices, sometimes also called Improvement Notices, must specify the steps necessary to ensure the supply is wholesome and sufficient. They can be served where non-health parameters are above the maximum allowable concentration and on single property supplies. The Drinking Water Inspectorate however advises caution as far as single property supplies are involved. It recommends that notices for these supplies only require improvements where the water poses a hazard to health (HMSO, 1991c).

In situations where there are several relevant persons, the authority must decide which ones to serve the notice on. The legislation allows plenty of scope but requires the local authority to have regard to any appropriate documentation covering the supply of the water. In many cases, the water supplied to a house arises on land owned by someone else. There is often no payment made for the water and no legal agreements drawn up identifying responsibility; even in these cases the landowner can still be served with a notice. It is important that authorities are careful about whom they serve notice upon and they should have written procedures to follow to help show that they have acted fairly if appeals are lodged.

Where a local authority considers serving a notice, it must evaluate the appropriate remedial measures. There are two schools of thought regarding the specification of works. The first is that when serving a notice, regard must be had to the possibility of having to default it (carrying out the works themselves and sending the responsible person the bill). Because a local authority has these powers, the recipient should be aware of exactly what action the council will take and the likely cost, if he or she fails to improve the supply. The council should therefore specify a single system of improvement and be prepared to defend that decision if the notice is appealed against. This is the line of thought that the Drinking Water Inspectorate appears to favour (Drury, 2003). The Inspectorate also expects that local authorities will always choose the cheapest option that will ensure compliance with the regulations. The other theory is that there may be several ways of improving a supply. Therefore, providing the supply is wholesome at the end of the day, as wide a choice of improvement methods as possible should be given to the recipient of the notice. If, for example, a supply failed the faecal coliform standard, adequate treatment may consist of source protection and either chlorination, reverse osmosis, ultraviolet irradiation or connection to the public supply. It could be considered unfair to insist that one particular system is installed. The wording of section 80 (3b) appears to confirm this. It states that the person who has had the notice served upon them must 'take one or more of the steps' specified by the notice. This indicates that the notice can include several (different) steps to improve the supply, only one of which needs to be carried out. It would seem only fair though, that the local authority tells the responsible person what treatment they intend to install if they have to default the notice.

The notice may include a requirement to connect to the public supply and the local authority can insist that the water undertaker for the area does this work. Connection to the mains sometimes involves a new supply pipe, which may have to be laid over land owned by a third party. The legislation does not allow the person served with the notice to connect to the public supply where the third party does not agree to this, but the local authority has powers under the Water Industries Act 1991 section 83 (2) to compulsorily purchase or acquire rights over the land. It can then dispose of the land or those rights to the person served with the notice. It also appears that if the supply owners serve a Connection Notice on the Water Undertaker under section 41 of the Act and consent by a third party is refused or unacceptable conditions are attached, the Undertaker can lay the pipe across the land anyway (section 46 (1) and (4)).

Legislation dealt with by environmental health departments normally involves appeals to the magistrates' court but appeals against Private Supply Notices are to the Secretary of State. There have only been a handful of appeals since the Act was introduced (Drury, 2003), but it is interesting to speculate why this change was included in the legislation. It does not appear to have been introduced to speed up the appeals process as the first appeal took several years to determine (Clapham, 1995). It may therefore have been thought that such a specialized subject was beyond the ability of magistrates to decide; this is patronizing and contrary to the natural laws of justice. Perhaps the reason is that the legislators felt the need to be kept aware of the number of appeals against section 80 notices. If this is the case, they could have included appeals within the normal reporting requirements of local authorities. If this system of appeal merely fitted in with the then current legislation procedures, experience has shown that it is less than satisfactory and any new legislation should return the appeal process back to the magistrates' court.

The Private Water Supplies Regulations

Following the 1998 Drinking-water Directive, the UK Government updated the requirements for drinking-water quality in public supplies, in the Water Supply (Water Quality) Regulations 2000. However, at the time of writing, the relevant legislation for private supplies remains the Private Water Supplies Regulations 1991 (HMSO, 1991b) and their Scottish equivalent of 1992. New legislation will be introduced for private water supplies in the near future and the regulations for Scotland will probably be updated before the English and Welsh ones. It is not known exactly how the legislation will change, but there will be much that is similar to the existing regulations. It is likely that there will still be a classification of supplies based on volume or usage numbers, with monitoring frequencies dependent on those classifications. The ability to give a derogation or relaxation where water quality is below the wholesome standard but is not injurious to health may also be in. However, because WHO and the European Union have begun to omit aesthetic contaminants in their guidelines and requirements, the relaxation process may soon no longer be necessary. This will be a useful reduction in the amount of bureaucracy. It will also encourage people to concentrate on health-related contaminants and their removal. The chemical, physical and microbiological parameters that need to be monitored should still be detailed in the new regulations and will follow the maximum allowable concentrations in the Directive.

It is expected that the regulations will reflect suggestions from local authority practitioners and scientists that a more risk-based approach is necessary, through the use of a sanitary survey. WHO is promoting this approach, particularly for small supplies that are not fully treated. It will be interesting to see how this will be written into the new regulations. One of the other interesting questions is – what size of domestic supply will no longer be required to meet the standards? Will it remain as a single property supply or will the government take advantage of the exemptions allowable in the Directive and increase it?

Because the legislation from 1991 is still in force, this book looks at the Private Water Supplies Regulations in some detail, but as they are due to change, concentrates on those aspects that are likely to be included in the new ones.

General requirements

The wholesomeness of water is defined in regulation 3 as

- Water that does not containing anything (element, organism or substance), other than a parameter listed in the appendices of the regulations, at a concentration that would be detrimental to health either on its own or in conjunction with anything else; and
- Water that does not contain any parameter above the value or concentration laid down in those appendices.

To ensure that the water is wholesome, sampling has to be carried out. Regulation 19 requires that sampling be done properly; that the water is not contaminated post-sampling and is representative of the water in the supply. The laboratory that is used must have adequate quality controls. When the results are received from the laboratory, the authority must inform all the relevant people of those results within twenty-eight days. Regulation 21 allows the local authority to enter into an agreement with another person to allow them to sample the water from a private water supply as long as they are competent, do not cost the authority anything and inform it of the results of the sampling. This sort of arrangement is most common where large manufacturers, such as breweries or soft drinks companies, have their own supply and can save money by carrying out the sampling themselves.

Relaxations

Both Drinking-water Directives allow member states to make derogations. The first one specified that they were to take account of situations arising from 'the nature and structure of the ground', 'exceptional meteorological conditions' and in the event of emergencies. The term 'derogation' becomes 'authorization' in regulations 4 to 7, but they are usually known as 'relaxations'. The relaxations for private water supplies are dependent on several factors:

- The relaxation cannot be for substances that would cause a risk to public health.
- In situations where emergencies or meteorological conditions apply, a time limit must be incorporated within the terms of the relaxation.
- Relaxations under the 'nature and structure of the ground' provisions, except for those to small private water supplies (categories C, D, E, F, 3, 4 and 5) must be notified to the Secretary of State.
- Relaxations for larger supplies must involve consultation with the Secretary of State who is then required to inform the European Union.
- Where the supply is used solely for food or drink production, it can be given a relaxation without reference to emergencies, meteorological conditions or the nature of the ground, but the water cannot make the food or drink unfit.

Relaxations have been a possible course of action for local authorities for nearly twenty years. Local authorities in the UK were initially informed of their ability to relax certain parameters in larger private supplies in 1984 (HMSO, 1984a). They did not appear to make use of these powers at that time, nor when they were repeated in the 1989 Water Act (HMSO, 1989). In 1991, many authorities were still unaware of their ability to exempt low-risk failures when an official circular on private water supplies (HMSO, 1991c) indicated that where there was a non-trivial breach of the standards for aesthetic parameters, local authorities should consider relaxation as the *first* possible course of action.

Because the 1991 regulations quantified the standards required for the first time and prescribed a sampling regime, many water supplies were suddenly found to be unwholesome. Aesthetic parameters caused some of these supplies to be unwholesome so local authorities were expected to deal with them. But some authorities had so many poor quality supplies that they needed to concentrate their resources on dealing with the health-related contaminants There are two opposing arguments about the use of these relaxation powers. The first is that, if the supply fails to reach the standards in the legislation, the enforcing authority should require them to be met. There is an ever-increasing interest in the quality of drinking-water and the public may consider the authority remiss in its duties if it does not ensure that water quality is kept as high as possible. Additionally, some parameters that are considered to be aesthetic, for example aluminium, are subject to occasional speculation about their safety. Until the research is complete (if that is actually possible) it may be unwise to allow a failure of the legal standard to be officially sanctioned. This is in line with the 'precautionary principle'. The converse argument is that private supplies are often located in areas where the quality of the groundwater is affected by the soil and geology to such an extent that it will constantly fail the drinking-water standards. It is therefore inappropriate for a local authority to use its scarce resources on formally forcing owners to improve a supply that has only failed for aesthetic parameters. For these contaminants it could be argued that the water becomes undrinkable before it is dangerous to drink and that decisions about them should be left to an individual householder, with the role of the council restricted to that of educator and advice giver.

In the majority of cases, relaxations allow local authorities to easily resolve the problem of non-health-related failures of the standards. Where a local authority feels that they are appropriate, or an owner approaches it with a request for one, it can either give blanket relaxations or appraise each supply separately. In order to relax particular parameters due to the 'nature and structure of the ground', most authorities normally only have to decide officially to relax them, up to a designated limit and the relaxations will apply. Interestingly enough, after the regulations came into force, several local authorities considered relaxations for parameters outside the strict interpretation of the legislation. For example, some relaxed the requirements for nitrates and in two instances, pesticides (Clapham, 1993). These parameters are not dependent on the nature of the ground so much as local farming methods. Moreover, they are usually considered to be health parameters.

Where relaxations are granted, the new maximum allowable concentrations have to be determined and specified by the local authority. Advice on this can be obtained from Circular 24/91 (HMSO, 1991c). The parameters listed in the guidance are aluminium, ammonium, calcium, chloride, colour, conductivity, iron, magnesium, manganese, nitrite, odour and taste, pH, sulphate, sodium, potassium and turbidity. It is unclear whether the guidance considers this to be a definitive list or not. It is also interesting that nitrites are included considering the questions about their involvement in stomach cancer and methaemoglobinaemia. The circular states that a permanent relaxation may be appropriate in the case of smaller supplies and normally it would be simpler not to fix a time limit on these supplies because the relaxations can always be revoked or altered in the light of subsequent information.

How is water quality affected by the nature and structure of the ground? Circular 24/91 discusses five general categories:

- *Upland surface catchments.* Low buffering and low pH cause greater amounts of colour and some metals to be present. This category is also associated with seasonal variation.
- *Evaporite deposits.* This is because of calcium sulphate, sodium chloride and sodium sulphate in the rocks cause elevated levels of sulphates and sodium.
- *Chalk, limestone and greensand.* Potassium is found in the potash deposits in greensand. Ammonium comes from clay minerals within

these aquifers and nitrite is a product of the oxidation of ammonium. Magnesium is dissolved from magnesium carbonate in the limestone or dolomitic cement in sandstone. Calcium is in hard waters from the calcium carbonate in limestone and elevated conductivity levels are due to high ionic concentrations in the rocks.

- *Underground waters with low dissolved oxygen and low pH.* Low buffering and low pH again cause greater amounts of iron and manganese to be taken into solution.
- *Other possible 'natural' problems.* Chlorine from saline intrusion, colour due to leaf fall in surface waters, odour and taste caused by microorganisms such as algae in surface waters and sodium from natural ion exchange mechanisms.

In the UK, water undertakers are required to produce annual information on their authorizations. Although they cover 'water supply zones' rather than catchment areas, it is useful to see which parameters attract relaxation because of the nature and structure of the ground. Iron, manganese and aluminium for example, have been regularly subject to relaxation for public water supplies. In some cases, the relaxations have been quite substantial. Some time ago, Yorkshire Water plc. had a small supply where an authorization was allowed down to pH 3.9 (Short, 1992). With this level of acidity the water would be able to eat your dinner for you!

Draft guidance when the regulations were first published advised consultation with public health advisors when deciding about relaxations. Most local authorities however preferred to consult with the Drinking Water Inspectorate when deciding what exemptions to give and by what amount (Clapham, 1993). Very few referred to WHO Guidelines (WHO, 1993a). Discussions with several Consultants in Communicable Disease Control (CCDCs) have shown that they usually have little involvement with private water supplies. When I discussed the matter with one a few years ago he was surprised by the number of people on private supplies and had not considered any potential health problems associated with them, either from diseases such as cryptosporidiosis or potentially carcinogenic contaminants. He was unaware, for example, that he regularly received bacteriological results of private supply sampling via the Public Health Laboratory Service. Because they were so bad, he had assumed they were river water sample results. In some areas, public health doctors have a deep interest in the safety of small water supplies and their advice can be very useful. However, because they are normally communicable disease specialists, they are not necessarily experts in the other aspects of drinking-water safety such as chemical contamination or radiological problems. It is important that a good dialogue is built up with them before any major problem occurs and gaps in the knowledge base identified and filled.

Relaxations for exceptional meteorological conditions only apply to non-health-related parameters. They can be applied, for example, where supplies become temporarily contaminated from high levels of rainfall or snowmelt. Relaxations may also be appropriate where an extended drought causes the normal source to dry up, meaning that supplies with poorer water quality have to be used. Another example is colour problems. The water from some peat moorlands in Yorkshire had consistently high colour levels for several years following a severe drought in 1976 (McDonald, 1991).

Health-related parameters, even microbiological ones, can be included in the final category of relaxation – water used in food production. Understandably, the water quality must not affect the fitness for human consumption of the final product and examples include the brewing and distilling industries where high temperatures will render microorganisms harmless. Each case must be considered on its merits and local authorities will wish to consult with their public health advisors and food specialists before granting such relaxations. It must be remembered that water may not only be used in the main food processes, but also for staff to drink and for cleaning. Because staff may be used to high quality mains water, where the supply is a private one, they may not consider the possibility that the tap water could be contaminated. Employees and consumers may therefore be put at risk if appropriate precautions are not taken to safeguard all the outlets of the water supply. It should also be remembered that between 1937 and 1984, up to twelve incidents of typhoid fever in the UK may have been caused by water-contaminated canned meat. In addition, eleven milk-borne outbreaks of disease (including typhoid, paratyphoid and dysentery) were also associated with contaminated water supplies. In the conclusion to their report on water and disease, Galbraith *et al.* (1987) emphasized the need for chlorination of *all* water used in food processing plants.

Categorization of supplies

The Private Water Supplies Regulations 1991 divide supplies into two categories. Category one contains supplies to domestic properties where water is used for drinking, cooking and washing; category two deals with supplies to commercial properties such as restaurants, food factories, hospitals, schools, hotels, campsites and holiday cottages. The two categories are then subdivided into five classes that depend on the number of people being served by the supply or the volume of water used. This is detailed in regulations 8 to 12. Once the supply had been classified, the regulations lay down the frequency of sampling and the parameters to be looked at. As the number of people served by a supply increases, the frequency of sampling goes up, as does the list of parameters to be tested. This is a sensible approach that attempts to prioritize the potential risks to the health of consumers. The classifications laid down in the regulations are summarized in Tables 5 and 6.

The main aim of classification was to separate the widely differing types of supply into similar risk groups that could be fairly and uniformly sampled

Table 5 Classification of domestic private water supplies

Class	People served on any one day	Average daily volume (m³/day)
A	>5,000	>1,000
B	501–5,000	101–1,000
C	101–1,000	21–100
D	25–100	5–20
E	<25 (except for supplies in category F)	<5
F	Those residing in a single domestic property	

Source: Private Water Supplies Regulations 1991 (HMSO, 1991b) London.

Table 6 Classification of private water supplies for food production

Class	Average daily volume of water supplied for domestic or food use (m³/day)
1	>1,000
2	101–1,000
3	21–100
4	2–20
5	<2

Source: Private Water Supplies Regulations 1991 (HMSO, 1991b) London.

(DoE, 1991). The risks associated with supplying water to over five thousand people are much greater than when supplying to five. This is because the greater the number served, the greater will be the possibility that 'at risk' people will come into contact with the water. 'At risk' groups include children under five, the elderly, the infirm and the immuno-compromised. Consequently, sampling regimes should become more arduous as the number of people served by a supply increases.

The first problem for local authorities when the regulations came in was to actually locate their private supplies and make an accurate database of the properties they served. Local authorities had collated information on private water supplies to a greater or lesser extent for many years. Some authorities had identified them by comparing the premises on the council's rating list with the ones that paid water rates. Others had their own records compiled from complaints about the quality of the water, enquiries from potential purchasers and previous surveys. This work was done many years ago and it is important that the records are kept up to date. Planning or building regulation departments should normally identify private water supplies in new buildings and arrangements must be made to inform the environmental health department when this happens. Section 25 of the

Building Act 1984 (HMSO, 1984b) requires a local authority to reject plans for a house unless they are satisfied that it is provided with a wholesome and sufficient supply of water. Where new domestic dwellings intend to use a private water supply, the building regulation section will usually ask for advice from the environmental health department. As far as properties that have ceased using a private supply are concerned, this will become known the next time they are sampled. Most people seem to enjoy telling the local council that they have got their records wrong, particularly if this also means that they do not get a bill for sampling charges.

It is often easier to obtain information on larger supplies, because the owners are more likely to monitor water usage. People usually have no notion of the volume of water used in smaller supplies. Should they still need to obtain this information, local authorities have several options:

- to ignore volume measurement and use the number of people as the classifier;
- to estimate the volume of water used by assigning a set amount to each individual. Regulation 10(6) allows local authorities to allot $1\,m^3$ of water per day for every five persons served by a category two supply;
- to contact the Environment Agency, which has information on abstraction licenses in their Public Register. This information is freely available and can be useful but it must be remembered that only maximum rates will be listed and these figures will not normally be the actual amounts used by the licensee;
- to measure the amount of water used. Brassington (1995) suggests a variety of simple ways in which this can be done and of course, there are many sophisticated and expensive flow-measurement devices available.

Discovering the number of people using a particular supply should be a relatively simple matter, occasionally however, there are problems. Many private water supplies are old and information about the premises they serve is hazy. It has been found, for example, that two neighbours on a single supply believe they are on separate supplies, arising in two different places. This does not apply to boreholes very often but many spring supplies are located away from houses, on land owned by other people, where access is remote or restricted. Also, populations change; households grow and decline, families move and builders connect new properties to private supplies. Regulation 12 requires that local authorities review their classifications at least once a year, but I do not imagine that many do. If a supply is feeding seven or eight properties, the number of occupants may fluctuate around the twenty-five person limit and will sometimes be a class D and at others a class E. It may prove difficult to convince householders that due to the birth of a baby in the house next door they will be expected to pay for additional sampling, even though the supply itself has not changed. There does not seem to be any answer to this other than to accept that once supplies are separated into various categories, there will be supplies close to the boundaries. The classification bands do not appear to be too narrow however, or the number of classes excessive.

If a local authority needs information about a particular water supply, there are three methods that can be used. Visiting the property, which is time consuming, writing to the householders asking them to complete a questionnaire or serving them with an Information Notice under section 85 of the Water Industries Act 1991. The notice can be served on any person and require such reasonable information as the local authority needs to carry out its duties. The penalty for failure to reply to the Information Notice or making false statements is a fine of up to £5,000. Local authorities will normally only use formal action after an informal approach has failed and in most cases, will not need to because they will be fairly confident that their classification is accurate. An Information Notice can be used for more than classifying supplies and would be appropriate if a local authority needed ownership details for a Private Supply Notice.

The regulations introduced a particular category of supply that has caused more concern and confusion than all the others. These are class F supplies, that is those that only serve one dwelling. They are by far the most common type found and are virtually exempt from the legislation. Section 3.11 of Circular 24/91 (HMSO, 1991c) stated that local authorities should monitor them with a view to ensuring that the supply does not 'pose a threat to public health'. It has been argued that if the only people involved all live in one house, they cannot be defined as 'public' in the wider sense of the term. But, if the government meant to exempt class F supplies from all health considerations, what is the point of discussing them in the Circular. It would also have meant that over 45,000 private water supply users were completely unprotected by the UK government or European legislation.

The exemption arose as a result of a 1989 judgment by the European Court of Justice (*Commission* v. *Kingdom of Belgium*, Case C-42/89) in which the Belgian government questioned the definition of 'supply'. The Directive referred to water 'supplies' and the Belgians successfully argued that this had to involve someone supplying water to someone else. The Court found that if only one property was involved, no one was technically being 'supplied' and thus it was exempt from the Directive. As the Directive led to the Private Water Supply Regulations, the judgment allowed the government to exempt single property supplies. The government also decided to include within class F, multi-occupied single dwellings, even where several households used the same water supply. It is perhaps regrettable that the European Court found in favour of the Belgian government, as the original Directive must have been intended to ensure that all water intended for human consumption was fit to drink. It has been suggested (Kunzlik, 1992) that the original Directive may have stepped outside the bounds of the 1960 Treaty of Rome. The Treaty required that all Directives must provide a 'level playing field' for inter-European trade and it could have been argued that private water supplies have nothing to do with this. Thus the 1980 Directive was *ultra vires* (outside their powers). Kunzlik proposed that the Belgian appeal might have been allowed because, if they lost their case, the Belgians might then question the correctness of the application of the Directive to any private supply.

It is unfortunate that local authorities were not formally required to identify unwholesome class F supplies, although many do so. It may be considered to be over zealous and unnecessarily intrusive to sample and require improvements to class F supplies but these premises may nevertheless contain 'at risk' people who could benefit from protective action by the local authority.

Most agricultural activities that use private water supplies are exempt from the regulations. 'Agriculture' includes activities such as horticulture, fruit growing and seed growing. The only exemption to this rule is a supply that is used for washing food where the quality of the water affects the final product.

When the regulations first came in, responsibility for monitoring private water supplies on dairy farms moved from what was then the Ministry of Agriculture Fisheries and Food (MAFF) to local authorities. Environmental health departments are responsible for the safety of milk from on-farm pasteurization and so also being responsible for the safety of the water used for cooling and cleaning was a sensible move. On-farm pasteurization often causes concern for environmental health officers (Petts, 2003) so extra precautions are needed if they use a private water supply. MAFF originally split dairy farms into two classes – class one water supplies did not need treatment and class two supplies were required to have some form of disinfection. If a supply ever failed a coliform test, it was put into class two but MAFF also carried out a sanitary survey and if the source were likely to become polluted, it was also placed in class two. This would happen even if the bacteriological analysis were satisfactory. The approach was ahead of its time as it involved a risk analysis of the supply that could over-ride the results of snapshot sampling. It is illegal for dairymen to add water to milk and chemical contamination of the milk from any extraneous water would normally be diluted to insignificant levels by the larger volume of milk. As long as there is no cross-contamination from cooling water at the end of the pasteurization process, pathogenic contamination from small amounts of accidentally added water from private supplies should be inactivated by the temperature in the pasteurizer. Unfortunately, small amounts of raw milk are still sold throughout England and Wales ('green top' milk). It has however, sensibly been banned in Scotland, as unpasteurized milk is generally considered to be a health hazard amongst public health professionals.

Monitoring

Schedules 3 and 4 of the regulations dictate the frequency of sampling for each parameter, dependent on the classification of the supply. The formal sampling requirements in the regulations replaced advisory recommendations from 1984 (HMSO, 1984a) and increased the range of parameters in line with the 1980 Drinking-water Directive. Schedule 5 of the regulations lays down the maximum charges a local authority can recover from a relevant person. The amounts are now out of date, as they have not been increased in over ten years. When the new regulations are published, they will raise these costs as it is unlikely that powers can be given to the Secretary of State to raise the charges without having to change the primary legislation. At present, the local authority can charge up to £50 towards the cost of collection but is not supposed to make a profit from this. Charges cannot be made for repeat sampling, class F supply monitoring or any pesticide sampling. Local authorities have therefore had to budget for some of the cost of compliance with these regulations and thus subsidize people who use private supplies.

The only monitoring decisions left to the discretion of a local authority are as listed here:

- the time of year and time of day the samples are taken;
- for smaller classes, which of the forty-three 'additional parameters' detailed in schedule 4 part II, should also be analysed for;
- whether to sample class F supplies;
- which parameters to monitor in class F supplies;
- whether to charge for sampling.

The best times to carry out sampling have been discussed already and local authorities should know which parameters they are interested in, over and above the 'basic' ones. The regulations refer to a parameter that in the opinion of the authority should be investigated because of any 'characteristics of the locality in which the supply is situated or of the supply's distribution system'. The list of additional parameters included most of the remaining ones in the original Drinking-water Directive and comprised of those

- that are likely to be found in water with a high or low pH;
- from different types of geological strata;
- found in water that has had contact with different sorts of pipework (this means copper and possibly asbestos, as lead is in the list of basic parameters);
- associated with treatment systems (such as sodium from water softeners) or surface abstraction (HMSO, 1991c).

Local authorities should be aware of the different characteristics of the geology of their area and the effects that treatment can have on water supplies. Where this is still a problem, advice can be obtained from the local water undertaker or the Drinking-Water Inspectorate.

Should single property supplies be tested and if so, for which parameters and how regularly? A local authority that chooses to sample for unnecessary parameters or too regularly could be accused of acting *ultra vires*. There are also financial restraints on local authorities; criticism can be levelled if auditors feel too much money is being spent on sampling class F supplies to the detriment of another service. It is important to balance the need for knowledge about the safety of water with ensuring appropriate levels of spending for supplies that have been given a low priority by the government. Circular 24/91 (HMSO, 1991c) did not encourage regular monitoring of class F supplies, but there is a moral obligation on local authorities to ensure that public health is not at risk and a statutory one to keep themselves aware of the quality of the water in their area. It would seem appropriate that some form of monitoring programme for single property supplies is in place, even if it is infrequent. Class E monitoring is required every five years, so it is hard to see how class F supplies could be examined more often. Most local authorities decided to sample class F supplies when the regulations first appeared (Clapham, 1993) and the majority either sampled five-yearly (28 per cent) or annually (33 per cent). Sixteen authorities sampled even more regularly than once a year but this seems a tad excessive. There is little point in sampling for aesthetic parameters in class F supplies, so sampling should only be for microbiological health-related parameters, lead and perhaps nitrates in arable areas.

A small number of authorities have taken the decision not to charge domestic properties for the cost of monitoring. This is because they consider that the amount of extra money spent by the authority will not compare with the lost goodwill and the costs involved in the recovery of bad debts. The vast majority, however, have been charging for sampling for ten years and do not have major problems in recovering the money.

Reduced and increased monitoring frequencies

The regulations introduced the concept of reducing or increasing the frequency of sampling for larger supplies based on previous results. If the samples contain concentrations of certain parameters at less than 50 per cent of the prescribed level and in the opinion of the local authority they are unlikely to increase significantly in the next year, the number of samples can be reduced. For some parameters, this can be 66 per cent, 50 per cent or 25 per cent of the normal rate. Some parameters cannot be sampled at a reduced rate. These include the following:

- the bacteriological category, containing faecal indicators and disinfectant residual levels;
- the annual wide-spectrum test that includes several substances which are known to be detrimental to health;
- some of the basic and additional parameters in schedule 3. These include quantitative odour and taste, turbidity, colour and the more commonly encountered contaminants. It is suggested that these were exempted because of the need to monitor their presence accurately over the course of the year.

Conversely, when a parameter fails a prescribed concentration in a larger supply, the frequency of sampling increases for the rest of the year and the following one. The sampling for all parameters increases to at least monthly and for some parameters to twenty-four times a year.

When the Regulations were introduced, the total costs for sampling private water supplies was calculated to be somewhere between £3 million and £5 million per annum (Clapham, 1993). The Government presumably considered the money well spent, otherwise the regulations would not have been written as they were. Whether this was because they were a good public health measure or to stop the UK being taken to court for failure to comply with the Drinking-water Directive is another matter. There did not appear to be a great clamour by the public, pressure groups or the media to introduce the regulation of private water supplies in the UK and in 1984 the Government felt that additional legislation was unnecessary (HMSO, 1984a). So, have the regulations achieved the object of providing an adequate monitoring regime for private supplies? With certain exceptions, yes, they have. The smaller supplies are tested for a sensible set of basic parameters and larger ones are required to conform to a standard similar to that of a public water supply. It must be repeated, however, that the timing of monitoring is vital.

Survey for pesticides

Regulation 13 (6) requires local authorities 'from time to time to ascertain and record the extent to which pesticides and related products are present' in their smaller supplies. Local authorities have had to decide which pesticides to look for, how many samples to take, and when and where to take them. The maximum allowable concentration for pesticides is 0.1 μg/l, above which the water ceases to be 'wholesome', so the local authority can take action to improve the supply if pesticides are found. The actual risk to health, of course, may be minimal or nil.

It is not possible to carry out a survey that looks for all possible pesticides and their metabolites (the 'related products' referred to in the regulation). This is because there are over 400 active ingredients approved for pesticide use in the UK (NRA, 1992). Additionally, the US has over 1,200 basic ingredients formulated into over 30,000 preparations and Western Europe is one of their main pesticide export areas (Hurst *et al.*, 1991). When the regulations were introduced, the Drinking Water Inspectorate advised authorities on setting up a programme to monitor pesticides that followed their guidance for water undertakers (HMSO, 1991c). This guidance included information on the following:

- the most commonly used substances at the time;
- those that had been found in water supplies;
- the usage, solubility and persistence characteristics of pesticides;

- advisory maximum safety concentrations;
- details of examination techniques.

The advice from the Drinking-Water Inspectorate was to carry out annual surveys but as finances have been squeezed, some local authorities have interpreted 'from time to time' more liberally. Where local authorities still carry out pesticide surveys, a fivefold approach is needed:

- Consultation with various organizations to see if any pesticides have been identified in the waters of the catchment areas. The local water undertaker is required to sample annually for a range of pesticides. The results of these surveys should be sent to local authorities. Public and private water supply catchment areas will often overlap and the substances identified by a water utility may also be present in private supplies. Nevertheless, it must be remembered that the water in a particular supply zone may not have originated in that area. The Environment Agency carries out surveys of ground and surface waters and organizations such as ADAS (the Agricultural and Dairy Advisory Service) may collate data on pesticide usage on farms. Water testing laboratories will also have developed expertise in sampling specific suites of pesticides that are considered important in the locality. It would make financial sense if the local authority also used these particular groups, rather than looking for its own unique set of parameters.
- Knowledge of the local farming community will identify the main types of pesticides used. The area may only carry out minor operations, such as sheep dipping or bracken removal. Alternatively, a whole range of agricultural chemicals can be used in arable areas.
- A basic knowledge of pesticides will indicate which ones are more likely to be highly soluble in water (such as MCPA) and which are likely to be quickly degraded before reaching water supplies (such as glyphosate). Advice may be obtained from pesticide manufacturers, their trade organizations and the Drinking-Water Inspectorate.
- Previous testing of water supplies for nitrates may indicate areas that have high agrochemical usage or groundwater that is susceptible to pollution. Pesticides are not solely used by the farming community and therefore will not automatically correlate with nitrates in water and conversely, nitrate levels are not exclusively correlated to fertilizer usage. In arable areas however, one method of organizing a pesticide survey would be to identify supplies where nitrates are highest.
- The survey may need to include a search for the more common pesticides used by non-agricultural organizations. This will include local authorities, railway companies, domestic gardeners, golf clubs and sports grounds.

Circular 24/91 suggested that supplies should be separated into two groups – spring/surface supplies and wells/boreholes. The recommended number of samples in each category ranges from two where there are few supplies to ten for those areas with many. Costs will range from £500 to over £5,000 depending on the number of supplies and the suites of pesticides being analysed for. The survey needs to be carried out at the time when the pesticides are being used and should ensure that sampling points are evenly spaced to give a general picture. Any particularly suspect supplies should also be included. The number of samples must be linked to a realistic expectation of the results. An area of high moorland, with little pesticide usage may only need a few strategically placed supplies to be sampled for a few obvious pesticides such as sheep dip and common herbicides. Areas with a higher pesticide usage will need to carry out a more detailed survey and perhaps devote more resources to this project.

It is debatable whether it is necessary to look for pesticides that are no longer in use, as there is be little that can be done about it. The benefits of such a discovery would be either to prove that the usage had not in fact finished, leading to the possible tracing of illegal use, or that the particular water supply needs treating or abandoning because of the potential risk to the consumer.

A lot of data on pesticides come from research on public supplies, which often use surface sources. The Drinking-Water Inspectorate report on pesticides (DWI, 1992) only used surface water supplies for its discussion on seasonal variation and it stated that the number of detections of pesticides was greater in surface supplies. This is because they are expected to be more vulnerable to pesticide contamination from spray drift, run-offs and field drains. Although spring supplies often have almost direct contamination pathways, private water supplies will rarely come from surface sources such as rivers and streams. Because groundwater supplies will have had more time to reduce the levels of pesticides by degradation, adsorption, etc., it would be expected that pesticides will be found in lesser amounts than in public surface water supplies. Nonetheless, there will always be the potential for accidental spillages or inappropriate disposal to cause significant problems to individual supplies.

Other legislation for securing improvement

The Housing Act 1985

Where the water supply to a dwelling is unwholesome, it may render the property unfit for human habitation. Section 604 of the Housing Act 1985 (HMSO, 1985) states that a house without a wholesome or sufficient supply of water fails the fitness standard. The local authority then has to decide whether in those circumstances, it is reasonably suitable for occupation and, if the water supply is unfit to drink, it would seem that it is not. They are then required by the Act to consider what is the most appropriate action to take. Powers available include closure and demolition. Mind you, demolishing someone's house because it had a contaminated private water supply may be considered a bit drastic, even if it encouraged others to comply with Private Supply Notices a bit more smartly!

Where the property is rented, local authorities should serve a repair notice on the owner, under section 109a of the Housing Act 1985. The notice can require source protection or the provision of water treatment. Failure to comply with this notice could result in prosecution or the work being carried out in default. For owner-occupied property, a notice is unlikely to be served, but it is possible in some circumstances, particularly if grant aid is dependent upon it. Owner-occupiers are normally expected to be responsible for their own property without being forced to improve it by the local authority; so councils could decide to take no action over the unfitness or serve a section 80 Private Supply Notice under the Water Industries Act as a more appropriate use of their powers.

Grant aid

The Water Industry Act 1991 allowed local authorities to assist with the cost of improving domestic private water supplies, if a notice under section 80 has been served. It is not known how many have taken up this option but it is likely to be few, because of the tight financial constraints they operate under. New legislation – The Regulatory Reform (Housing Assistance) (England and Wales) Order (HMSO, 2002) – required local authorities in these countries to produce a formal plan by July 2003 that detailed what they will and what they will not give domestic grant aid for. This allows them a lot more leeway than previously and they can decide therefore to give grants to improve private supplies that have bacteriological or serious chemical problems. They can means-test the grant if they wish, and can also choose to give a grant where a Repair Notice under the Housing Act 1985 has been served. Previously, grant aid was usually only given for works within the curtilage of a property. If works are necessary to the source of a private supply, which is located on land away from the curtilage and not owned by the applicant, local authorities can now consider them as part of a grant application.

The solution to the problem of lead in small water supplies is primarily the replacement of the piping, which can be expensive. Local authorities can decide whether they wish to encourage the replacement of lead piping by giving grants. Sampling the water will enable them to analyse their private supplies for lead content and rank them in order of significance of failure. It is a simple process to contact the persons consuming the water to discuss grant aid. The grading of supplies that fail the lead standard and contacting the householders achieves several useful objectives:

- The worst problems are dealt with first.
- If the grant is means-tested, money is channelled to those most in need of financial assistance and incapable of correcting the problem themselves.

- When discussing the problems of lead, people who will not qualify for a grant will be informed of the dangers and may be persuaded to improve their supply without the need for formal action.
- The intervention could be honed to households containing young children or pregnant women. This has the dual purpose of dealing with those most at risk and contacting parents who may be more inclined to remove lead piping when their children are small than at any other time.

Food Safety Act 1990

Powers are available to environmental health staff if a small water supply is used for food production. Where a food business does not have a potable water supply for cleaning, use in food production and hand washing, the owner of the business can be prosecuted under part VII of the Food Safety (General Food Hygiene) Regulations 1995 (HMSO, 1995) made under the Food Safety Act 1990 (HMSO, 1990). Environmental health officers may also serve an Improvement or Prohibition Notice requiring improvements to be made or for the supply to be discontinued and an alternative one provided. If there is a serious risk to the health of its customers or the water will render food unfit to eat, the business may be closed by order of a Magistrates Court using an Emergency Prohibition Notice. As well as closing the whole premises, the notice can simply require the use of the water supply to be suspended. During food safety inspections, it is important to ensure that food producers are aware of their duties with regard to the wholesomeness of the water. It is not sufficient, for example, to rely on the local authority's monitoring programme to ensure that the water used in the premises is safe for use in food.

Local authority schemes

One interesting alternative to the use of notices or grant aid is the raising of additional revenue from the Council Tax to pay for schemes to connect private supplies with the public mains. I am only aware of this being done once, by Staffordshire Moorlands District Council, but they should be commended for being so innovative. They began a programme in the 1980s that involved an additional two-penny rate to fund first-time water and sewage schemes. Private water supplies in the area were reduced from over 1,200 to 653 within eight years (Taylor, 1993). Money was diverted from mere sampling towards positive improvements, arguably a much more sensible way of spending it. The initial connection programme was targeted, by catchment area, towards supplies that had constantly failed the faecal coliform standard. All schemes were carried out with the full agreement of the householders involved. Smaller schemes were paid for during the year of completion, with larger ones financed over a twelve-year period. The scheme was a forward-looking undertaking, which took a significant health problem and tried to resolve it. It may be considered by many local authorities that schemes like these are a thing of the past and not possible in today's tight financial climate, but it should be remembered that major new health schemes have always occasioned criticism. A *Times* editorial of 3 August 1854 stated, 'A certain amount of opposition may doubtless be anticipated in all measures of sanitary reform. Cleanliness is highly becoming and supremely beneficial, but it involves some trouble and, what is worse, some expense. There will always be found people to vote against a rate, for whatever purpose levied, but in the long run commonsense prevails, and when the reforms are once accomplished everyone testifies to the advantage of the result.'

20 UK governmental control

The main government body with responsibility for drinking-water quality and safety in England and Wales is the Drinking Water Inspectorate (DWI). The DWI is part of the Department for the Environment, Food and Rural Affairs (DEFRA) and is based in London with about thirty-six members of staff. There is a similar department called the Drinking Water Regulation Team in the Scottish Executive based in Edinburgh, and for Northern Ireland there is a department called The Water Service and is in the Northern Ireland Office in Belfast. Their main statutory functions are to check that the public water supply is safe to drink and meets the standards laid down in water quality regulations. These organizations also provide scientific and technical guidance for Ministers and officials of the Government and the regional Assemblies and contribute to debates with international and European bodies on drinking-water-related issues. Each of the three government departments produces an annual report that gives an overview of water quality in their area and how it has changed over the years, plus discussions on any topics of general interest that have recently arisen. In addition, they produce guidance on various topics related to water, including technical advice, general information to the public on water safety issues and official direction accompanying legislation.

Water undertakers have to report all events to these bodies that are likely to have affected water quality. They will then assess each event to see what caused it, whether the water became unfit for human consumption, what the undertaker did about it, if it was avoidable and if any regulations were breached. They then decide whether to take any further action to prevent a recurrence of the problem or to punish the perpetrator and if they do, what form that will take. They can prosecute for breaches of legislation, but do not do so for every minor problem. The inspectors also investigate complaints from the public, and study outbreaks of waterborne disease.

The departments run various expert committees that look into particular problems of drinking-water. The DWI for example, has expert groups for *Cryptosporidium* and disinfection by-products. It also has money to invest in research to find out about these problems and to help solve them. This is called the Water Distribution, Conservation and Quality (WDCQ) research programme. This programme helps the DWI assist in formulating policy on the quality of water and whether additional regulation or advice is needed to deal with newly identified problems or changes in legislation from Europe. Recent research topics include *Mycobacterium avium*, a study of the distribution of Crohn's disease, analytical methodology for detecting steroids in water, mutagenicity testing for various chemicals and the efficacy of membrane treatment technology. The DWI has also initiated useful research into the quality of private water supplies, particularly *Cryptosporidium* incidence.

These bodies are the overall guardians of private water supplies. Each year, for instance, the DWI audits a percentage of local authorities to ensure that they are complying with the regulations. This is a small programme and in the year 2000 only one local authority was inspected. It is unlikely that many more will be inspected until the new regulations are introduced. A senior inspector also coordinates responses to requests for guidance from local authorities and takes part in and coordinates finance for seminars on various aspects of quality and safety in small water systems.

These bodies can also commission guidance on private water supplies. The DWI has recently helped produce a revised *Manual on Treatment for Small Water Supply Systems* (WRc-NSF, 2002). This is a useful guide to the different treatment systems available and provides a good background to the problems of private water supplies. A final responsibility for the DWI is to collate information from local authorities on the quality of water in private supplies and how they are meeting their regulatory responsibilities. Unfortunately, this is not a statutory requirement and not every local authority provides this information, therefore a complete picture has never been produced.

21 USA

Like the UK, water supplies in the US are either public or private but the definitions are different and more complex. In the US, a public supply has to serve at least fifteen permanent connections or twenty-five people for at least sixty days. It can serve anything from a large metropolis to a small trailer park and may be provided by a city, organization or a private company.

Public supplies are then either community or non-community supplies. A community supply has to have at least fifteen permanent connections or twenty-five residents on a year-round basis. A non-community supply serves at least twenty-five individuals for at least sixty days in the year. To make it more confusing, if the water serves an average of twenty-five different people daily for at least six months of the year, it is a transient supply. If they are the same people, for example, at a school or factory then it is a non-transient supply. If the supply is none of these, it is a private supply and usually serves a single domestic property or business. Many of the community and non-community supplies in the US would therefore be classified in the UK as private water supplies.

Generally, small water supplies in the US are boreholes (invariably called wells) but there are a few exceptions such as on the Southern coast of Louisiana. Here supplies are subject to saline intrusion and alternative sources such as surface water or rainwater harvesting have to be employed. In more mountainous states, the sources of water are split between boreholes and surface and spring supplies. In Tennessee, for example, this is approximately 50 : 50 (Foster, 2003).

A report in 1997 (USGAO), estimated that 15 million households obtained their water from private supplies. At that time, coliform bacteria contaminated up to 42 per cent, and 18 per cent had excessive levels of nitrates. Between 1971 and 1985, 45 per cent of all reported waterborne outbreaks were associated with non-community water systems. In 1985 for example, there were nine outbreaks (60 per cent) from non-community supplies and six of those were from seasonal supplies – that is campsites, parks or resorts. At that time, there were 180 million community, 20 million non-community and 30 million individual water supply users in the US (St Louis, 1988).

Public supplies in the US are closely regulated and whenever water is provided to more than twenty-five people, there are clear state or federal standards and requirements (Louisiana DTE, 1985). All water utilities must be licensed, with an annual inspection of the water treatment plant and a yearly report sent to consumers detailing water quality (Watson, 2003). The regulation of private water supplies is less clear-cut and several agencies can be involved. The US is, of course, a collection of states with their own legislation based on federal law. As with European water laws, individual states adopt overarching legislation and can include tighter standards but not laxer ones. It is therefore difficult to discuss general water issues in absolutely definitive terms across the whole US. The information considered here may not therefore apply to every single state, but represents the situation in most of them.

The authority for enforcing drinking-water legislation depends on whether a state or Tribal Council has formally applied for what is called 'primacy'. If it has and meets the conditions required by the legislation, it and not the Environmental Protection Agency enforces the 'Safe Drinking Water' statutes. Most states go for primacy and enforcing authorities have wide-ranging powers to control water quality. These include closing systems, seizing records, serving notices and prosecution. Most can also take action based on poor results from a sanitary survey. Nevertheless, they help and advise where possible and try to maintain a supportive relationship with the water industry. But, if there is a risk to public health, they are usually happy to use their legal powers.

The direct influence of surface water

The US has a fundamental split between types of water supply and that is whether they are 'under the direct influence of surface water' or not. Borehole or well water in the US is normally expected to come from a deep aquifer, free from microbiological contamination and not subject to sudden changes in turbidity, conductivity, temperature or pH. The boreholes are usually protected against any contamination from the surface and can be relied upon to provide reasonably safe water. Any other supply, where a continuously good quality of water cannot be guaranteed, is considered to be 'under the direct influence of surface water'. A good definition of this is a supply characterized either by

- a significant occurrence of insects or other macro-organisms in the water;
- the presence of algae or organic debris;
- large diameter pathogens in the water such as *Giardia lamblia*; or
- a supply that is subject to significant water quality changes following precipitation (TDEC, 1991).

Each public supply normally has to be assessed to see whether it is or is not under the direct influence of surface water. It should not be assumed that boreholes are never under the influence and spring supplies always are. A borehole for example, may be in a karst limestone region and easily contaminated. The supply to a spring on the other hand, may be well protected, both by surrounding impermeable layers of rock and its collection chamber construction (although that is less likely). Most states will have a set of guidance rules to assist engineers or health officials in deciding whether a particular supply falls within this category (TDEC, 1991). There will, for example, be a requirement to test the raw water for faecal coliforms on a regular basis for a particular period of time. There will also be an expectation to test the system after heavy rain.

If a supply is not found to be under the direct influence of surface water it will only require chlorination but if it is, a full filtration system is needed. This will, for example, require a treatment system that provides a 3-log removal of *Giardia* (99.9 per cent). Therefore, being classed as under the direct influence of surface water brings some significant expenses to small water providers. All the same, this appears to be a very sensible way of dealing with potential problems from small water supplies. If the supply can be relied upon to provide safe water, it only has to be chlorinated. In fact, in a few states, if the water to a small supply has proved to be free from coliform indicator organisms over a long period of time, chlorination is not strictly necessary; but this is rare and the possibility of being sued for damages tends to ensure all public and small non-community systems are chlorinated. If the source cannot be relied upon however, not only does it need to be chlorinated but it also must have a good standard of treatment and be regularly monitored, at least monthly. This will improve the water quality to an acceptable standard for many parameters, not just bacteria. Of course, this does not apply to individual private wells and in most states it only applies to public water supplies. However, some states, Michigan for example, have adopted a more stringent approach to the safety of smaller supplies and the treatment rule applies to all those providing water to two or more households (Brown, 2003).

The UK does not presently have a risk-based approach to small water supplies and the use of residual disinfection is not obligatory even when private supplies provide water to a great number of people. This is unfortunate. Worries about disinfection by-products perhaps affect this and the protective nature of chlorination, both for accidental contamination of the source and subsequent problems with storage or pipework, is not fully appreciated by supply owners and some local authorities. The UK reliance on infrequent sampling to identify contamination and then asking for treatment to be provided appears to be a much less effective approach than that of the US.

Borehole protection

As well as the general rules about supplies under the direct influence of surface water, the approach to the safety of individual supplies is based on a prior approval system, which appears appropriate to the needs of most areas (Carlson, 2003). The prior approval systems require that the following are adhered to.

- Boreholes are only drilled by licensed operators. An operator will be visited every time they drill a borehole until the inspecting engineer is satisfied they are competent and conscientious. Private water supplies can be exempted from this requirement, if the householder drills the borehole themselves. They must however comply with the same strict rules and regulations as the licensed drillers and, as this is a skilled and technical operation, it is unlikely that many people would not use a licensed operator.

- Drilling operations have to comply with a comprehensive set of rules. Most states have a manual that details everything that needs to be done, from deciding where to drill to post-completion water testing. The rules include carrying out a survey of the area to identify potential contaminating sites, details of construction, including casing, screening and filling in the annular space (between the hole and the borehole casing), post-drilling chlorination and wellhead design (Louisiana DTE, 1985).

- All new boreholes are formally registered. This has been a legal requirement for many years and ensures the authorities are aware of the supplies in their area and that the water has been tested and is satisfactory. In addition when a property is purchased, the buyer can check if it is registered. In some states there is no legal duty to register, but it is often carried out to protect the owner in case someone else tries to claim the land or the water.

- Boreholes cannot be sited within a certain distance of a number of potential pollution sources, such as, for example, septic tanks, streams and sewers (50 ft – 15.24 m), cesspools and outdoor privies (100 ft – 30.48 m, but for single domestic supplies it can be 50 ft – 15.24 m) and sanitary landfill sites, feed lots, manure piles and similar (100 ft – 30.48 m) (Louisiana Register, 2002). Some states also have a 'two for one' rule. For every extra two feet of casing, a pollution source can be one foot nearer (LaBarbera, 2003).

- When the drilling is complete, the well has to be shock-chlorinated and then tested bacteriologically. The water can only be drunk when the tests are consistently negative.

Licensed drillers quickly become experienced in the geology of the area where they operate. In Brazos, Texas for example, they know that they must drill through one aquifer with high iron and manganese concentrations into a better quality one below it (Alonso, 2003). When the well is being drilled, an experienced driller can tell when the aquifer has been reached as the tailings being brought to the surface usually change consistency. Boreholes are

Plate IX A picture of a borehole being drilled in Loiuisiana. (See also Colour Plate IX.)

normally drilled using a rig attached to the rear of a specially adapted vehicle, although, where access is difficult or the ground is particularly wet, a separate, towed rig can be used. Some unscrupulous drillers have been known to 'deep drill' (go deeper than necessary) to raise costs. In some areas, if they go too deep (to below sea-level) the water becomes saline. When drilling is complete, the casing is added and carefully jointed. Casings are either metal (usually steel) or plastic (PVC). Depending on the use of the borehole, the backfilled grouting filling the annular space has to be provided to a specified depth. In private domestic boreholes, this is usually 10 ft (3.04 m) below ground level. Public or community supplies need backfilling down to the screen (this is the metal grill at the bottom of the borehole, where the water enters). Vacuum pumps are usually fitted to US boreholes. Vacuum pumps have to be kept primed with water to enable them to work and the maximum depth they can operate at is about 200 ft (60.96 m) (Ham, 2003). If it is necessary to lift water to a greater height than this, a submersible pump must be used.

In some states, private water supply treatment installers also have to be licensed. Sometimes installers over-prescribe remedial measures or install inappropriate treatment. They can make a lot of money doing this, but also risk losing their license.

Other than to ensure there are no new contamination sources in the vicinity, a regular sanitary survey is inappropriate once most boreholes are completed; they are protected from surface contamination and the weather has little impact on water quality. Of course, if a major hurricane passes over and protection is ripped out of the ground, this may cause a problem. Similarly, where flooding is common, measures have to be taken to prevent floodwater getting in.

When a public supply is operational, it needs a minimum number of licensed operatives, depending on size. They must take daily measurements of free chlorine and report the results monthly. Utilities need a 'water quality monitoring plan' and a 'wellhead protection plan', which has to be updated every three years. In some states, schools also need a 'water emergency plan' (which usually involves a plan to go out and buy bottled water).

Where there is a problem with a public water supply, the electronic media and local press have to be informed. If a campsite or trailer park has water problems it has to post notices until the problem is corrected. However, there is no requirement to report previous sampling results to new arrivals. If the problem is so bad as to require immediate action, a 'boil water advisory' can be issued. Some states issue these as a matter of course where there is a bacteriological failure; others are more cautious and only issue them for very serious problems. An example from Tennessee involved flooding from heavy rainfall and snowmelt that caused muddy, turbid water to enter a small town's water treatment works, blocking it completely. A boil water advisory was in place for the three weeks it took to put the problem right (Foster, 2003). Occasionally a water utility expects the enforcing authority to stop them supplying water completely when there is a really serious problem, but because of the obvious need for water for toilet flushing and fire safety, it is usually considered to be a better idea to issue a boil water advisory.

Where water is in short supply and there are problems with over-abstraction, some states have identified areas classed as Groundwater Conservation Districts (or similar). In these control zones, water use is limited and maximum volumes are laid down for abstraction. Normally, people are not allowed to use up water from adjoining properties, but in the majority of Texas this rule does not apply. This has been case law since 1904; if someone moves next to you and drills an enormous borehole and the drawdown renders your well dry, then that is just your bad luck. This is locally known as the 'law of the biggest pump' (LaBarbera, 2003).

Another interesting difference between the UK and the US is the litigious approach of many American consumers (although the UK appears to be rapidly catching up). It has been suggested (Wood, 2003) that in the US environmental legislation is often irrelevant because of the strength of the civil liability lobby. In other words, companies and individuals are more frightened of being sued by someone who has been injured by the water they supply, than they are of falling foul of the more formal legal system.

Most companies take precautions to prevent themselves from being sued and this is their main driver, far outweighing any requirements of the Environmental Protection Agency or state enforcement body. When properties change hands for example, the company giving the loan (the loan agent) requires a certificate of potability. This is to prevent the possibility of being sued by the new occupant, if they subsequently become ill from the water. The seller has to get the water sampled and pays for a certificate. The result is that the water supply is checked for safety whenever the property changes hand. A similar certificate is also required by employer liability insurance companies for small public or private systems at commercial premises such as restaurants or garages. They do not want to have to pay money because their client is being sued for drinking-water related ill-health problems.

In some states chlorination is rare in single domestic borehole supplies, nevertheless bacteriological quality is usually good if the borehole has been drilled within the last fifteen to twenty years. Out of 3,319 private and public borehole sampled by Texas A&M University only 7.5 per cent were contaminated with faecal material (Dozier, 2003). In mountainous areas, where there are more spring supplies, the quality is more variable. In Tennessee, where 240,000 families have their own private borehole, water quality ranges from good to very poor (Foster, 2003). The majority of the water from these private supplies is unchlorinated and they are not regulated at all if the supply is not from a borehole. As I was told (Hall, 2003), 'If someone wants to drink out of a muddy hole and its a private supply, then they can – we still have few freedoms left!' Some of the problems encountered in Tennessee have included a cross-connection with an irrigation borehole at a country club (to save money); dead raccoons and skunks being found in the system (with fur coming out of the tap) and a cross-connection with a sewer pipe at a prison, where a third of the staff and prisoners subsequently became ill.

Education and advice

Many authorities have education programmes for the public, license holders and regulators. The one in Tennessee for example, has been very successful, with microbiological failure rates dropping dramatically year on year (Foster, 2003). This programme consists of simple, well-explained, graphically descriptive teaching methods that try to give an understanding of the processes and problems involved. One example is the way water operatives are warned about the problems caused by falsifying records. If there is a problem with the water and they are worried and need help, operatives are told that if they let the authorities know, they will get it – but if they lie about the results then they may lose their job or get sent to jail.

Non-regulatory advice and assistance comes from several sources including the 'Farm A Syst' programme, which encourages farmers to get their borehole tested annually. It can also give financial support for health-related improvements. Rural Water Associations are independent organizations helping owners of small water supplies. Some get state funding but they are mainly membership organizations that provide practical advice and advocacy to members (LRWA, 2002). Membership is open to anyone and there is no minimum number of properties that they will not help. The Associations have advisory leaflets and organize seminars on particular topics for members. To attract participants, the associations always provide freshly cooked food at their meetings.

Sanitarians in County Environmental Health Departments have some responsibility for drinking-water safety. In many states, the owners of all non-private supplies have to send samples to them on a monthly basis (Alonso, 2003). The samples can be posted and as long as they are received within 30 hours, have not leaked and the water is not too turbid, are examined for the presence of coliforms. Those that fail the standard are notified to the person responsible for the supply and the state. If a failure occurs, more samples are taken and advice is given on potential causes of contamination. Sampling for a larger suite of chemical contaminants is less regular and depends for frequency on the size of the supply and the particular substances being looked for. The owners of the supply can take the water sample themselves but must have attended a training course to be allowed to do it. This does not appear to cause any major problems other than the occasional failure due to bad practice. Sampling is carried out at the source and not the kitchen tap, as is done in the UK and a sampling tap is usually incorporated in the wellhead design. Water is run for some time prior to sampling but the tap is not usually sterilized (Pontiff, 2003). Sometimes however, chorine solution is run over the tap to prevent cross-contamination from a leaking washer. The water is collected in specially produced small plastic bags that contain sodium thiosulphate. Where the owner of a property is concerned about their water, sanitarians will visit, sample the water and advise the householder about treatment and best practice. Laboratories also provide simple advice to well owners on potential reasons for contamination, what the results of water testing means and how to chlorinate and flush the system if there is a microbiological problem.

Some universities provide independent water-testing facilities for both chemical and microbiological parameters. These can be for drinking-water or water used for irrigation. They tend to be easier to contact for some rural people than a state regulatory department, mainly because of worries about enforcement action, real or imaginary. University laboratories can be used by anyone and they often receive posted samples from a large geographical area. Plastic containers (such as a baby's drinking bottle) are sometimes recommended for people sending in a sample where the local water has a boron problem (Pitt and Provin, 2003). Boron can restrict plant growth, thus if a supply has a high boron content, it may have to be abandoned for irrigation purposes. As most glass contains boron, if it is leached into the sample, the laboratory result can be prejudiced and will make a borderline water supply seem worse than it is. The normal solution to the problem of high boron in water is to try and blend it with water containing lesser amounts. Another chemical causing irrigation problems is sodium. In Texas for example, levels of up to 700 µg/l occur. In soil, this produces a pH of 10 and can turn the crust to 'concrete'.

Indian land

In the US, Indian land is often subject to a different set of rules, with tribal leaders forming the local council. A good example of the way this works is the Tohono O'odham Indian Nation, in southern Arizona. Water, electricity, gas and telephone services are all provided by the central Tohono O'odham Utility Authority. It is the federal Environmental Protection Agency that checks that legal water responsibilities are met, rather than the Arizona authorities. An interesting aspect of the work of the Utility's water laboratory is that it is much more hands-on than those in the UK (Natividad, 2003). As well as being responsible for a certified microbiology laboratory, the manager collects water samples, does field-testing, cleans out and disinfects water tanks and carries out minor repairs to pumps and treatment plant. To maintain this additional skill base, there are twice-monthly training sessions and some repairs are helped by mobile phone conversations with engineers back at head office. The manager believes this makes it a much more interesting job, as well as making sure that minor repairs are carried out straight away. At one village we visited, she carried out on-site testing for free residual chlorine. It was too low, so she went back to the treatment plant, checked through the chlorination system, found a fault, repaired it and reconnected the system – all within the space of an hour. The water was re-tested the next day to ensure the repair had been carried out correctly. This is a very efficient way of maintaining drinking-water safety.

There are two boreholes at each Tohono O'odham village site so that supplies can be maintained if one of them fails. The water from the borehole passes through a chlorinator and then to a storage tank. Disinfection units are in waterproof cabinets and a sturdy fence surrounds the whole site with barbed wire around the top. Most of the pumps are above ground but a few are sunk in covered compartments. These are not popular with staff as there is a problem with illegal immigrants from Mexico and these underground areas have been used as hideouts. Other on-site dangers include rattlesnakes and rats. The area around the water stations is open to grazing, but the fence ensures that there is an adequate protection zone.

Problems

Chemical contamination problems include leaching (arsenic), agricultural pollution (nitrates and pesticides), copper and lead from pipework and high levels of radiological parameters (uranium or radon). Iron and manganese are also commonly found but because aluminium is so ubiquitous and not considered to be a health-related parameter, it is not analysed for in most laboratories. These problems are comparatively infrequent and, with the possible exception of arsenic in some areas, most small regulatory authorities only have to deal with a handful of failures at any one time.

One of the problems that seems to be encountered more often in the US than in the UK is contamination from pollution incidents. These are usually a result of inappropriate industrial waste disposal or accidental spillage. One example is the past practice of pouring used dry cleaning fluid (trichlorethane, etc.) on the ground to dispose of it cheaply; another is leakage from defective underground gasoline storage tanks. In rural areas the herbicides atrazine and simazine can also be found in small drinking-water supplies during the spring spraying season. In Louisiana, the Department of Agriculture has about 150 test boreholes around the state to monitor the groundwater. There are also many 'hits' for pesticides in surface water samples after the crops are sprayed and education programmes have been designed to teach farmers how to use the chemicals responsibly (Carlson, 2003). Nitrates can also be a problem if water supplies are near fertilizer usage areas or septic tanks.

In several states, particularly in the South and West, such as Arizona or California, naturally occurring arsenic can be a serious problem in groundwater. Even in central states such as Michigan, the main driver for change at the moment is the reduction in the amount of arsenic allowed in drinking-water (Brown, 2003). The US Environmental Protection Agency has a new standard of $10\,\mu g/l$ for arsenic and many boreholes will fail this standard – some have up to $200\,\mu g/l$. The majority have less however and it is considered by some authorities that $25\,\mu g/l$ would be a more suitable standard. The reason is that at this level, there is a sufficient safety factor to prevent illness but far fewer boreholes would need remedial treatment. Many systems are presently completely closed and deliver pristine groundwater, but if treatment is installed, they will have to be opened up, allowing a possible entry route for contamination (Brown, 2003).

Of course, once chemicals contaminate an aquifer it is difficult to remove them. Depending on the chemical and the extent of the problem some boreholes have to be abandoned and new ones drilled or expensive treatment installed. Sometimes, for minor problems, the treatment will only involve increased pumping, so that the pollution is physically removed, with the contaminated water run to waste. It is important that the water does not re-contaminate another part of the aquifer, so advice has to be sought on its disposal. Another remediation method involves pumping hydrogen peroxide into the borehole. This feeds oxygen to naturally occurring bacteria in the borehole that then breakdown the contaminants in the water.

Some older systems have regular coliform and nitrate failures. Where problems are encountered with a small village supply, the usual option is to form a local committee and ask the nearest town or city to connect it to the public supply. Very often a way can be found to achieve this using a variety of grants and easements (Foster, 2003). The second choice is to drill another borehole, away from the problem. Many enforcement authorities are not keen on recommending treatment for individual boreholes and see this as a last option; they much prefer to ensure that uncontaminated groundwater is used. As with arsenic removal systems, whenever microbiological treatment is put in, a potential path for contamination is introduced. In addition, many treatment systems are not maintained and if this is the case why put one in only to have it fail later on?

Effluent from small sewage treatment plants often causes these problems. When the septic system blocks up, a safety valve opens allowing raw sewage to flow overland. In other areas there are septic tanks that merely discharge into shallow field drains. Where heavy clay soils do not allow the sewage to soak into the ground, it ponds on the surface and shallow boreholes can easily become contaminated. A community of about one hundred people in Tennessee was recently subject to a series of hepatitis A outbreaks because of this (Foster, 2003). The community was poorly educated and did not appreciate that the overflowing sewage in their village was making them ill. Eventually however, they were given federal financial assistance to sort the problem out.

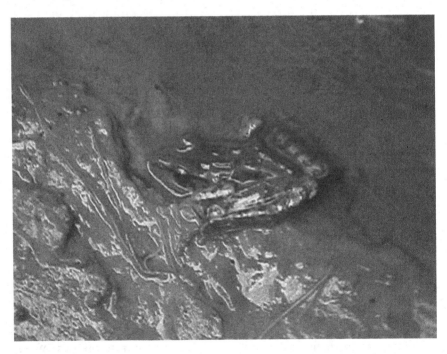

Plate X A frog in a well in Nashville, USA. (See also Colour Plate X.)

Health Departments and other enforcement authorities are normally strict in their control of the rural sewage treatment systems they are aware of; all septic systems should be visually checked three times a year by a licensed contractor and sampled annually. Another legal requirement for most states is that any sewage discharged onto the surface must be chlorinated. Nevertheless, it is generally accepted that there are many problems, in the septic systems still to be found.

Abandoned boreholes are another problem in some states. Usually defined as a well that has not been used for more than six months, abandoned, unprotected boreholes can contaminate an aquifer by short-circuiting any protective overlying geology. There are rules for capping abandoned boreholes in order to prevent this problem but this can cause major logistic headaches. There are, for example, approximately 800,000 uncapped abandoned boreholes in Texas (Mayhew, 2003).

In the mountain areas of Tennessee, there is also a problem with the quality of water at many rural churches. They have their own community water supplies, but because they are not commercial organizations, there is little money for repairs or improvements. The local courts do not look favourably on state agencies taking formal action against them, and the pastor is often unwilling to be told what to do by 'government bureaucrats'. If the congregation are advised about the water and still chose to use it, then it is considered that they have made an informed choice and will often be left to their own devices.

In rural areas therefore, a dual approach of enforcement and education has to be carefully planned to ensure that the water people have drunk for years gets improved. One village in Tennessee had a very poor spring supply, with a filter system so old that any treatment media had been washed away years before. The village, particularly its water utility manager, was very unwilling to spend any money on the supply and considered the state's Water Supply Division an irritation when they suggested improvements could be made. Because of the strength of the village's opposition, a public meeting was organized to discuss the situation. Previously, the Division had taken samples of the water and photographed the creatures they had found in it. At the meeting they showed large slides of them, particularly the nematodes, which were 'big ugly critters'. As part of the presentation the nematode's anus and genitalia were pointed out to the crowd. Some people in the audience were then physically sick. One of the people who attended on behalf of the Division said that towards the end of the meeting he thought that the utility manager was going to be lynched (Foster, 2003). The water supply was quickly improved.

References

ACSH (2002) American Council on Science and Health ASCH Holiday Dinner Menu 2002. Online article http//www.acsh.org/publications/booklets/menu02.html accessed 6 May 2003.

Alegre, M. (1999) Oficina de Aesesoria y Consultoria Ambiental, Lima, Peru. Personal communication.

Alexander, A. and Heaven, A. (1992) Its in the pipeline: the continuing hazard of lead in domestic drinking water. *Paper for Conference on Housing and Health, Manchester GMB College.*

Allott, K. (1991) Nitrate, nitrate everywhere. *Process Engineering Environmental Protection Supplement* 1991, 23–4.

Alonso, M. (2003) Brazos County Health Department, Brazos, Texas. Personal communication.

Altvater, F. (2001) Personal communication.

Amin, A.B. *et al.* (1979) Incidence of helminthiasis and protozoal infections in Bombay. *Journal of the Indian Medical Association* 72(10), 225–7.

Anderson, I. (1992) When contaminated food can be good for you. *New Scientist* 1840, 6.

Anon. (1997) Drinking water – a source of exposure to oestrogenic substances? *Water and Health* North West Water (20), 1–2.

Anon. (1999) Aluminium study puts drinking water in the clear. *Water and Health* North West Water (29), 3–5.

Anon. (2000) Cholera gives up its secrets. *New Scientist* 167(2250), 25.

Anon. (2001a) Turkish women boycott sex. *Cape Cod Times*, 15 August 2001 C9.

Anon. (2001b) EPA to implement 10ppb standard for arsenic in drinking water. *Environmental Protection Agency Office of Water, Ground Water and Drinking Water Arsenic Factsheet EPA 815-F-01-010* October 2001.

Anon. (2001c) Feedback. *New Scientist* 2294, 104.

Anon. (2002a) There's water on Mars, but you wouldn't want to drink it. *New Scientist* 174 (2346), 11.

Anon. (2002b) Trends in selected gastrointestinal infections 2001. *Communicable Disease Review Weekly* 12(7), 3.

Anon. (2002c) Food and waterborne diseases associated with travel part two: hepatitis a, cholera and typhoid. *Travel Health CDR Weekly*, 4 July 2002.

Anon. (2003) Outbreak of Verocytotoxin-producing *Escherichi Coli* O157 (VTEC O157) and *Campylobacter* spp. associated with a campsite in North Wales, *CDR Weekly News*. Online article http://www.phls.org.uk/publications/cdr/archieve02/news/news3402/html accessed 12 May 2003.

Aquafine (1982) Aquafine Corporation. *Ultraviolet Sterilizers Owners Manual.* Aquafine UV Water Treatment System.

Armstrong, C.W., Stroube, R.B., Rubio, T., Sidudyla, E.A. and Millar, G.B. (1981) Outbreak of fatal arsenic poisoning caused by contaminated drinking water. *Archives of Environmental Health* 39(4), 276–9.

Attorney General (2003) Report of the Wakerton *E. coli* O157 Outbreak. Online report http://www.attorneygeneral.jus.gov.on.ca/english/about/pubs/walkerton/part1/ accessed 12 May 2003.

AWWA (1999) American Water Works Association – Microbiological Contaminants Research Committee, *Journal of American Water Works Association* 91(9), 101–9.

Baker, E.L., Landrigan, P.J., Field, P.H., Basteyns, B.J., Bertozzi, P.E. and Skinner, H.G. (1978) Phenol poisoning due to contaminated drinking water. *Archives of Environmental Health* 33(2), 89–94.

Ball, K. (1993) Personal communication.

Barraclough, J., Collinge, R. and Horan, N.J. (1988) The quality of private drinking water supplies. *Environmental Health* 96(11), 12–6.

Barrell, R.A.E. (1989) The microbiology of treated and untreated small private water supplies in relation to the European Community guidelines. *Environmental Health* 97(7), 171–3.

Bassett, W.H. (Editor) (1999) *Clay's Handbook of Environmental Health Eighteenth Edition*, E&FN Spon (London and New York).

Bell, C. and Kyriakides, A. (1998) *E. coli: A Practical Approach to the Organism and its Control in Food*, Blackie Academic and Professional (London, Weinheim, New York, Tokyo, Melbourne, Madras).

Bell, F.A. (1991) Review of effects of silver-impregnated carbon filters on microbial water quality. *Journal of American Water Works Association* 83(8), 74–6.

Beller, M., Ellis, A., Lee, S.H., Drebot, M.A., Jenkerson, S.A., Funk, E., Sobsey, M.D., Simmons, O.D. 3rd, Monroe, S.S., Ando, T., Noel, J., Petric, M., Middaugh, J.P. and Spika, J.S. (1997) Outbreak of viral gastroenteritis due to a contaminated well: international consequences. *Journal of the American Medical Association* 278(7), 563–8.

Bergeisen, G.H., Hinds, M.W. and Skaggs, J.W. (1985) A waterborne outbreak of hepatitis A in Meade County, Kentucky. *American Journal of Public Health* 75, 57–63.

Birchall, J.D. (1991) The role of silicon in aluminium toxicity. In *Alzheimer's Disease and the Environment*. Lord Walton of Detchant (Editor) Royal Society of Medicine Services (London).

Boccia, D. Eugenio Tozzi, A., Cotter, B., Rizzo, C., Russo, T., Buttinelli, G., Caprioli, A., Luisa Marziano, M. and Ruggeri, F.M. (2002) Waterborne outbreak of norwalk-like virus gastroenteritis at a tourist resort, Italy *Emerging Infectious Diseases* 8(6). Online article htpp//www.cdc.gov/ncidod/EID/pastcon.htm accessed 29 April 2003.

Brady, P.G. and Wolfe, J.C. (1974) Waterborne giardiasis. *Annals of Internal Medicine* 81, 498–499.

Brassington, R. (1995) *Finding Water (Second Edition)*, John Wiley and Sons Ltd (Chichester, New York, Brisbane, Toronto and Singapore).

Bray, D.T. (1990) Performance and application of RO systems. In *Point of Use/Point of Entry. Pollution Technology Review 188*, US Environmental Protection Agency/American Water Works Association.

Bremer, P.J., Webster, B.J. and Wells, D.B. (2001) Biocorrosion of copper in potable water. *Journal of the American Water Works Association* 93(8), 82–91.

Bridges, J.W., Wheeler, D. and Bridges, O. (1991) Water quality – does 'don't know' mean 'safe'. In *New Health Considerations in Water Treatment.* Holdsworth, R. (Editor) Avebury Technical (Aldershot, Brookfield, Hong Kong, Singapore, Sydney).

British Water (2003) Website address http//www.britishwater.co.uk/contact_us/contact_us.html accessed 30 April 2003.

Brown, E. (2003) Michigan Department of Environmental Quality, Ann Arbor, MI. Personal communication.

Brown, M. (1992) Radon and geology. *Environmental Health* 100(4), 96–8.

Brugha, R., Vipond, I.B., Evans, M.R., Sandifer, Q.D., Roberts, R.J., Salmon, R.L., Caul, E.O. and Mukerjee, A.K. (1999) A community outbreak of food-borne small round-structured virus gastroenteritis caused by a contaminated water supply. *Epidemiology and Infection* 122(1), 145–54.

Bryant, E.A., Fulton, G.P. and Budd, G.C. (1992) *Disinfection Alternatives for Safe Drinking Water.* Van Nostrand Reinhold (New York).

Bryson, B. (2000) *Down Under.* Doubleday (London).

Buchanan, S.D., Diseker, R.A. 3rd, Sinks, T., Olson, D.R., Daniel, J. and Flodman, T. (1995) Copper in drinking water, Nebraska, 1994 *International Journal of Occupational Environmental Health* 5(4), 256–61.

Burfitt, C.K. (1997) A critical assessment of the effectiveness of private water supply monitoring regimes in terms of microbiological parameters. MSc Thesis, University of Leeds.

Burke, I. (1991) Eliminating nitrate from drinking water. *Water and Waste Treatment* November 1991, 40–52.

Butterworth, W. (2002) Eliminating nitrate pollution, *Water and Effluent Treatment News* April 2002, 8–9.

Button, J. (1988) *Green Pages.* MacDonald & Co (Publishers) Ltd (London).

Campbell, A.T., Nicholls, P.H. and Bromilow, R.H. (1991) Factors influencing rates of degradation of an arylamide in subsoils. In *BCPC MONO No. 47 Pesticides in Soils and Water.* British Crop Protection Council.

Candy, J.M., Oakley, A.E. and Klinowski, J. (1986) Aluminosilicates and senile plaque formation in Alzheimer's disease. *The Lancet* 340, 354–7.

Carlile, P.R. (1994) Further studies to investigate microcystin-lr and anatoxin-a removal from water. *Foundation for Water Research* Report no FR0458 – April 1994.

Carlson, C. (2003) Louisiana Department of Health and Hospitals, Office of Public Health, New Orleans, Louisiana. Personal communication.

Casemore, D.P. (1991a) Cryptosporidium (effects, detection and elimination). *seminar Cryptosporidium (Effects, Detection and Elimination) Wakefield.*

Casemore, D.P. (1991b) Cryptosporidium and protozoan parasites and the water route of infection. *Health and Hygiene* 12, 84–9.

Castle, R.G. (1988) Radioactivity in water supplies. *Journal of Institute of Water and Environmental Management* 2(4), 275–84.

Chadwick, E. (1842) *The Report on the Sanitary Condition of the Labouring Population.* Report for the Poor Law Commissioners.

Chalmers, R. (2003) Head of the Cryptosporidium Reference Laboratory, Wales. Personal Communication, Royal Institute of Public Health seminar on Cryptosporidium and Swimming Pools, London, November 2003.

Chorus, I. and Bartram, J. (Editors) (1999) *Toxic Cyanobacteria in Water: A Guide to Their Public Health Consequences, Monitoring and Management*. E&FN Spon on behalf of WHO (Geneva).

Clancy, Jennifer L., Bukhari, Zia, Hargy, Thomas M., Bolton, James R., Dussert, Bertrand W. and Marshall, Marilyn M. (2000) Using UV to inactivate Cryptosporidium. *Journal of the American Water Works Association* 92(9), 97–104.

Clapham, D. (1993) *The Implications of the Private Water Supplies Regulations 1993 for Local Authorities*. MSc Thesis, University of Leeds.

Clapham, D. (1995) Unfit private water supplies. *Water Law* 6(2), 46–9.

Clapham, D. (1997) Cryptosporidium incidence in private water supplies and correlatory indicators. *Institute of Biology Conference*, University of Warwick (Warwick).

Clark, S.W. (1990) Point of entry and point of use devices for meeting drinking water standards. In *Point of Use/Point of Entry. Pollution Technology Review*, US Environmental Protection Agency/American Water Works Association.

Cohen, B.L. (1990) An experimental test of the linear no threshold theory of radiation carcinogenesis. In *Radon, Radium and Uranium in Drinking Water*, Cothern, C.R. and Rebers, P.A. (Editors), Lewis Publishers Inc. (Chelsea, MI).

Coker, A.O., Isokepehl, R.D., Thomas, B.N., Amisu, K.O. and Obi, C.L. (2002) Human campylobacteriosis in developing countries. *Emerging Infectious Diseases* 8(3). Online article http://www.cdc.gov/ncidod/eid/vol8no3/01-0233.htm accessed 29 April 2003.

Combs, R.F. and McGuire, P.J. (1988) The use of ultra violet light for microbial control in industrial water treatment applications. *Paper presented at The Industrial Treatment of Water Seminar, Dublin*.

Corso, P.S., Kramer, M.H., Blair, K.A., Addiss, D.G., Davis, J.P. and Haddix, A.C. (2003) Cost of illness in the 1993 waterborne *Cryptosporidium* outbreak, Milwaukee, WI. *Emerging Infectious Diseases* 9(4). Online article http://www.cdc.gov/ncidod/eid/vol9no4/02-0417.htm accessed 7 May 2003.

Costa, P.T. (1991) The role of aluminium in cognitive functioning. In *Alzheimer's Disease and the Environment*, Lord Walton of Detchant (Editor) 1991 Royal Society of Medicine Services (London).

Craig, F. and Craig, P. (1989) *Britain's Poisoned Water*. Penguin (London, New York, Victoria, Ontario, Aukland).

Crapper McLachan, D.R. (1991) The possible relationship between aluminium and Alzheimer's disease and the mechanisms of cellular pathology. In *Alzheimer's Disease and the Environment*. Lord Walton of Detchant (Editor) Royal Society of Medicine Services (London).

Craun, G.F. (1984) Health aspects of groundwater pollution. In *Groundwater Pollution Microbiology*, Wiley and Sons (New York, London, Sydney).

Craun, G.F., Hubbs, S.A., Frost, F., Calderon, R. and Via, S. (1998) Waterborne outbreaks of cryptosporidiosis. *Journal of American Waterworks Association* 90(9), 81–91.

Crawford, T. and Crawford, M.D. (1967) Prevalence and pathological changes of ischemic heart-disease in a hard-water and in a soft-water area. *The Lancet* 321, 229–32.

Crawford-Brown, D.J. (1992) Cancer risk from radon. *Journal of American Waterworks Association* 84(3), 77–81.

Curriero, F., Patz, J., Rose, J. and Lele, S. (2001) The association between extreme precipitation and waterborne disease outbreaks in the United States, 1984–1994. *American Journal of Public Health* 91, 1194–9.

Dadswell, J.V. (1990) Microbiological aspects of water quality and health. *Journal of Institute of Water and Environmental Management* 4(6), 515–9.

Denny, S., Whitmore, T. and Jago, P. (1991) Efficiency of UV radiation as a water disinfectant. *WRc report FR 0209 March 1992, Foundation for Water Research*.

DETR (2001) *Cryptosporidium in Water Supplies. Third report of the group of experts.* Appendix 3A (8.1) Department for Environment, Food and Rural Affairs (London).

DETR (2002) *Farm Waste Grant (Nitrate Vulnerable Zones) (England) (No. 2) Scheme 2000* HMSO (London).

Dixon, K.L., Lee, R.G., Smith, J. and Zielinski, R. (1991) Evaluating aeration technology for radon removal. *Journal of American Water Works Association* 83(4), 141–8.

DoE (1986) Nitrate in water, *Pollution Paper No 26. Report of the Nitrate Coordination Group.* HMSO (London).

DoE (1988) Acidity in United Kingdom freshwaters: UK acid waters Review Group Second Report. *Acidity in United Kingdom Fresh Waters.* DoE, HMSO (London).

DoE (1989) *Guidance on the Safeguarding of Public Water Supplies.* HMSO (London).

DoE (1991) *Private Water Supplies: A Consultation Paper.* HMSO (London).

DoE (1992) (Warden, J., Jukes, G. and Green, R.) Treatment of Private Water Supplies (DWQ 9028) Draft Consultation Paper. *Draft consultation document "Treatment of Private Water Supplies".* HMSO (London).

DoE (1996) Department of the Environment Report No. DWI0754, 1996 *Removal of Oocysts of Cryptosporidium from Private Water Supplies – Assessment of Point-of-Use Filters,* UK Drinking Water Inspectorate (London).

Dougan, H. (1996) *An Investigation into Self-reported Gastrointestinal Illness in People on Private Water Supplies in Shropshire.* Unpublished report.

Dozier, M. (2003) Texas A&M University. Personal communication.

Drury, D. (1993) Drinking Water Inspectorate. Personal communication.

Drury, D. (2003) Drinking Water Inspectorate. Personal communication.

Dudley, N. (1990) *Nitrates The Threat to Food and Water* Green Print. The Merlin Press (London).

Dufour, A.P. (1990) Health studies of aerobic heterotrophic bacteria colonising GAC systems. In *Point of Use/Point of Entry. Pollution Technology Review 188.* US Environmental Protection Agency/ American Water Works Association.

Duke, L.A., Breathnach, A.S., Jenkins, D.R., Harkis, B.A. and Codd, A.W. (1996) A mixed outbreak of *Cryptosporidium* and *Campylobacter* infection associated with a private water supply. *Epidemiology and Infection* 116, 303–8.

DuPont, H.L., Chappell, C.L., Sterling, C.R., Okhuysen, P.C., Rose, J.B. and Jakubowski, W. (1995) The infectivity of *Cryptosporidium parvum* in healthy volunteers. *New England Journal of Medicine* 332, 855–9.

DWI (1987) Drinking Water Inspectorate Report no DWI0002. *The Removal of Volatile and Non-volatile Organic Compounds from Small Water Supplies.* Final report 1330/RD9/10.87.

DWI (1992) *Nitrate, Pesticides and Lead 1989 and 1990.* Drinking Water Inspectorate, Department of the Environment, HMSO (London).

DWI (1999) *Nitrate in Drinking-water and Childhood-onset Insulin-dependent Diabetes Mellitus in Scotland and Central England.* Drinking Water Inspectorate Report DWI0801. Drinking Water Inspectorate (London).

DWI (2001a) *Mycrobacterium avian paratuberculosis*. Online article htpp// www.dwi.gov.uk/consumer/consumer/mapcron.htm accessed 29 April 2003. Drinking Water Inspectorate (London).

DWI (2001b) *MTBE*. Online article htpp//www.dwi.gov.uk/consumer/consumer/ mtbe.htm accessed 29 April 2003. Drinking Water Inspectorate (London).

DWI (2002) *Endocrine Disrupters*. Online article htpp//www.dwi.gov.uk/consumer/ consumer/Drugs%20and%20EDs2.htm accessed 29 April 2003. Drinking Water Inspectorate (London).

DWI (2003a) *Nitrates*. Online article htpp//www.dwi.gov.uk/consumer/consumer/ nitrate.htm accessed 15 May 2003. Drinking Water Inspectorate (London).

DWI (2003b) *Nickel*. Online article htpp//www.dwi.gov.uk/consumer/pr0302.htm accessed 29 April 2003. Drinking Water Inspectorate (London).

Edwards, M. and McNeill, L.S. (2002) Effect of phosphate inhibitors on lead release from pipes. *Journal of the American Water Works Association* **94**(1), 79–90.

Edwardson, J.A. (1991) The pathogenesis of cerebral beta amyloid deposition and the possible role of aluminium. In *Alzheimer's Disease and the Environment Royal Society of Medicine Services*. Lord Walton of Detchant (Editor) 1991 Royal Society of Medicine Services (London).

EEC (1980) *Council Directive Relating To The Quality of Water Intended For Human Consumption*. European Economic Community. Council Directive 80/778/EEC.

EPA (1999) *Proposed Radon in Drinking Water Rule*. US Environmental Protection Agency. Technical Fact Sheet: EPA 815-F-99-006 October 1999.

European Union (1991) *Council Directive Concerning the Protection of Waters against Pollution Caused by Nitrates from Agricultural Sources* European Union. Council Directive 91/676/EC.

European Union (1996) *Council Directive on the Approximation of the Laws of Member States Relating to the Exploitation and Marketing of Natural Mineral Waters*. European Union, Council Directive 96/70/EC.

European Union (1998) *Council Directive on the Quality of Water Intended for Human Consumption*. European Union, Council Directive 98/83/EC.

Farthing, M. Professor (1999) Expert Group on Cryptosporidium in Drinking Water. Personal communication.

Fawell, J.K. (1991) Pesticide residues in water – imaginary threat or imminent disaster. In *BCPC MONO No. 47 Pesticides in Soils and Water*. British Crop Protection Council.

Fawell, J.K. (2000) Pharmaceuticals in the drinking water – issue or scare story. *Water and Health* North West Water 29(1).

Fawell J.K. (2003) quoted in DWI. Online article htpp//www.dwi.gov.uk/ consumer/pr0302.htm accessed 29 April 2003.

Fawell, J.K. and Miller, D.G. (1992) Drinking water quality and the consumer. *Journal of Institute of Water and Environmental Management* **6**(6), 726–33.

Fawell, J.K., Fielding, M., Horth, H., James, H., Lacey, R.F., Ridgway, J.W., Wilcox, P. and Wilson, I. (1986) *Health Aspects of Organics in Drinking Water*. WRc Environment Paper TR 231.

Fewtrell, L. and Kay, D., (1996) *Health Risks from Private Water Supplies* CREH Report EPG 1/9/79, Centre for Research on Environment and Health, University of Leeds, Leeds.

Fewtrell, L., Kay, D., Jones, F., Baker, A. and Mowat, A. (1996) Copper in drinking water – an investigation into possible health effects. *Public Health* **110**, 175–7.

Fewtrell, L., Kay, D., Wyer, M., Godfree, A. and O'Neill, G. (1997) Microbiological quality of bottled water. *Water Science Technology* 35(11–12), 47–53.

Flaten, T.P. (1990) Geographical associations between aluminium in drinking water and death rates with dementia (including Alzheimer's Disease), Parkinson's Disease and Amyotrophic Lateral Sclerosis in Norway. *Environmental Geochemical Health* 12, 152–68.

Food Standards Agency (2001) A UK-wide survey of *Salmonella* and *Campylobacter* contamination of fresh and frozen chicken on retail sale. *FSA News* August 2001, Food Standards Agency (London).

Foster, R. (2003) Tennessee Department of Environment and Conservation – Division of Water Supply, Personal communication.

Foster, S.D. and Chilton, P.J. (1991) Pesticides in groundwater: some preliminary observations on behaviour and transport. In *British Crop Protection Council MONO No. 47, Pesticides in Soils and Water*. British Crop Protection Council.

Foust, C. (1990) Performance and application of ultraviolet light systems. In *Point of Use/Point of Entry. Pollution Technology Review 188*. US Environmental Protection Agency/American Water Works Association.

Frost, J.A., Gillespie, I.A. and O'Brien, S.J. (2000) Combining typing data and epidemiological information in *Campylobacter* surveillance – new opportunities. *The Increasing Incidence of Human Camplylobacteriosis*. Report and Proceedings of a WHO Consultation of Experts.

Furtado, C., Adak, G.K., Stuart, J.M., Evans, H.S. and Casemore, D.P. (1998) Outbreaks of waterborne infectious intestinal disease in England and Wales 1992–5. *Epidemiology and Infection* 121, 109–19.

FWR (1982) Foundation for Water Research Asbestos in Drinking-water. *FWR Report no DWI0073*, November 1982.

Galbraith, N.S., Barrett, N.J. and Stanwell-Smith, R. (1987) Water and disease after Croydon: a review of water-borne and water associated disease in the UK 1937–86. *Journal of Institute of Water and Environmental Management* 1(1), 7–21.

Gale, P. (2001) Developments in microbiological risk assessment for drinking-water. *Journal of Applied Microbiology* 91, 199–205.

Gale, P. and Stanfield, G. (2000) *Cryptosporidium* during a simulated outbreak. *Journal of American Water Works Association* 92(9), 105–16.

Gjessing, E.T., Alexander, J. and Rosseland, B.O. (1989) Acidification and aluminium – contamination of drinking water. In *Watershed 89 The Future of Water Quality in Europe*. Pergammon Press (Oxford, New York, Beijing, Frankfurt, Sao Paulo, Sydney, Tokyo, Toronto).

Gleeson, C. and Gray, N. (1997) *The Coliform Index and Waterborne Disease: Problems of Microbial Drinking Water Assessment*. E&FN Spon (London, Weinheim, New York, Tokyo, Melbourne, Madras).

Glicker, J.L. (1992) Convincing the public that drinking water is safe. *Journal of American Water Works Association* 84(1), 46–51.

Gluber, D.J. and Clark, G.C. (1995) Dengue/dengue hemorrhagic fever: the emergence of a global health problem. *Emerging Infectious Diseases* 2(1). Online article http://www.cdc.gov/ncidod/EID/pastcon.htm accessed 29 April 2003.

Graham, R.D., Hannam, R.J. and Wren, N.C. (Editors) (1988) *Manganese in Soils and Plants*. Kluwer Academic (Dordrecht, Boston, London).

Graham-Bryce, I.J. (1983) Pesticide research for the improvement of human welfare. In *Pesticide Chemistry – Human Welfare and the Environment, vol. 1.* Miyamoto, J. and Kearney, P.C. (Editors) Pergammon Press (Oxford, New York, Beijing, Frankfurt, Sao Paulo, Sydney, Tokyo, Toronto).

Gray, N.F. (1994) *Drinking Water Quality: Problems and Solutions.* John Wiley and Sons (Chichester, New York, Brisbane, Toronto, Singapore).

Green, B. (1992) Personal communication.

Green, B.M.R., Cliff, K.D., Miley, J.C.H. and Lomas, P.R. (1991) Radon studies in UK homes. In *Fifth International Symposium on the Natural Radiation Environment,* Salzburg, Austria on 22–28 September 1991.

Green, B.M.R., Lomas, P.R. and O'Riordan, M.C. (1992) *Radon in Dwellings in England.* National Radiological Protection Board (Chilton).

Grimason, A.M., Smith, H.V., Smith, P.G., Jackson, M.E. and Girdwood, R.W.A. (1990) Waterborne cryptosporidiosis and environmental health. *Environmental Health* 98(9), 228–33.

Hall, R. (2003) Tennessee Department of Environmental Conservation, Division of Water Supply. Personal communication.

Ham, D. (2003) Texas Parks and Wildlife Department, Austin, Texas. Personal communication.

Handysides, S. (1999) Underascertainment of infectious intestinal diseases. *Communicable Disease and Public Health* 2(2), 78–9.

Hardie, R.M., Wall, P.G., Gott, P., Bardhan, M. and Bartlett, C.L.R. (1999) Infectious diarrhea in tourists staying in a resort hotel. *Emerging Infectious Diseases* 5(1). Online article http://www.cdc.gov/ncidod/EID/pastcon.htm accessed 29 April 2003.

Hardy, J. (1991) The Genetics of Alzheimer's disease. In *Alzheimer's Disease and the Environment.* Lord Walton of Detchant (Editor), 1991 Royal Society of Medicine Services (London).

Harold, E., Hampton, E. and McKirdy, K. (1990) Lead levels in drinking water from water boilers in Glasgow district council premises. *Environmental Health* 98(2), 32–3.

Healey, M.G. (1991) *Drinking Water 1990: A Report by the Chief Inspector Drinking Water Inspectorate.* HMSO (London).

Health and Welfare Canada (1977) *Survey and test protocols for point of use water purifiers.* Environmental Health Directorate, Health and Welfare Canada.

Heath, M.J. (1991) Radon in the surface waters of south west England and its bearing on uranium distribution, faults and fracture systems and human health. *Quarterly Journal of Engineering Geology* 24, 183–9.

Hegarty, J.P., Dowd, M.T. and Baker, K.H. (1999) Occurrence of *Heliobacter pylori* in surface water in the United States. *Journal of Applied Microbiology* 87, 697–701.

Heseltine, M. (1992) quoted in: Britain in court over nitrates. *Guardian* 22, January 1992.

Hesketh, G.E. (1982) Natural radioactivity in water. *Journal of the Society for Radiological Protection* 2(3), 233–41.

Hess, C.T., Michel, J., Horton, T.R. *et al.* (1985) The occurrence of radioactivity in public water supplies in the United States. *Health Physics* 48(5), 553–86.

Hiagins, D.A. *et al.* (1984) Human intestinal parasitism in three areas of Indonesia, a survey. *Annals of Tropical Medicine and Parasitology* 78(6), 637–48.

HMSO (1936) *The Public Health Act 1936*. HMSO (London).

HMSO (1977) *The Rent Act 1977*. HMSO (London).

HMSO (1984a) EC Directive Relating to the Quality of Water Intended For Human Consumption (80/778/EEC). *Circular 25/84 EC Directive Relating to the Quality of Water Intended For Human Consumption*. HMSO (London).

HMSO (1984b) *The Building Act 1984*. HMSO (London).

HMSO (1985) *The Housing Act 1985*. HMSO (London).

HMSO (1988) *The Public Health (Infectious Diseases) Regulations 1988*. HMSO (London).

HMSO (1989) *The Water Act 1989*. HMSO (London).

HMSO (1990) *The Food Safety Act 1990*. HMSO (London).

HMSO (1991a) *The Water Industries Act 1991*. HMSO (London).

HMSO (1991b) *The Private Water Supplies Regulations 1991*. HMSO (London).

HMSO (1991c) *Private Water Supplies Circular 24/91*. *Private Water Supplies 1991*. HMSO (London).

HMSO (1992) *The Private Water Supplies (Scotland) Regulations 1992*. HMSO (London).

HMSO (1995) *Food Safety (General Food Hygiene) Regulations 1995*. HMSO (London).

HMSO (2000) *The Water Supply (Water Quality) Regulations 2000*. HMSO (London).

HMSO (2002) *The Regulatory Reform (Housing Assistance) (England and Wales) Order 2002*. HMSO (London).

HOLSCOTEC (1989) *Nitrate in Water (16th Report with Evidence)*. House of Lords Report of Select Committee on the European Communities (London).

Horan, N.J. (1994) University of Leeds. Personal communication.

HSE (1998) Health and Safety Executive. *Agriculture Information sheet No 23 Avoiding Ill Health at Open Farms – Advice to Farmers (with Teacher's Supplement)* HSE (London).

Humphrey, T.J. and Cruickshank, J.G. (1985) The potability of rural water supplies – a pilot study. *Community Medicine* 7, 43–7.

Hunter, P. (1997) *Waterborne Disease: Epidemiology and Ecology*. John Wiley and Sons (Chichester, New York, Weinheim, Brisbane, Singapore, Toronto).

Hunter, P. (2001) University of East Anglia. Personal communication.

Hunter, P. (2002) University of East Anglia. Personal communication.

Hurst, P., Hay, A. and Dudley, N. (1991) *The Pesticide Handbook*. Journeyman Press (London, Concorde, MA).

Inglis, T.J.J., Garrow, S.C., Henderson, M., Clair, A., Sampson, J., O'Reilly, L. and Cameron, R. (2000) *Burkholderia pseudomalleii* traced to water treatment plant in Australia. *Emerging Infectious Diseases* 6(1). Online article http://www.cdc.gov/ncidod/EID/pastcon.htm accessed 29 April 2003.

Ives, K.J. (1991) Cryptosporidium (effects, detection and elimination). Seminar paper *Cryptosporidium (Effects, Detection and Elimination)* Wakefield.

Ives, K.J. (1993) Personal communication.

Jackson, M.H., Morris, G.P., Smith, P.G. and Crawford, J.F. (1989) *Environmental Health Reference Book*. Butterworths (London, Boston, Singapore, Sydney, Toronto, Wellington).

Jackson, P. (1992) Point of use devices. *Environmental Health* 100(7), 192–6.

Jackson, S.G., Goodbrand, R.B., Johnson, R.P., Odorico, V.G., Alves, D., Rahn, K., Wilson, J.B., Welch, M.K. and Khakhria, R. (1998) *Escherichia coli O157* diarrhoea associated with well water and infected cattle on an Ontario farm. *Epidemiology and Infection* **120**, 17–20.

Jarvis, S.N., Staube, R.C., Williams, A.L. and Bartlett, C.L. (1985) Illness associated with contamination of drinking-water supplies with phenol. *British Medical Journal (Clinical Research Edition)* **290**, 800–1802.

Jones, I.G. and Roworth, M. (1996) An outbreak of *Escherichia coli* O157 and campylobacteriosis associated with contamination of a drinking water supply. *Public Health* **110**, 277–82.

Jones, K. (2003) Lancaster University. Personal communication.

Karanis, P., Maier, W.A., Seitz, H.M. and Schoenen, B. (1992) UV sensitivity to protozoan parasites. *Aqua* **41**(2), 95–100.

Kay, D. (2003) Risk assessment. Presentation at *Private Water Supply – Microbial Risk Assessment Reporting and Surveillance Workshop*, Peebles, Scotland. April 2003.

Khan, L.K., Li, R., Gootnick, D. and the Peace Corps Thyroid Investigation Group (1998) Thyroid abnormalities related to iodine excess from water purification units. *The Lancet* **352**, 1519.

Kim, D.H., Lee, S.K., Chun, B.Y., Lee, D.H., Hong, S.C. and Jang, B.K. (1994) Illness associated with contamination of drinking-water supplies with phenol. *Journal of Korean Medical science* **9**(3), 218–23.

King, A. (1991) Pesticides in water – what is going on? *Water and Waste Treatment* **83**(12), 18–9.

Kool, H.J. (1990) Health risk in relation to drinking water treatment. In *Biohazards of Drinking Water Treatment*. Larson, R.A. (Editor) Lewis Publishers (Michigan).

Kuennen, R.W., Taylor, R.M., Van Dyke, K. and Groenevelt, K. (1992) Removing lead from drinking water with a point-of-use gac fixed-bed adsorber. *Journal of American Water Works Association* **84**(2), 91–101.

Kukkula, M., Maunula, L., Silvennoinrn, E. and von Bonsdorff, C.H. (1999) Outbreak of viral gastroenteritis due to drinking-water contaminated by norwalk-like viruses. *Journal of Infectious Disease* **180**, 1771–6.

Kunzlik, P. (1992) Personal communication.

LaBarbera, J. (2003) Resource Development and Conservation Department, State of Texas. Personal communication.

Lacey, R.F. (1981) Changes in water hardness and cardio-vascular death rates. *Technical Report 171*, WRc (Medmenham).

Lamb, A.J., Reid, D.C., Lilly, A., Gauld, J.H., McGaw, B.A. and Curnow, J. (1998) *Improved Source Protection for Private Water Supplies: Report on the Development of a Microbiological Risk Assessment Approach*. School of Applied Sciences, Robert Gordon University, Aberdeen.

Landsberg, J. and Watt, F. (1992) Absence of aluminium in neuritic plaque cores in Alzheimer's disease. *Nature* **360**, 65–8.

Lao, K.Q. (1990) *Controlling Indoor Radon: Measurement, Mitigation and Prevention*. Van Nostrand Reinhold (New York).

Lee, D. (1992) University of Leeds. Personal communication.

Lees, A. (1991) Government faces prospect of ten high court cases over illegal H2O cocktails. *Friends of the Earth Press Release*.

Lerner, D.N. and Tellam, J.H. (1992) The protection of urban groundwater from pollution. *Journal of Institute of Water and Environmental Management* **6**(1), 28–37.

Lezcano, I., Perez Rey, R., Baluja, C. and Sanchez, E. (1999) Ozone inactivation of *Pseudonmonas aeruginosa*, *Escherichia coli*, *Shigella sonnei* and *Salmonella typhimurium* in water. *Ozone Science and Engineering* **21**, 293–300.

License, K., Oates, K.R., Synge, B.A. and Reid, T.M. (2001) An outbreak of *E. coli* O157 infection with evidence of spread from animals to man through contamination of a private water supply. *Epidemiology and Infection* **126**(1), 135–8.

Lichtenstein, S., Sloric, P., Fischoff, B., Layman, M. and Combs, D. (1978) Judged frequency of lethal events. *Journal of Experimental Psychology: Human Learning and Memory* **4**, 551–78.

Lilly, A. (2003) Microbial risk assessment for private water supplies. Presentation at *Private Water Supply – Microbial Risk Assessment Reporting and Surveillance Workshop*, Peebles, Scotland. April 2003.

Lloyd, B.J. and Helmer, R. (1991) *Surveillance of Drinking Water Quality in Rural Areas*. Longman Scientific and Technical (Harlow).

Lloyd, P. (1999) City of Bradford Metropolitan District Council. Personal communication.

Longmate, N. (1966) *King Cholera: The Biography of a Disease*. Hamish Hamilton (London).

Louisiana DTE (1985) *Water Well Rules, Regulations and Standards State of Louisiana*. Louisiana Department of Transportation and Development, Public Works and Flood Control Directorate, Water Resources Section, November 1985.

Louisiana Register (2002) Water Supplies. *Louisiana Register* Promulgated 20 June 2002. 28(6), 318–1342.

Lowry, J.D. and Lowry, S.B. (1988) Radionuclides in drinking water. *Journal of American Water Works Association* **80**(7), 51–64.

LRWA (2002) *Louisiana Rural Water Association's Information Brochure: Serving the Smaller Water and Wastewater Systems*. Louisiana Rural Water Association, Kinder (Louisiana).

Luhr, J.F. (Editor) (2003) *Earth*. Dorling Kindersley (London).

McDonald, A. (1991) *Discoloured Water Investigations: Final Report to Yorkshire Water plc* Yorkshire Water plc.

McFetters, G.A., Camper, A.K., Davies, D.A., Broardaway, S.C. and Le Chevalier, M.W. (1990) Microbiology of GAC used in the treatment of drinking water. In *Biohazards of Drinking Water Treatment*. Larson, R.A. (Editor), Lewis Publishers (MI).

MacKenzie, D. (2002) Anthrax, *New Scientist* 9/2/2002 8–10.

McLauchlin, J.S., Pedraza-Díaz, S., Amar-Hoetzeneder, C. and Nichols, G.L. (1999) Genetic characterization of *Cryptosporidium* strains from 218 patients with diarrhea diagnosed as having sporadic cryptosporidiosis. *Journal of Clinical Microbiology* 37, 3153–8.

Maddison, T. (2000) *Cryptosporidiosis and Private Water Supplies*. Thesis for Masters degree in Public Health Medicine. University of Birmingham, August 2000.

MAFF (1992) Government announces restrictions on use of atrazine and simazine. *MAFF Food Safety Directorate News Release FSD 26/92*, 12 May 1992. Ministry of Agriculture, Fisheries and Food (London).

MAFF (1995) *Assessment of Environmental Oestrogens. Assessment A1. Consequences to Human Health and Wildlife*. Institute for Environmental Health. Ministry of Agriculture, Fisheries and Food (London).

Mara, D.D. (2003) Personal communication.

Mara, D.D. and Oragui, J. (1985) Bacteriological methods for distinguishing between human and animal faecal pollution of water: results of fieldwork in Nigeria and Zimbabwe. *Bulletin of the World Health Organization* 63(4) 773–83.

Mara, D.D. and Clapham, D. (1997) Water-related carcinomas: environmental classification. *Journal of Environmental Engineering* 123(5), 416–22.

Martyn, C.N., Barker, D.J.P. and Osmond, C. (1989) Geographical relation between Alzheimer's disease and aluminium in drinking water. *Lancet* 345, 59–62.

Mason, B. (1996) *Principals of Geochemistry*, Third Edition. John Wiley and Sons (New York, London, Sydney).

Matthews, R. (1997) Wacky water. *New Scientist* 21 June 1997, 40–3.

Mayhew, M. (2003) Texas Parks and Wildlife Department, Austin, Texas. Personal communication.

Meara, J.R. (1989) An investigation of health and lifestyles in people who have private water supplies at home. *Community Medicine* 11(2), 131–9.

Merritt, A., Miles, R. and Bates, J. (1999) An outbreak of *Campylobacter* enteritis on an island resort, North Queensland. *Communicable Disease Intelligence* 23(8), 215–9.

Michel, J. (1990) Relationship of radium and radon with geological formations. In *Radon, Radium and Uranium In Drinking Water*. Cothern, C. and Rebers, P. (Editors), Lewis Publishers (Michigan).

Michel, P., Commenges, D. and Dartigues, J.F. (1991) Study of the relationship between aluminium concentration in drinking water and risk of Alzheimer's disease. In *Alzheimer's Disease: Basic Mechanisms, Diagnosis and Therapeutic Strategies*. Iqbal, K., McLachlan, D., Winblad, B. (Editors), Wiley (Chichester), pp. 387–91.

Mills, W.A. (1990) Risk assessment and control management of radon in drinking water. In *Radon, Radium and Uranium in Drinking Water*. Cothern, C. and Rebers, P. (Editors), Lewis Publishers (Michigan).

Milvy, P. and Cothern, R. (1990) Scientific background for the development of regulations for radionuclides in drinking water. In *Radon, Radium and Uranium In Drinking Water*. Cothern, C. and Rebers, P. (Editors), Lewis Publishers (Michigan).

Morgan, D.R. (Editor) (1990) *The BMA Guide to Living With Risk* BMA. Penguin Books (London, New York, Victoria, Ontario, Auckland).

Morgan, D.R. (Editor) (1992) *The BMA Guide to Pesticides Chemicals and Health* BMA. Edward Arnold (London, Melbourne, Auckland).

Morgan, D., Allaby, M., Crook, S., Casemore, D., Healing, T.D., Soltanpoor, N., Hill, S. and Hooper, W. (1995) Waterborne cryptosporidiosis associated with a borehole supply. *Communicable Disease Report and Review*, 5, 93–7.

Morris, B.L. (2001) Practical implications of the use of groundwater-protection tools in water supply risk assessment. *Journal Chartered Institute of Water and Environmental Management* 15(4), 265–70.

Morris, G. (1997) Environment Agency. Personal Communication.

Mullins, J. (2003) Sink or swim. *New Scientist* 178(2396), 28–31.

Murphy, J. (2003) Marshall's Pumps. Personal communication.

National Safety Associates (1991) *Publicity leaflet for Bacteriostatic Water Systems*. National Safety Associates of America (UK) Ltd, NSA House, Queen Street, Maidenhead, Berkshire, UK.

National Trust (1996) *Private Water Supplies: A Guide to Design, Management and Maintenance*. The National Trust, Estates Advisory Office (Cirencester).

Natividad, N. (2003) Laboratory Manager Tohono O'odham Utility Authority, Water Department, Sells, Arizona. Personal communication.

Nazaroff, W.W., Doyle, S.M., Nero, A.V. and Sexton, R.G. (1987) Potable water as a source of airborne Rn-222 in US dwellings: a review and assessment. *Health Physics* 52(2), 281–95.

NRA (1992) *Policy and Practice for the Protection of Groundwater*. National Rivers Authority (London).

NRPB (1990) *National Radiological Protection Board. Radon: Questions and Answers*. National Radiological Protection Board Booklet Radon Questions and Answers. NRPB (Didcot, Oxon).

NRPB (2003) *Radon in Private Drinking Water Supplies*. National Radiological Protection Board. Online article http://www.nrpb.org/radon/radon_in_drinking_water.htm accessed 19 May 2003. NRPB (Didcot, Oxon).

Oakes, D. (1992) Risk assessment and catchment protection. *Institute of Water and Environmental Management Symposium: Pesticides and the Water Industry* (Loughborough).

O'Connor, D.R. (2002) *Report of the Walkerton Inquiry*. Ontario Ministry of the Attorney General (Ontario).

O'Donnell, M., Platt, C. and Aston, R. (2000) Effects of a boil water notice on behaviour in the management of a water contamination incident. *Communicable Diseases and Public Health* 3, 56–9.

Olsen, S.J., Miller, G., Breuer, T., Kennedy, M., Higgins, C., Walford, J., McKee, G., Fox, K., Bibb, W. and Mead, P. (2002) A waterborne outbreak of *Escherichia coli* O157:H7 infections and hemolitic uremic syndrome: implications for rural water. *Systems Emerging Infectious Diseases* 8(4). Online article http://www.cdc.gov/ncidod/EID/pastcon.htm accessed 29 April 2003.

O'Riordan, M.C. and O'Riordan, C.N. (1991) The matter of radon. *Paper presented at Nuclear Inter Jura*, Bath, UK. September 1991.

Ozonia Ltd. (2000) *Information brochure Compact Ozone Generators for Oxygen or Air Feed* (Duebendorf, Switzerland).

Packham, R.F. (1990) Chemical aspects of water quality and health. *Journal of Institute of Water and Environmental Management* 4(5), 484–8.

Paragon (2003) Paragon Water Systems. Website http://www.paragonwater.com/fataltap.shtml accessed 30 April 2003.

Patel, T. (1992) Water without a whiff of chlorine. *New Scientist* 1846, 19.

Payment, P., Franco, E., Richardson, L. and Siemiatycki, J. (1991) Gastrointestinal health effects associated with the consumption of drinking water produced by point-of-use reverse-osmosis filtration units. *Applied and Environmental Microbiology* 57(4), 945–8.

PCI-Memtech. (2003) Choosing the right membrane. *Water* 189, 10.

Pearce, F. (1992) Lower limits for lead in the pipeline. *New Scientist* 1839, 4.

Peng, M.M., Xiao, L., Freeman, A.R., Arrowood, M.J., Escalante, A.A., Weltman, A.C., Ong, C.S.L., Mac Kenzie, W.R., Lal, A.A. and Beard, C.B. (1997) Genetic polymorphism among *Cryptosporidium parvum* isolates: evidence of two distinct human transmission cycles. *Emerging Infectious Diseases* 3(4). Online article http://www.cdc.gov/ncidod/EID/pastcon.htm accessed 29 April 2003.

Perl, D.P. and Good, P.F. (1991) The relationship between aluminium and neurofibrillary tangle formation. In *Alzheimer's Disease and the Environment* Lord Walton of Detchant (Editor), Royal Society of Medicine Services (London).

Petts, M. (2003) City of Bradford Metropolitan District Council. Personal communication.

Pickford, J. (1991) *The Worth of Water: Technical Briefs on Health, Water and Sanitation*. Water, Engineering and Development Centre (Loughborough).

Pinfold, J. (1993) University of Leeds. Personal communication.

Pitt, J. and Provin, A. (2003) Texas A&M University, College Station, Texas. Personal communication.

Pontiff, D. (2003) Louisiana Department of Health and Hospitals, Office of Public Health, Lafayette, Louisiana. Personal communication.

Pound, B.R.E. (1992) Monitoring strategy for pesticides – a water company approach. *Institute of Water and Environmental Management Symposium: Pesticides and the Water Industry* (Loughborough).

Pouria, S., de Andrade, A., Barbosa, J., Cavalcanti, R.L., Barreto, V.T.S., Ward, C.J., Preiser, W., Grace Poon, K., Neild, G.H. and Codd, G.A. (1998) Fatal microcystin intoxication in haemodialysis unit in Caruaru, Brazil. *The Lancet* 352, 9121.

Powell, R., Packham, R.F., Lacey, R.F. and Russell, P.F. (1982) Water *Quality and Cardiovascular Disease in British Towns*. Technical Report 178 WRC (Medmenham).

Price, M. (1985) *Introducing Groundwater*. George Allen and Unwin (London, Boston, Sydney).

Raina, P.S., Pollari, F.L., Teare, G.F., Goss, M.J., Barry, D.A. and Wilson, J.B. (1999) The relationship between *E. coli* indicator bacteria in well-water and gastrointestinal illness in rural families. *Canadian Journal of Public Health* 90(3), 172–5.

Reacher, M., Ludlam, H., Irish, N., Buttery, R. and Murray, V. (1999) Outbreak of gastroenteritis associated with contamination of a private borehole water supply. *Communicable Disease and Public Health* 2(1), 27–31.

Reasoner, D.J. (1990) Microbiological studies of GAC point of use devices. In *Point of Use/Point of Entry. Pollution Technology Review 188*, US Environmental Protection Agency/American Water Works Association.

Regunathan, P., Beauman, W.H. and Jarog, D.J. (1990) Precoat carbon filters as barriers to incidental microbial contamination. In *Point of Use/Point of Entry. Pollution Technology Review 188*, US Environmental Protection Agency/American Water Works Association.

Reily, W. (2001) *Final Report Task Force on E. Coil 0157*. Ministry of Health and Community Care, Food Standards Agency (Scotland) and Scottish Executive Health Department (Edinburgh).

Reintjes, R., Dedushaj, I., Gjini, A., Jorgensen, T.R., Cotter, B., Lieftucht, A., D'Ancona, F., Dennis, D.T., Kosoy, M.A., Mulliqi-Osmani, G., Grunow, R., Kalaveshi, A., Gashi, L. and Humolli, I. (2002) Tularemia outbreak investigation in Kosovo: case control and environmental studies. *Emerging Infectious Diseases* 8(1), 69–73.

Reiter, P. (2000) From Shakespeare to Defoe: malaria in England in the Little Ice Age. *Emerging Infectious Diseases* 6(1). Online article http://www.cdc.gov/ncidod/EID/pastcon.htm accessed 29 April 2003.

Rice, R.G. (1990) Ozone oxidation products: implications for drinking water treatment. In *Biohazards of Drinking Water Treatment*. Larson, R.A. (Editor) Lewis Publishers (Chelsea, MI).

Richard, Y. (1989) Nitrates and drinking water. In *Watershed 89 The Future of Water Quality in Europe*. Pergammon Press (Oxford, New York, Beijing, Frankfurt, Sao Paulo, Sydney, Tokyo, Toronto).

Rivett, M.O., Lerner, D.N., Lloyd, J.W. and Clerk, L. (1990a) Organic contamination of the Birmingham aquifer. *Journal of Hydrology* 113, 307–23.

Rivett, M.O., Lerner, D.N. and Lloyd, J.W. (1990b) Chlorinated solvents in UK aquifers. *Journal of Institute of Water and Environmental Management* 4(3), 242–50.

Robert, D.R., Laughlin, L.L., Hsheih, P. and Legters, L.J. (1997) DDT, global strategies, and a malaria control crisis in South America. *Emerging Infectious Diseases* 3(3). Online article http://www.cdc.gov/ncidod/EID/pastcon.htm accessed 29 April 2003.

Rogella, F. (2002) Technically speaking. *Water and Waste Treatment* 45(4), 30–1.

Rose, J.B. (1990) Occurrence and control of *Cryptosporidium* in drinking water. In *Drinking Water Microbiology*. Mc Fetters, G.A. (Editor), Springer-verlag (New York).

Rose, J.B., Gerba, C.P. and Jakubowski, W. (1991) Survey of potable water supplies for *Cryptosporidium* and *Giardia*. *Environmental Science and Technology* 25(8), 1393–1400.

Rowland, A., Grainger, R., Stanwell-Smith, R., Hicks, N. and Hughes, A. (1990) Water contamination in North Cornwall: a retrospective cohort study into the acute and short-term health effects of the aluminium sulphate incident in July 1988. *Journal of the Royal Society of Health* 110, 66–172.

Rubenowitz, E., Axelsson, G. and Rylander, R. (1999) Magnesium and calcium in drinking-water and death from acute myocardial infarction in women. *Epidemiology* 10(1), 31–6.

Ruiz-Tiben, E., Hopkins, D.R., Ruebush, T.K. and Kaiser, R.L. (1995) Progress toward the eradication of dracunculiasis (guinea worm disease) *Emerging Infectious Diseases* 1(2). Online article http://www.cdc.gov/ncidod/EID/pastcon.htm accessed 29 April 2003.

Rutter, M., Nichols, G.L., Swan, A. and de Louivois, J. (2000) A survey of the microbiological quality of private water supplies in England. *Epidemiological Infection* 124, 417–25.

Sacks, J.J., Leib, S., Baldy, L.M., Berta, S., Patton, C.M., White, M.C., Bigler, W.J. and Witte, J.J. (1986) Epidemic campylobacteriosis associated with a community water supply. *American Journal of Public Health* 76(4), 424–8.

St Louis, M.E. (1988) Water-related disease outbreaks 1985. *Morbidity and Mortality Weekly Report*. Online article http://www.cdc.gov/mmwr/preview/mmwrhtml/00001765.htm accessed 29 April 2003.

Sartory, D. (2003) Severn Trent Water Ltd. Personal communication.

Schnare, D.W. (1990) Raid on sanity: policy and economic analysis of radionuclide removal from drinking water. In *Radon, Radium and Uranium in Drinking Water*. Cothern, C. and Rebers, P. (Editors), Lewis Publishers (Michigan).

Seegers, H. (1992) 7,300 years old well is the world's oldest wooden building. *European Water Pollution Control* 2(5), 9–12.

Shaw, J.P., Malley, J.P. and Willoughby, S.A. (2000) Effects of UV irradiation on organic matter. *Journal of the American Water Works Association* 92(4), 157–67.

Short, C. (1992) Yorkshire Water plc. Personal communication.

Skirrow, M.B. (1977) *Campylobacter entiritis:* A "new" disease. *British Medical Journal* ii, 9–11.

Smith, A.H., Lingas, E.O. and Rahman, M. (2000) Contamination of drinking-water by arsenic in Bangladesh: a public health emergency. *Bulletin of the World Health Organization* 78(9), 1093–103 WHO (Geneva).

Smith, D.M. and Welham, D. (1989) *Radiological Assessment of Private Water Supplies in North West England.* DoE Commissioned Research on Radioactive Waste Management 1990.

Smith, H. (2003) Scottish Parasitic Diagnostic Laboratory. Personal communication.

Snow, J. (1854) The cholera near Golden-square, and at Deptford. *Medical Times Gazette* 9, 321–2.

SOA (1997) Prevention of Environmental Pollution from Agricultural Activities. *Scottish Office Agriculture*, Environment and Fisheries Development (Edinburgh).

Solt, G. (1991) Special uses for ion exchange resin 1: nitrate removal. *Dewplan Technical Bulletin No. 34 Dewplan (WT) Ltd.*

Spitalny, K.C., Brondum, J., Vogt, R.L., Sargent, H.E. and Kappel, S. (1984) Drinking-water-induced copper intoxication in a Vermont family. *Paediatrics* 74(6), 1103–6.

Starr, S. (1969) Social benefit versus technological risk. *Science* 165, 1232–8.

Steadman, L. (2002) Taking the pain out of arsenic. *Water* 158, 10–1.

Stirling, C.R. and Ortega, Y.R. (1999) *Cyclospora:* an enigma worth unravelling. *Emerging Infectious Diseases* 5(1). Online article http://www.cdc.gov/ncidod/EID/pastcon.htm accessed 29 April 2003.

Stupples, L. (1991) *Friends of the Earth Forces Government Action on Water – But Polluter Should Pay for Drinking Water Clean Up says FoE.* Friends of the Earth Press Release.

Swerdlow, D.L., Woodruff, B.A., Brady, R.C., Griffin, P.M., Tippen, S., Donnell, H.D. Jr, Geldreich, E., Payne, B.J., Meyer, A. Jr., Wells, J.G., Greene, K.D., Bright, M., Bean, N.H. and Blake, P.A. (1992) A waterborne outbreak in Missouri of *Escherichia coli* O157:H7 associated with bloody diarrhoea and death. *Annals of Internal Medicine* 117, 812–9.

Tauxe, R.V. (2000) Major risk factors for human campylobacteriosis – an overview. *The Increasing Incidence of Human Camplylobacteriosis.* Report and Proceedings of a WHO Consultation of Experts.

Taylor, G. (1993) Personal communication.

Taylor, P. (2003) Agriculture and water. *Water* (189), 6.

TDEC (1991) *Guidance for Determining if a Ground Water Source is Under the Direct Influence of Surface Water.* Tennessee Department of Environment and Conservation, Division of Water Supply, August 1991.

Toft, B. (1992) *Reduction in WHO Guideline for Lead.* WHO Press Statement.

Tompkins, J. (2003) Oh NO_3! nitrate levels are rising. *Water* 174, 7.

Toyoda, A. (2002) History of *Shistosoma japonica.* Online article http//www2.ttcn.ne.jp/~akky/parasite/history.htm accessed 6 January 2003.

USEPA (1989) United States Environmental Protection Agency Pesticides and Food. *Environmental Backgrounder* USEPA.

USGAO (1997) United States General Accounting Office, June 1997 *Drinking Water: Information on the Quality of Water Found at Community Water Systems and Private Wells.* Report GAO/RCED-97-123.

Van Dyke, K. and Kuennen, R.W. (1990) Performance and applications of GAC point of use systems. In *Point of Use/Point of Entry. Pollution Technology Review 188*, US Environmental Protection Agency/American Water Works Association.

van Maanen, J.M.S., Albering, H.J., de Kok, T.M.C.M., van Breda, S.G.J., Curfs, D.M.J., Vermeer, I.T.M., Ambergen, A.W., Wolffenbuttel, B.H.R., Kleinjans, J.C.S. and Maarten Reeser, H. (2000) Does the risk of childhood diabetes mellitus require revision of the guideline values for nitrate in drinking water? *Environmental Health Perspectives* 108, 457–61.

Victoria, C.G., Smith, P.G., Vaughan, J.P. and Noble, L.C. (1988) Water supply, sanitation and housing in relation to the risk of infant mortality from diarrhoea. *International Journal of Epidemiology* 17(3), 651–4.

Vogt, R.L., Sours, H.E., Barrett, T., Feldman, R.A., Dickinson, R.J. and Witherell, L. (1982) *Campylobacter enteritis* associated with contaminated water. *Annals of Internal Medicine* 96, 292–6.

Vogtmann, H. and Biederman, R. (1989) *The Nitrate Story: No End in Sight*. Elm Farm Research Centre (Newbury).

Walbran, S. (2001) City of Bradford Metropolitan District Council. Personal communication.

Walton (1991) Alzheimer's disease and the environment: chairman's conclusions. In *Alzheimer's Disease and the Environment*. Lord Walton of Detchant (Editor), 1991 Royal Society of Medicine Services (London).

Watkins, J., Francis, C., Kay, D. and Fewtrell, L. (2001) *Report on the Incidence of Cryptosporidium in Private Water Supplies*. Report for the Drinking-water Inspectorate Contract DWI/70/2/129.

Watson, K. (2003) Wickson Creek Utility Company, Brazos, TX. Personnal communication.

West, P. (1998) North West Water. Personal communication.

West, P. (1999) North West Water. Personal communication.

West Devon Borough Council (2001) *Natural Radiation in West Devon's Private Water Supplies* Joint Press Release – West Devon Borough Council and South and West Devon NHS Health Authority, 9 January 2001.

White, G.F., Bradley, D.J. and White, A.V. (1972) *Drawers of Water: Domestic Water Use in East Africa*. Chicago University Press (Chicago, IL)

Whitehead, R. (1992) Pesticides in agribusiness – where would we be without them. *Institute of Water and Environmental Management Symposium: Pesticides and the Water Industry*. Loughborough.

WHO (1970) *European Standards For Drinking Water*. World Health Organization (Geneva).

WHO (1984) *Guidelines For Drinking-water Quality Volumes 1–3*. World Health Organization (Geneva).

WHO (1987) *Drinking-water Quality and Health-related Risks*. World Health Organization (Geneva).

WHO (1992) *Our Planet Our Health*. World Health Organization (Geneva).

WHO (1993a) *Guidelines for Drinking-water Quality, Volume 1*, Second Edition. World Health Organization (Geneva).

WHO (1993b) World Health Organization. *The Control of Shistosomiasis*: Second Report of the WHO Expert Committee. Technical Report Series 830. World Health Organization (Geneva).

WHO (1995) *Lead and Health (local authorities and environment briefing pamphlet series; 1)*. World Health Organization (Geneva).

WHO (1996a) *WHO Guidelines for Drinking-water Quality, Volume 2*, Second Edition, Health Criteria and Other Supporting Information. World Health Organization (Geneva).

WHO (1996b) Epidemic dysentery. *Fact Sheet Number 108*. World Health Organization (Geneva).

WHO (1996c) Shistosomiasis. *Fact Sheet Number 115*. World Health Organization (Geneva).

WHO (1996d) *The World Health Report 1996: Fighting disease fostering development*. World Health Organization (Geneva).

WHO (1998a) Zoonotic Non-O157 Shiga Toxin-producing *Escherichia coli* (STEC). *Report of a WHO Scientific Working Group Meeting. Department of Communicable Disease Surveillance and Response WHO/CSR/APH 98.8*, World Health Organization (Geneva).

WHO (1998b) *Guidelines for Drinking-water Quality – Boron*. WHO Online article http://www.who.int/docstore/water_sanitation_health/GDWQ/Chemicals/boronfull.htm accessed 31 December. World Health Organization (Geneva).

WHO (2000a) The increasing incidence of human Camplylobacteriosis. *Report and Proceedings of a WHO Consultation of Experts*. World Health Organization (Geneva).

WHO (2000b) *Fact Sheet – hepatitis A*, WHO Website http//www.who.int.emc accessed 6 June 2001 (no longer available). World Health Organization (Geneva).

WHO (2001) *Fact Sheet – hepatitis E*, WHO Website http//www.who.int.emc accessed 6 June 2001 (no longer available). World Health Organization (Geneva).

WHO (2002a) *Dengue and Dengue Fever. Fact Sheet Number 117* Revised April 2002. http://www.who.int/inf-fs/en/fact117.html accessed 29 April 2003. World Health Organization (Geneva).

WHO (2002b) *Guidelines for Drinking-water Quality – Copper*. WHO Online article http://www.who.int/water_sanitation_health/GDWQ/Chemicals/coppersum.html accessed 29 April 2003. World Health Organization (Geneva).

WHO (2002c) *Guidelines for Drinking-water Quality: Nitrate and Nitrite*. Online article http://www.who.int/water_sanitation_health/GDWQ/Chemicals/nitratenitritesum.htm accessed 29 April 2003. World Health Organization (Geneva).

WHO (2003) *Emerging Issues in Water and Infectious Disease*. Online article http://www.who.int/water_sanitation_health/emerging/emerging_issues/en/ accessed 30 December 2003. World Health Organization (Geneva).

Wilcock, G.K. (1991) Alzheimer's disease and the environment: a clinician's viewpoint. In *Alzheimer's Disease and the Environment*. Lord Walton of Detchant (Editor), Royal Society of Medicine Services (London).

Willocks, L.J., Sufi, F., Wall, R., Seng, C. and Swan, A.V. for the Outbreak Investigation Team (2000) Compliance with advice to boil water during an outbreak of cryptosporidiosis. *Communicable Diseases and Public Health*, 3, 137–8.

Wilkins, M. (1992) Going potty on pesticides. *Water Bulletin* **493**, 6–7.

Wisniewski, H.M. (1991) Neuropathology and biochemistry of Alzheimer's disease and aluminium encephalopathy. In *Alzheimer's Disease and the Environment*. Lord Walton of Detchant (Editor), 1991 Royal Society of Medicine Services (London).

Wolff, S.J. (1993) A story of diminishing returns. *New Scientist* 16 January 1993, 42.

Wood, P. (2003) Tetra Tech E.M. Inc. Memphis, TN. Personal communication.

WRc-NSF (2002) WRc-NSF Ltd *Manual on Treatment for Small Water Supply Systems*. Drinking Water Inspectorate and WRc-NSF Ltd (London and Medmenham).

Zoetman, B.C. and Brinkmann, F.J. (1976) *Hardness of Drinking Water and Public Health*. Pergammon Press (Oxford).

Index

pump 28, 29, 34, 35; Broad Street 51, 77, 114; electric 24, 31; hand 24; hydraulic ram 31; law of the biggest (US) 283; submersible 29, 282; vacuum 282

Queen Victoria 192

raccoons 284
radiation: effects of 210
radioactive contamination 210; natural versus man-made 210
radioactivity: units of 211
radium 231
radon 28, 167, 210, 211, 219, 235, 245, 286; aeration 214; airborne 213, 215; health problems 213; risks from 211; in small supplies 212; and smoking 215; standards 215; treatment 242; in water 212; waterborne 214
rainwater collection 15, 142, 169, 278
rainwater jars 32
raspberries 142
rattlesnakes 286
relaxations 256, 259; for food production 263
Rent Act 1977 250
Report on the Sanitary Condition of the Labouring Population 113
resampling 65, 66, 245
reservoir 36, 51, 109, 140, 142, 157
residual effect 68
retention 203
reverse osmosis 15, 33, 51, 108, 122, 169, 208, 219, 232; problems with 234
Rhodococcus coprophilus 89
risk assessment 10; *see* sanitary survey
river blindness 152
River Dee 208
River Nakdong 209
River Thames 50, 77
Rochepot 22
rock: basalt 6; chalk 45, 48, 50, 261; granite 6, 45, 178, 213; gravel 9, 22, 31; igneous 6, 9, 50, 178; impermeable 6, 7, 8, 16, 26, 28; impervious *see* impermeable; limestone 6, 45, 48, 103, 106, 213, 240, 261; mafic 178; marble 6; metamorphic 6, 9, 178; millstone grit 41, 45, 48; permeable 20, 25; Quaternary 11; sandstone 6, 9, 48,

103, 167, 178, 262; sedimentary 6, 45, 178; shale 6, 22, 178; slate 6
rodding eyes 19
Rotavirus 86, 102, 123, 135
run-off 30, 110, 144, 157, 184, 189
rural churches: US 288
Rural Water Associations: US 284
Rutherford, Ernest 179, 211

safety zone *see* protection zone
saline intrusion 9, 186, 189, 262, 278
Salmonella 94, 120, 135
sampling 13, 30, 34, 62, 259; charges 268, 269; chemical 69; frequency of 267, 269; from a tank 72; microbiological 69; procedural document 246; programme 66; snapshot 12, 267; US 284
sand 16, 31, 156
sand filters 236; rapid 237, 238; slow 51, 236
sanitarians 13, 74, 284
sanitary assessment *see* sanitary survey
sanitary survey 10, 21, 30, 34, 40, 62, 69, 73, 95, 97, 119, 139, 140, 184, 198, 267, 279, 282
saturated zone 6, 61
Saudi Arabia 122
scabies 86
scaling 45
schistosomiasis 86, 132, 153; bilharzia 86
schmutzdecke 236
Scotland 14, 45, 68, 79, 98, 108, 113, 115, 117, 118, 119, 144, 172, 267
scratching posts 20
seagulls 32
sedimentation 206, 238
seep hole 8
septic tanks 74, 103, 106, 123, 125, 184, 281
settlement 35, 37
sewer 123
sheathing 28
shed 20
Sheet Memoir 11
Shigella 94, 122
shoved-up-pipe method 18, 20
sievert 211
silicon 163, 165
silver 230
single property supplies 245, 247, 256, 266, 268
skunks 284

Printed in the United States
by Baker & Taylor Publisher Services